Calcium Metabolism, Bone and Metabolic Bone Diseases

Editors
Friedrich Kuhlencordt and Hans-Peter Kruse

With 122 Figures

Springer-Verlag Berlin · Heidelberg · New York 1975

Proceedings of the
X. European Symposium on Calcified Tissues
Hamburg (Germany), 16–21 September 1973

Friedrich Kuhlencordt, Dr. med.
Professor of Internal Medicine
Director of the Department Clinical Osteology
University of Hamburg
D – 2000 Hamburg 20, Martinistraße 52

Hans-Peter Kruse, Dr. med.
I. Medical Clinic
University of Hamburg
D – 2000 Hamburg 20, Martinistraße 52

ISBN-13:978-3-642-80877-7 e-ISBN-13:978-3-642-80875-3
DOI: 10.1007/978-3-642-80875-3

 Library of Congress Cataloging in Publication Data: European Symposium on Calcified Tissues, 10th, Hamburg, 1973. Calcium metabolism, bone and metabolic bone diseases. Bibliography: p. Includes index. 1. Bones — Diseases — Congresses. 2. Calcium metabolism disorders — Congresses. I. Kuhlencordt, Friedrich, 1917 – ed. II. Kruse, Hans-Peter, 1943 – ed. III. Title. [DNLM: 1. Bone and bones — Metabolism — Congresses. 2. Bone Diseases — Therapy — Congresses. 3. Calcification, Physiologic — Congresses. 4. Fluorides — Pharmacodynamics — Congresses. 5. Vitamin D — Metabolism — Congresses. W3 EU938 1973c / WE200 E89 1973c]. RC930.E9 1973 616.7'1'071. 74–20630.
Softcover reprint of the hardcover 1st edition 1975

Organizing Committee
X. European Symposium on Calcified Tissues

Chairman

F. Kuhlencordt, Hamburg

H. Czitober, Wien
J. Eschberger, Wien
H. J. Dulce, Berlin
F. Heuck, Stuttgart
H. J. Höhling, Münster
K. H. Knese, Stuttgart

Preface

The *X. European Symposium on Calcified Tissues* took place in Hamburg from 16th to 21st September 1973. The financial backing came from the Gesundheitsbehörde der Freien und Hansestadt Hamburg (President Dr. Zylmann), the Bundesministerium für Jugend, Familie und Gesundheit, and from industry. This made it possible to carry out the scientific program, to invite several European and non-European scientists, and to organize a social program designed to establish personal contact between delegates and to make visitors acquainted with our country. In the name of the organizing committee, we herewith express our warmest appreciation of all the help given to us.

Professor Bartelheimer, as the representative of the Medical Faculty and on behalf of the President of the University of Hamburg, welcomed the Participants in the symposium to our city.

The history of these meetings began in Oxford in 1963, with the *First European Bone and Tooth Symposium,* organized by H.J.J. Blackwood, B.E.C. Nordin, and Dame Janet Vaughan. The idea was to found in Europe an institution similar to the American *Gordon Research Conferences of Bone and Tooth.* After the U.K., the host countries were Belgium, Switzerland, the Netherlands, France, Sweden, Italy, Israel and Austria. The second symposium in Liège already bore the present name. The original approach has persisted until today, that is, to have a limited number of participants, and to group the main topics with an introduction and single papers and with sufficient time allowed for discussion. It was always the intention to accommodate the delegates under one roof in order to encourage personal contact, and this we accomplished in the hotel Loews Hamburg-Plaza and the Congress Centrum. Thus, the *Symposium on Calcified Tissues* developed its own style, largely thanks to a group of enthusiastic colleagues. We are very lucky to have found an international coordinator in the person of Professor *Gaillard.*

The increasing research in the field of calcified tissues made it necessary to restrict the papers to the main topics proposed by us and accepted at the symposia in Israel and Austria. Preference was given to those adhering to the main clinical topics with which this symposium is principally concerned. It is hoped that this emphasis on clinical effects may stimulate clinicians in their work at the bedside, while scientists will recognize the problems present in the clinical aspects and see where the theoretical foundations are inadequate.

The Organizing Committee therefore chose *Methods for Analysing Bone Metabolism* as the first topic. These methods form the background to the diagnosis and differential diagnosis of the various bone diseases. Although research on *Vitamin D* has been in progress for some 50 years, recent advances in this field have been so great that, so soon after the *IX. European Symposium* in Baden in 1972, it seemed expedient to include this theme again. In the meantime, other fundamental aspects have been studied in connection with the pathogenesis of a number of diseases, and a start has been made in treatment with various *Vitamin D Metabolites*. In the light of expected progress, it is to be hoped that the dangerous and not always effective treatment with vitamin D_3 will be replaced by that with vitamin D metabolites.

Now that the efficacy of these metabolites has been established, it no longer seems appropriate to look upon D_3 as a vitamin. Today vitamin D is classed as a hormone.

For many participants representing widely different disciplines, *Fluoride* continues to be of vital interest. In internal medicine, its effects on bone metabolism and the resulting therapeutic applications, especially its use in osteoporosis, are most remarkable. However, long-term observation of patients with clearly defined osteoporosis will be necessary before a true assessment can be made of the value of this treatment.

In the field of *Collagen Research,* aspects of contemporary importance concern the mechanism of the invasion of the collagen structures by calcium and phosphate ions, as well as the question of active germ centers. Here the link with clinical work consists essentially in the investigation of the pathological forms of collagen associated with pathological bone formation. We may hope that this work will help to explain the pathogenesis of a number of diseases of bone that puzzle us today. Of course, this research is closely related to the *Biochemistry and Histochemistry of Bone Diseases,* the clinical consequences of which have yet to be realized. As the physician's goal is always successful treatment, we scheduled *Therapeutical Aspects of Bone Diseases* for the last day of the 1973 symposium. The meeting ended with the presentation of research results from areas not covered elsewhere, first and foremost the parathyroid hormone, which plays a central role in the whole complex story of calcified tissues.

For the help given in preparing for and carrying out the symposium, we should like to thank the members of the Department of Clinical Osteology of the First Medical University Clinic, Hamburg, especially CORINNA BORGARDT, MARIE-LUISE DANZ, EVA-MARIA HOBBJE, VERA HOHLWEG, FRANCES JOHNSON, Dipl. Chem. ELDA SOMMER, Dr. C. LOZANO-TONKIN, G. MAASS, Dr. J.-D. RINGE and Dr. A. v. ROTH. We express our thanks especially to INGEBORG KRUSE, who helped in the revision of the manuscripts for this book.

Hamburg, January 1975

FRIEDRICH KUHLENCORDT
HANS-PETER KRUSE

Table of Contents

List of Contributors

Aaron, J.E. MRC Mineral Metabolism Unit, The General Infirmary Leeds, (GB)

Aitken, J.M. Medical Professional Unit, St. Bartholomew's Hosp. London EC1A 7BE (GB)

Alevizaki, C.C. Department of Endocrinology and Radioisotopes, Evangelismos Hospital, Athen 140 (G)

Anderson, J.B. see Smith, D.A.

Armstrong, W.D. University of Minnesota, Department of Biochemistry, Medical School, 227 Millard Hall, Minneapolis, Minnesota 55455 (USA)

Arnaud, C.D. Mayo Clinic, Rochester, Minnesota 55901, (USA)

Ascenzi, A. Istituto di Anatomia e Istologia Patologica, Ospedale Policlinico Umberto I, Viale Regina Elena 324, 00161 Rom, (I)

Ashton, B.A. see Höhling, H.J.

Bachra, B.N. Laboratory for Physiological Chemistry, Wassenaarseweg 62, Leiden (NL)

Backer Dirks, O. see Sluys Veer, J. v. d.

Bärenholdt, O. Department of Clinical Physiology, Glostrup Hospital, 2600 Glostrup, Copenhagen (DK)

Balsan, S. Hôpital des Enfants malades, 149 Rue de Sèvres, 75730 Paris Cedex 15, (F)

Baud, C.A. Institut de Morphologie, Ecole de Médicine, 1211 Genf 4, (CH)

Bauer, G.C.H. Lund University, Department of Orthopaedic Surgery, S-22185 Lund, (S)

Bauze, R.J. see Francis, M.J.O.

Baxter, L.A. see Bonjour, J.-P.

Bellwinkel, S. see Ziegler, R.

Bijvoet, O.L.M. Afdeling Inwendige Geneeskunde, Academisch Ziekenhuis, Rijnsburgerweg 10, Nijmegen, (NL)

Blumenthal, N.C. see Posner, A.S.

Bohr H. Heslegardsvej 25, 2900 Hellerup Copenhagen, (DK)

Bonjour, J.-P., Pathophysiologisches Institut der Universität, Hügelweg 2, 3012 Bern, (CH)

Bon-Nijssen, K.L.M. see Rösingh, G.E.

Bonucci, E. Istituto Anatomia Patologica, Policlinico Umberto I, Viale Regina Elena 324, 00161 Rom, (I)

Bordier, Ph. Hôpital Lariboisière, Centre du Metabolisme Phosphocalcique, 6 Rue Guy Patin, 75 Paris 10e, (F)

Boskey, A.L. see Posner, A.S.

Bouillon, R. Rega Instituut, Minderbroederstraat 10, 3000 Leuven, (B)

Boyde, A. see Nicholson, W.A.P.

Brewer, B. National Institutes of Health, Bethesda, Maryland 20014 (USA)
Brezillon, P. see Perault-Staub, A. M.
Bronner, F. The University of Connecticut, Health Center, School of Dental Medicine, Department of Oral Biology, Farmington, Connecticut 06032 (USA)
Büttner, W. Universitäts-Klinik, Robert-Koch-Straße 27a, 44 Münster, (D)
Burckhardt, P. Clinique Médicale Universitaire (Hôpital Nestlé), 1000 Lausanne, (CH)

Copp, D.H. Department of Physiology, University of British Columbia, Vancouver 8, B.C. (CDN)
Czitober, H. Wiener Allgem. Poliklinik, I. Med. Abteilung, Mariannengasse 10, 1090 Wien (A)

Dambacher, M.A. Kantonsspital, 4000 Basel, (CH)
Delling, G. Pathologisches Institut der Universität, Martinistr. 52, 2 Hamburg 20, (D)
DeLuca, H.F. University of Wisconsin, Department of Biochemistry, Madison, Wisconsin 53706 (USA)
Deutsch, D. see Weatherell, J.A.
Doyle, F.H. see Greenberg, P.B.
Dryburgh, F.J. see Morgan, H.G.
Dulce, H.J. Freie Universität Berlin, Institut für Klinische Chemie und Biochemie, Hindenburgdamm 30, 1 Berlin 45 (D)
Dymling, J.-F. Department of Endocrinology, General Hospital, 21401 Malmö 8, (S)

Eastwood, J.B. Department of Medicine, New Charing Cross Hospital, London W 6, (GB)
Eilon, G. Research Laboratory, Shaare Zedek Hospital, Jerusalem (IL)
Ericsson, Y. Karolinska Institut, Odont. Fak. Holländargatan 17, 10364 Stockholm (S)
Eyre, D.R. Kennedy Institute of Rheumatology, Bute Gardens, Hammersmith, London W 6, (GB)

Felber, J.-P. see Burckhardt, P.
Ferguson, H.W. Department of Dental Sciences, School of Dental Surgery, The University of Liverpool, Pembroke Place, POBox 147, Liverpool L69 3BX (GB)
Fisher, M.T. see Greenberg, P.B.
Fleisch, H. Universität Bern, Pathophysiologisches Institut, Hügelweg 2, 3012 Bern (CH)
Foster, G.V. Department of Physiology, Rm 211 B.S.B., Johns Hopkins University, School of Medicine 725 N. Wolfe Street, Baltimore, Md. 21205, (USA)
Francis, M.J.O. Nuffield Department of Orthopaedic Surgery, Nuffield Orthopaedic Centre, Headington, Oxford OX 3 7LD (GB)
Freund, T. see Bronner, F.

Gaillard, P.J. Laboratory for Cell Biology and Histology, 10 Rijnsburgerweg, Leiden (NL)
Gallagher, J.C. see Nordin, B.E.C.

Garabedian, M. Hôpital des Enfants malades, 149 Rue de Sèvres, 75730 Paris Cedex 15, (F)
Gildersleeve, D.L. see Foster, G.V.
Glimcher, M.L. see Eyre, D.R.
Gore, L.F. Department of Dental Sciences, School of Dental Surgery, The University of Liverpool, Pembroke Place P.O. Box 147, Liverpool L69 3BX (GB)
Greenberg, P. Department of Endocrine Chemistry, Royal Postgraduate Medical School, Ducane Rd. London W12 OHS (GB)
Guncaga, J. Kantonsspital Basel, I. Med. Univ. Klinik, Stoffwechselabteilung, 4004 Basel (CH)

Haas, H.G. Kantonsspital Basel, I. Med. Univ. Klinik, Stoffwechselabteilung, 4004 Basel (CH)
Hagen, A.R. University of Oslo, Fak. Odontology, Geitmyrsvn 71, Oslo 4 (N)
Hart, G. see Rösingh, G.E.
Hartles, R.L. see Gore, L.F.
Hayashi, Y. see Igarashi, M.
Heitz, P. see Remagen, W.
Henderson, R.G. Renal Unit, Churchill Hospital, Oxford (GB)
Henschler, D. Institut für Pharmakologie und Toxikologie der Universität, 8700 Würzburg, Koellikerstr. 2 (D)
Hermsdorf, C.L. Department of Oral Biology, School of Dental Medicine, University of Connecticut, Health Center, Farmington, CT. 06032 (USA)
Hill, L.F. Department of Medicine, The Royal Infirmary, Oxford Road, Manchester (GB)
Hillyard, C. Endocrine Unit, Royal Postgraduate Medical School, Ducane Road, London W 12 OHS (GB)
Hioco, D. Service de Rhumatologie, Hôpital Henri-Mondor, 94010 Créteil (F)
Höhling, H.J. Westfälische Wilhelms-Universität, Institut für Medizinische Physik, Hüfferstr. 68, 44 Münster (D)
Horsman, A. see Nordin, B.E.C.
Houwink, B. see Sluys Veer, J. van der
Howell, D.S. University of Miami, School of Medicine, P.O. Box 875, Biscayne Annex, Miami, Florida 33152 (USA)

Igarashi, M. Tokyo Metropolitan Geriatrics Hospital, Department of Orthopedic Surgery, 35-2 Sakaecho, Itabashiku, Tokyo 173 (J)
Ikkos, D.G. Department of Endocrinology, Evangelismos Hospital, Athen 140 (G)

Johnson, R.L. see Norman, A.W.
Joplin, G.F. Department of Medicine, Royal Postgraduate Medical School, Ducane Road, London W 12 OHS (GB)
Jowsey, J. Mayo Clinic, Rochester, Minnesota 55901 (USA)

Kelly, D.T. see Foster, G.V.
Kelly, P.J. see Jowsey, J.
Kivisaari, J. see Niinikoski, J.
Krempien, B. Pathologisches Institut der Universität, Berliner Str. 5, 69 Heidelberg (D)

Kuettner, K. see Howell, D.S.
Kuo, Chi-Li, see Foster, G.V.

Larkins, R.G. see MacAuley, S.J.
Lauffenburger, T. Med. Univ. Klinik, Bürgerspital, 4000 Basel, (CH)
Ledingham, J.G.G. see Henderson, R.G.
Lefevre, C. see Perault-Staub, A.M.
Lemperg, R. see Rosenquist, J.
Lentner, C. see Haas, H.G.
Lippiello, L. see Wasserman, R.H.
Lüthy, H. see Remagen, W.

MacAuley, S.T. Endocrine Unit, Royal Postgraduate Medical School, Hammersmith Hospital, Ducane Road, London W 12 (GB)
MacIntyre, I. Department of Chemical Pathology, Endocrine Unit, Royal Postgraduate Medical School, Hammersmith Hospital, Ducane Road, London W 12 OHS (GB)
Martin, T.J. see MacIntyre, I.
Maruna, R.F.L. Universität Wien, Hygiene-Institut, Kinderspitalgasse 15, 1095 Wien (A)
Matsuura, M. Tokyo Metropolitan Institute of Gerontology, Department of Chim. Biochemistry 35-2 Sakaecho, Itabashiku, Tokyo 173, (J)
Mawer, E.B. Division of Metabolism, Department of Medicine, Manchester Royal Infirmary, Oxford Road, Manchester M 12 9WL (GB)
McLoughlin, J.L. see Greenberg, P.B.
Mechanic, G.L. Dental Research Center, University of North Carolina, Chapel Hill, N.C. 27514 (USA)
Melman, A.P.N. see Sluys Veer, J.v.d.
Menczel, J. Shaare Zedek Hospital, Department of Medicine, Jerusalem, (IL)
Messer, H. see Armstrong, W.D.
Milhaud, G. Université de Paris, Faculté de Médecine Saint-Antoine, Service de Biophysique, 27 Rue de Chaligny, Paris 12ᵉ (F)
Mills, E.A. see Morgan, H.G.
Minne, H. see Ziegler, R.
Mitra, M.N. see Norman, A.W.
Morgan, H.G. Department of Pathology and Biochemistry, The Royal Infirmary, Glasgow, G4 OSF (GB)
Munck, O. see Nielsen, S.P.

Nichols, W.W. see Foster, G.V.
Nicholson, W.A.P. see Höhling, H.J.
Nielsen, S.P. Glostrup Hospital, Department of Clinical Physiology, 2600 Glostrup, (DK)
Niinikoski, J. Department of Medical Chemistry, University of Turku, 20520 Turku 52 (SF)
Nikolaou, A. see Ikkos, D.G.
Nordin, B.E.C. General Infirmary Leeds, Great George Street, Leeds 1 (GB)
Norman, A.W. Department of Biochemistry, University of California, Riverside, Ca. 92502 (USA)

Okamura, W.H. see Norman, A.W.
Olah, A.J. Universität Bern, Anatomisches Institut, Bühlstraße 26, 3012 Bern (CH)

Oliver, D.O. see Henderson, R.G.
Otto, H.F. see Schäfer, H.-J.

Parsons, J.A. National Institute for Medical Research, Mill Hill, London NW 7 (GB)
Patz, J. see Büttner, W.
Pautard, F.G.E. MRC, Mineral Metabolisme Unit, The General Infirmary, Great George Street, Leeds LS1 3EX (GB)
Pavlovitch, H. see Garabedian, M.
Pearson, T.A. see Foster, G.V.
Pennock, J. see Greenberg, P.B.
Perault-Staub, A.M. Laboratoire de Isotopes, C.H.U. Saint-Antoine, 27 Rue Chaligny, 75 Paris 12e (F)
Pita, J.C. see Howell, D.S.
Poli, S. see Burckhardt, P.
Posner, A.S. Cornell University, Hospital for Special Surgery, College of Medicine, 535 East, 70th Street, New York, New York 10021 (USA)
Pot, T. see Sluys Veer, J. v. d.
Potts, Jr., J.T. Massachusetts General Hospital, Endocrine Unit, Boston, Massachusetts 02114 (USA)
Pouëzat, J.A. Institut de Morphologie, Ecole de Médecine, 20 Rue de l'Ecole de Médecine, 1211 Genf 4 (CH)
Pozo, E. del, Sandoz AG. Abt. Experimentelle Therapie, 4002 Basel (CH)
Preston, C. see Russell, R.G.G.

Rafferty, B. see Reit, B.
Rapoport, A. see MacAuley, S.J.
Reit, B. The National Institute for Biological Standards and Control, London N.W. 3 (GB)
Remagen, W. Institut für Pathologie der Universität, Schönbeinstr. 40, 4056 Basel (CH)
Reutter, F.W. Medizinische Klinik, Kantonsspital, 9006 St. Gallen (CH)
Riggs, C. Mayo Medical School, Mayo Clinic, Rochester, Minnesota 55901 (USA)
Robinson, C. see Weatherell, J.A.
Rösingh, G.E. University of Amsterdam, Department of Orthopaedics, Amsterdam (NL)
Rosenquist, J. Department of Oral Surgery, University of Umea, 90187 Umea (S)
Roycroft, J.H. see Talmage, R.V.
Russell, R.G.G. Nuffield Orthopaedic Centre, Headington, Oxford OX 3 7LD (GB)

Schäfer, A. Zentrum für Innere Medizin, Universitäts-Klinik, Steinhövelstraße 9, Safranberg, 79 Ulm (D)
Schäfer, H.-J. Pathologisches Institut der Universität, Martinistraße 52, 2 Hamburg 20 (D)
Schenk, R.K. see Haas, H.G.
Schoot, J.B.v.d. see Rösingh, G.E.
Schulz, A. Pathologisches Institut der Universität, Martinistr. 52, 2 Hamburg 20 (D)
Schuppler, J. see Remagen, W.

Segre, V. Massachusetts General Hospital and Harvard Medical School, Endocrine Unit, Department of Medicine, Boston, Massachusetts 02114, (USA)
Shimmins, J. see Smith, D.A.S.
Singer, L. see Armstrong, W.D.
Sluys Veer, J.v.d. Department of Internal Medicine, University Hospital, Catharijnesingel 101, Utrecht (NL)
Smith, D.A.S. Bone Metabolism Research Unit, Western Infirmary, Glasgow GII 6NT, (GB)
Smith, R. Nuffield Orthopaedic Centre, Headington, Oxford OX3 7LD (GB)
Sognnaes, R.F. University of California, School of Dentristry, Centre Health Sciences, Los Angeles, California 90024 (USA)
Speirs, R.L. London Hospital Medical College, Turner Street, London E1 2AD (GB)
Staub, J.F. see Perault-Staub, A.M.
Sugár, E. see Tóth, K.

Talmage, R.V. 327 Swing Building, UNC School of Medicine, Chapel Hill, N.C. 27514 (USA)
Tannenbaum, P.J. see Posner, A.S.
Taylor, C.M. Division of Metabolism, Department of Medicine, Manchester Royal Infirmary, Oxford Road, Manchester M 13 9WL (GB)
Tóth, K. University Medicine School, Department of Dentristry and Oral Surgery, Lenin Körut 64, 6720 Szeged, (H)
Trechsel, U. Pathologisches Institut der Universität, Hügelweg 2, 3012 Bern (CH)
Tröhler, U. see Bonjour, J.-P.
Trojan, E. Unfallchirurgie I, Alserstraße 4, 1097 Wien (A)

Verberckmoes, R. see Bouillon, R.

Walton, R.J. see Russell, R.G.G.
Ward, M.K. see Greenberg, P.B.
Wardener, H.E. de, see Eastwood, J.B.
Warner, G.T. see Henderson, R.G.
Wasserman, R.H. Cornell University, Department of Physiology and Biology, New York State Vet. College, Ithaca, New York 14850 (USA)
Weatherell, J.A. University of Leeds, Dental School, Biological Research Unit, Leeds (GB)
Weidmann, D. see Remagen, W.
Woodhouse, J.J.Y. see Greenberg, P.B.
Woods, C.G. see Russell, R.G.G.

Yoshino, S. Tokyo Metropolitan Ohtsuka Hospital, Department of Rheumatological Diseases, 2-8 Minami Ohtsuka, Toshimaku, Tokyo, (J)

Ziegler, R. Universität Ulm, Zentrum für Innere Medizin und Kinderheilkunde, Steinhövelstraße 9, 79 Ulm (D)

I. Methods for Analyzing Bone Metabolism — Progress and Critical Remarks

Chairmen: O. L. M. Bijvoet and B. E. C. Nordin

Morphological and Crystallographic Analysis of Bone Mineral*

C.A. Baud and J.A. Pouëzat

In the past ten years, a series of physical methods have been applied to the quantitative study of bone mineral. The amount of mineral in bone, its topographical distribution and fine structure, are important parameters which are of interest for many reasons, and may have a significant place in the study of human bone physiology and pathophysiology. This data could be used, for example, to diagnose and follow up various bone diseases. Summarized briefly in this paper are some of the significant results obtained with these methods.

I. Measurements at Macroscopic Level

An improved technique for accurately measuring bone mineral content directly *in vivo*, using photon absorptiometry, was developed by SORENSON and CAMERON (1967). The principal advantages of this method as compared with X-ray densitometry are the following: the radiation originates in a radioactive source and is monoenergetic, the beam is well collimated and diffused radiation is eliminated, the beam intensity is measured by directly counting transmitted photons.

Using this method, MAZESS (1970) measured the bone mineral content of the radius shaft in a control group of over 1.000 human males and females of different ages. There was an increase in the bone mineral mass up to age 30. A decline in bone mineral began in general at about 35 to 40 years of age in both males and females, but the decrease was much steeper in the latter.

According to the observations of GOLDSMITH et al. (1971) the measurement of average mineral content in the distal part of the left radius separates control group women from the osteoporotics with sufficient distinctness for diagnostic purposes.

The most interesting and useful application of this simple procedure is the study, by repeated measurements, of the changes of the bone mineral mass in the same patient over a period of weeks, months or even years.

In this way, changes in mineral content in os calcis with increasing times of bed rest were studied by RAMBAUT et al. (1972); the measurements showed a progressive decrease with a rate differing in different individuals. Bone loss during pregnancy and lactation, as well as bone loss reversal following removal of parathyroid adenoma were measured by SORENSON and CAMERON (1967).

+ The author's work described in this report was supported by grants from the Fonds National Suisse de la Recherche Scientifique and by the Fondation Sandoz pour l'Avancement des Sciences Médicales et Biologiques.

Tibial fractures, as observed by NILSSON (1966), result in a loss of about 25% of the bone mineral in the knee region, which even after a number of years, is not recovered. Similarly, after lesions of the semilunar cartilage, NILSSON and WESTLIN (1969) noted that the bone mineral mass in the distal end of the femur is decreased by approximately 9%, and there is no tendency toward restoration of the bone mass over the years.

This method is of somewhat limited interest, however, as it gives information only on the bone mineral mass at the macroscopic level, and not on the degree of mineralization of the bone tissue.

II. Measurements at Microscopic Level

1. Degree of Mineralization of the Bone Tissue

In order to obtain reliable quantitative data concerning the overall distribution of the mineral content in a bone tissue sample, it is necessary to measure a significant number of areas. To do this, a combined microradiographic and microdensitometric technique is used. The microradiograph of the undecalcified bone section alongside an aluminium step-wedge is performed. The thickness of the bone section must be uniform to 100 ± 1 micron. The X-ray beam must be very homogeneously distributed, which can be done by means of a rotating camera. A specially designed automatic scanning microdensitometer with pulse height analyzer (BAUD, 1969), is capable of analyzing the distribution and area of blackenings on the microradiographic film. After appropriate calibration, 10.000 points are measured in less than a minute. The digitized results appear directly as a histogram which shows the degrees of mineralization of bone in grams of mineral substance per cubic-centimeter, expressed as a percentage of the total sample volume. An average value of the degree of mineralization can easily be calculated from the histogram.

This technique has been used to measure changes in mineral density of bone tissue in control subjects of varying ages. The mean value of mineralization of compact bone tissue in the iliac crest of women increases slightly with age. The coefficient of variation for bone mineral content at specific ages is 10 per cent, except in those of over 45 years of age, in whom it is 12-15 per cent (Fig.1). The histograms which represent the percentage of bone volume occupied by the more or less mineralized tissue show that a larger part of the bone volume is only slightly mineralized at the two extremes of the age scale than in the middle range. In the case of the young, this can be explained by a very active remodeling of the bone, and in that of elderly women, by a slower mineralization process.

A comparison of bone from patients with postmenopausal and senile osteoporosis with that of control subjects of the same age revealed that the mean value of bone mineralization of most osteoporotic patients, although within the control range, is closer to the upper limits. In a small number of cases, the mineralization is below the control range for the age group (Fig.2); this was also observed by JOWSEY and GERSHON-COHEN (1964) in two cases, as well as by BOHR (1964) in one case. However, in a group of osteoporotic patients between 56 and 65 years of age, several cases were observed, where the mineralization was above

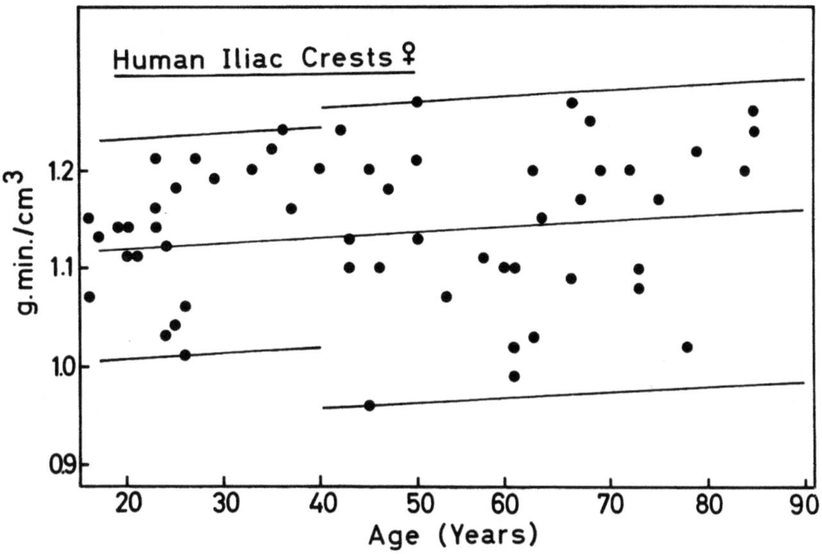

Fig. 1. Changes with age of the average mineral density of compact bone tissue in the iliac crest of control group women

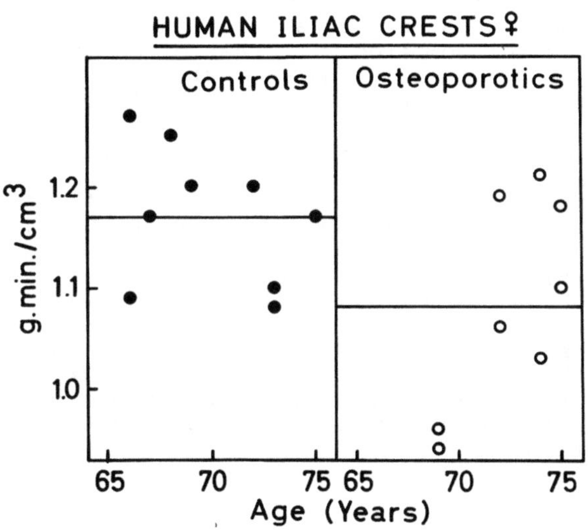

Fig. 2. Average mineral density of compact bone tissue in the iliac crest of women with clinically diagnosed osteoporosis (o) compared with that of a control group of women of the same age (●)

the limits of the control range. These latter results should be considered along with those of VOSE (1962) whose studies have shown that in several cases the mean degree of mineralization of extra-haversian bone tissue is higher in osteoporotic than in nonpathologic bone. These findings suggest that there are, in fact, different types of osteoporosis which might perhaps be distinguishable on the basis of differences in mineralization of bone tissue.

The changes with time in the mineralization of bone tissue in osteoporosis under treatment can be observed by quantitative microradiographic study of successive biopsies of the same patient. In this way, LANGER et al. (1971) were able to show that in osteoporotics, there is an increase of 3 to 6.5% in the degree of mineralization of bone tissue after a prolonged treatment with calcitonin.

The low degree of mineralization of bone tissue observed in certain experimental and pathological conditions, has, in fact, been found to be due to a loss of mineral from once normally mineralized bone. This important phenomenon of diffuse demineralization without concomitant resorption of the organic matrix, can be explained by the instability of the mineral substance, and for that reason is called "halastasis". It should be taken into consideration along with osteoclastic resorption and osteolytic or perilacunar resorption as a third aspect of bone loss. This process of demineralization was first observed under experimental conditions, as, for example, where there is hormone imbalance induced in eels by long-term administration of total pituitary extract (LOPEZ, LEE and BAUD, 1970).

As to pathological conditions in man, in post-traumatic osteoporosis or Sudeck's disease, the bone which existed prior to the lesion becomes demineralized. The measurements of BAUD and POUËZAT (1973) show that the mean degree of mineralization of this bone tissue is 6 to 7 per cent less than in control bone. The histogram of the distribution of bone substance as a function of its degree of mineralization is modified; compared with that of control bone, it has shifted toward the lower mineralizations (Fig.3). The heavily mineralized areas have disappeared, demonstrating an actual loss of mineral.

In patients suffering from industrial fluorosis, the degree of mineralization of bone can be different from that of control subjects of corresponding age groups. When the fluorine in the bone substance is at a level ranging between 0.4 and 0.8%, the mean degree of mineralization is higher than that in control subjects. In cases of severe fluorosis, however, when the fluorine level is above 0.8%, the mineralization is not significantly increased (Fig. 4).

These findings clearly show that the quantitative microradiography method, by providing data on the degree of mineralization of the tissue at the microscopic level, is capable of giving us important information regarding differences of mineralization in different pathological conditions.

2. Chemical Composition

Microradiography only indicates the level of bone mineralization, and gives no information about chemical composition. Since bone is a composite tissue, it is important to know the chemical composition at the level of microscopic structures. Such an investigation has been made possible by the use of the electron probe X-ray microanalyzer. With this instrument, a precise non-destructive elemental analysis can be performed on localized regions with diameters as small as 1 micron. The detectable concentration of elements is very low, less than 0.01%.

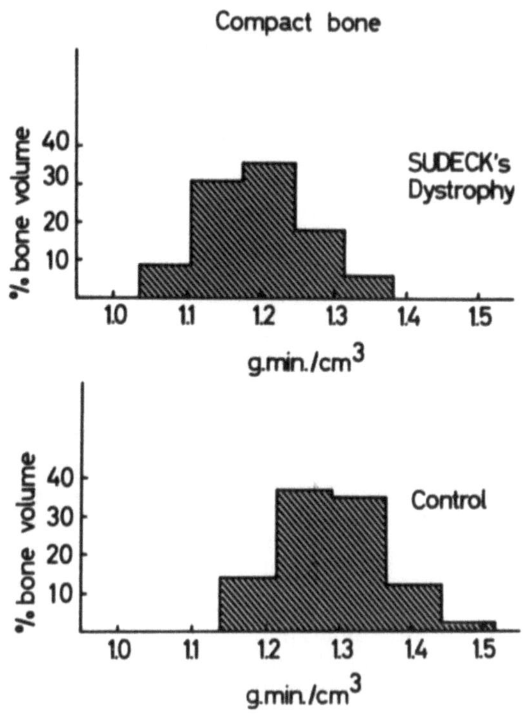

Fig. 3. Histograms showing distribution of mineral substance in percentage volume of compact bone tissue of a patient with Sudeck's dystrophy (top) and that of a control subject of the same age group (bottom)

The results concerning the topographical distribution of calcium and phosphorus, the principal components of bone mineral, have confirmed the microradiographic data. In particular, the variations in the concentration of calcium in the lamellar systems of the osteon have been observed by BAUD, KIMOTO and HASHIMOTO (1963).

The Ca:P molar ratio is far below that of the apatite during the early stages of mineralization, approaching it only in fully mineralized areas, according to the measurements of HÖHLING et al. (1967). The mean mass concentration of sodium and magnesium in the mineral phase shows a difference in distribution between osteons and interstitial bone, the latter having a lower content according to WOLLAST and BURNY (1971).

The distribution pattern of strontium deposited in the skeletons of mice on strontium-rich diets was observed by BANG and BAUD (1972). It was found to be almost identical with that of bone-seeking radioisotopes detected in autoradiographs as "hot spots" and "diffuse component".

Even the light elements such as fluorine can be detected with this method. Topographical distribution of fluorine in the bones of fluoride-treated osteoporotic patients was observed in this way by BAUD and BANG (1970). The bone layers formed during the treatment showed a high

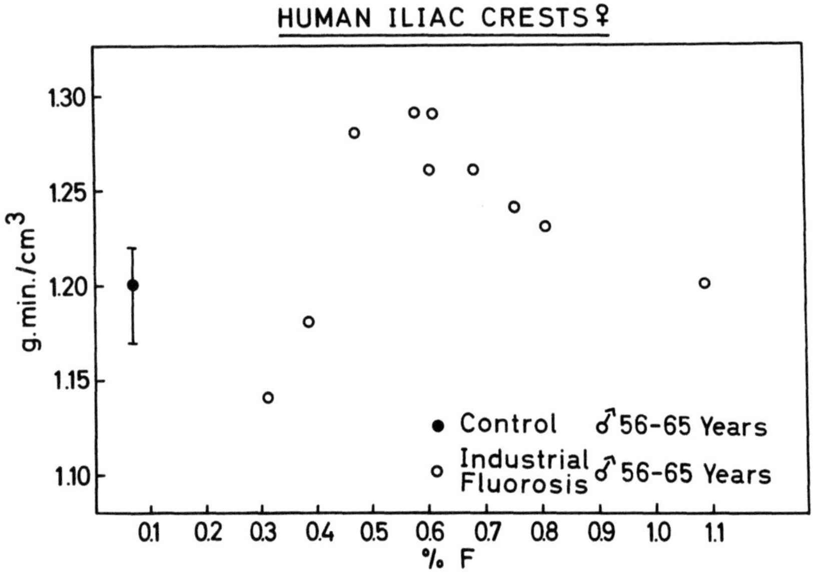

Fig. 4. Relationship between average mineral density and fluorine content of compact bone tissue in the iliac crest of control group men (•) and that of male patients of the same age group suffering from industrial fluorosis (o)

level of fluoride, whereas that existing before the beginning of the fluoride therapy showed only a small amount. Similar results were found by BAUD and BANG (1970) observing experimentally induced fluorosis in mice. In industrial fluorosis there is a striking correlation between the lesions of mottled osteons and the distribution of high fluoride zones, as observed by BAUD and BANG (1972).

The simultaneous detection of several elements can make certain interesting correlations evident. In fluorotic rabbit bone, semi-quantitative analyses by BANG , WEST and BAUD (1974) indicated an increasing Ca:P ratio with increase in fluoride concentration.

III. Measurements at Submicroscopic Level

1. Amorphous and Crystalline Material

At the submicroscopic level, it has recently become obvious that the bone mineral substance consists of two phases: an amorphous calcium phosphate and an apatite in small crystals.

According to TERMINE and POSNER (1967) a quantitative evaluation of the amorphous phase may be obtained from the magnitude of discrepancy between the observed and expected intensity values of the X-ray diffraction lines of the apatite phase, or from the measurement of the splitting of the P-O antisymmetric bending mode at 550-600 cm^{-1} in the infrared absorption pattern. By using these procedures, young bone was shown to be richer in amorphous calcium phosphate than in apatitic, and mature bone richer in crystalline phase than in amorphous.

To date, the amorphous phase in human bone has not been quantitatively evaluated in various age groups, since it becomes unstable after death, and necroptic material may therefore not be used. However, in the para-articular ossifications of paraplegics, BAUD, POUËZAT and VERY (1973) have shown that the amorphous calcium phosphate represents a very large part of the mineral. This noncrystalline phase hardly changes during the first 12 months of the evolution of the lesion and represents on the average 58% of the mineral. The percentage decreases slowly thereafter, and even after 30 months of evolution it is still at 40%, which is high compared with control adult bone.

Several changes in the mineral substance have been induced experimentally. In the rat, for example, as observed by TERMINE and POSNER (1967), after hypophysectomy, the relative proportion of amorphous mineral is high: this can be explained by a maturation defect. Similar observations were made by RUSSEL, TERMINE and AVIOLI (1973) studying rats with experimentally induced chronic renal insufficiency.

In the eel, in cases of hormone imbalance induced by injections of total pituitary extract, LOPEZ, LEE and BAUD (1970) have shown that the slow and progressive demineralization of the skeleton is essentially due to a loss of the amorphous component, while the crystalline component remains unchanged. This diffuse demineralization at the expense of the amorphous component characterises the phenomenon of "halastasis".

Preliminary studies of the nature of bone mineral in experimental conditions of massive and rapid resorption as induced by the parathyroid hormone (POSNER, 1967) or calciphylaxis (unpublished results of our laboratory) show that there is no preferential loss of either the amorphous or the crystalline portion of the mineral; the bone loss resulting from these conditions is non-selective with regard to the two calcium phosphate phases.

2. Crystal Size

Another field of investigation at the submicroscopic level with regard to bone mineral is that of the size of apatite crystals. The crystal size can be evaluated by measuring the width of the X-ray diffraction lines at half intensity or, more accurately, by means of the template method of POSNER et al.(1963) and the small-angle diffraction analysis of EANES et al.(1965).

The results show that the average apatite crystal size increases as the bone tissue becomes older. This explains why the average apatite crystal size is very small in para-articular ossifications of paraplegics studied by BAUD, POUËZAT and VERY (1973).

When the maturation of the organic matrix of bone is experimentally impaired, as in dietary deficiencies (TERMINE and POSNER, 1967) and chronic uremic state (RUSSEL, TERMINE and AVIOLI, 1973), the maturational sequence of bone apatite is also altered, and the crystal size remains small.

The crystallites in the untreated patients suffering from various types of osteoporosis and from osteogenesis imperfecta are normal in size according to the observations of GRØN, McCANN and BERNSTEIN (1966).

The ingestion of fluoride causes an increase in the average size of the bone mineral particles. As the fluoride content of the bone increases from 0 to 1.2%, there is a progressive increase in crystal size. When the fluoride level rises from 1.2 to 2%, however, the rate of increase becomes very slight (BAUD and MOGHISSI-BUCHS, 1965a). This phenomenon has been observed in endemic fluorosis in man (POSNER et al., 1963; EANES et al., 1965; BERNSTEIN and COHEN, 1967), and can be experimentally reproduced in animals.

In various pathological conditions, such as osteoporosis and Paget's disease, fluoride therapy brings about a significant increase in crystal size, as observed by BERNSTEIN and COHEN (1967), which is related to increased fluoride concentration in the bone mineral. This increase results in an improved chemical stability of the mineral system, due to the fact that with an increase in crystal size there is corresponding decrease in specific surface and that the rate of reaction between a solid and a solution is affected by the specific surface.

IV. Measurements at Atomic Level

Although electron probe X-ray microanalysis can detect the presence of elements, and can indicate with precision their concentration and their localization at the level of microscopic structures, it is unable to determine whether these elements are incorporated in the crystaline apatitic phase, adsorbed on the surface, or constitute a separate phase.

The X-ray diffraction method, however, allows for this determination since the progressive replacement of one species of ion by another causes a change in the dimensions of the unit cell proportionate with the concentration of the substituent. The order of magnitude of this change is between one thousandth and one hundredth of an angström. For the measurements, it was therefore necessary to develop a refined diffractometry method (BAUD et al., 1968), including step by step recording of diffraction lines and calculations of the peak positions by means of a computer programmed for the determination of the curves best fitting the experimental points. The measurement of the _a_ length of the apatite unit cell of the bone mineral was obtained in this way with a precision of about 0.001 Å.

It has been possible to demonstrate that the ingested fluoride which settles in the skeleton is incorporated in the apatite crystal lattice where it replaces the hydroxyl ions. The _a_ parameter of the bone mineral substance diminishes in proportion to the increase in fluorine content. When the fluorine content reaches 2 %, _a_ is shortened by 0.047 Å. These observations were made on the bone mineral substance of bovines (BAUD and SLATKINE, 1962; BAUD and BANG, 1972), of mice (BAUD and BANG, 1973), of humans with endemic or industrial fluorosis (POSNER et al. 1963; BAUD and BANG, 1971), as well as that of fluoride-treated osteoporotic patients (THIEBAUD et al., 1970). The effect of this bonding of fluoride in the crystal lattice is to increase the stability of the crystals and to decrease their solubility by a diffusion-blocking mechanism, as suggested by YOUNG, van der LUGT and ELLIOT (1969).

The strontium, administered orally or by injection, is also incorporated in the apatite crystal lattice and replaces the calcium ions. In this case, the _a_ parameter of the bone mineral substance increases in length along with the increase in the quantity of strontium. It increases by 0.05 Å when the percentage of calcium atoms replaced by strontium atoms is 9.5, as observed by BAUD and MOGHISSI- BUCHS (1965b), BAUD et al. (1968), BANG and BAUD (1972). The incorporation of strontium in the bone mineral lattice is of great interest for two reasons. It explains, first of all, the possibility of using strontium as a tracer element for calcium in metabolic studies, and secondly, the impossibility of removing the radioactive strontium bonded in the skeleton without dissolving the crystals.

Most of the carbonate groups present in the bone mineral crystals are substituted for phosphate ions, because the a parameter of the unit cell becomes shorter to the extent that the carbonate content increases, according to the observations of BAUD and LEE (1969, 1970). The amount of carbonate thus incorporated in the crystal lattice depends on humoral conditions at the time of crystallization. The highest levels are observed in animal species like the turtle where the CO_2 pressure and bicarbonate concentration are extremely high in the body fluids.

The possible presence, even only transitory, of other crystalline phases in the bone mineral, as that of calcite observed in the medullary bone of laying hens by LÖRCHER and NEWESELY (1969), of octacalcium phosphate or other acid phosphates such as brushite and monetite (LENART, BIDLO and PINTER, 1972; MUENZENBERG and GEBHARDT, 1973) requires further investigation.

Summary

The quantitative study of the bone mineral substance by biophysical methods provides much information. The bone mineral mass at the macroscopic level, the degree of mineralization of the bone tissue and the topographical distribution of the elements at the microscopic level, the amorphous-crystalline relationships and the crystal size at the submicroscopic level, and the ionic substitutions at atomic level, can all be evaluated. The variations of these parameters in different pathological or experimental conditions make it possible to distinguish significant characteristics which would be useful in the diagnosis of various metabolic bone diseases, and in the monitoring of their treatments.

References

BANG, S., BAUD, C.A.: Topographic distribution of strontium and its incorporation into bone mineral substance in vivo. In: Proceedings of the Sixth International Conference on X-Ray Optics and Microanalysis (ed. G. SHINODA, K. KOHRA and T. ICHINOKAWA), pp. 841-845. University of Tokyo Press 1972.

BANG, S., WEST, V.C., BAUD, C.A.: An electron microprobe analysis of fluorotic rabbit bone. J.dent.Res. 53, 738-739 (1974).

BAUD, C.A., BANG, S.: Fluoride and bone mineral substance. In: Fluoride in Medicine (ed. T.L. VISCHER), pp. 27-36. Bern: Huber 1970.

BAUD, C.A., BANG, S.: Microscopic and submicroscopic localization of F in fluorotic bones. IADR Abstracts 49, 72 (1971).

BAUD, C.A., BANG, S.: Fluoride in bone tissue: electron probe and X-ray diffraction microanalysis. In: Proceedings of the Sixth International Conference on X-ray Optics and Microanalysis (ed. G. SHINODA, K. KOHRA and T. ICHINOKAWA), pp. 835-840. University of Tokyo Press 1972.

BAUD, C.A., BANG, S.: Biophysical studies of the bone tissue of fluoride-treated mice. J.dent.Res. 52, 589 (1973).

BAUD, C.A., BANG, S., LEE, H.S., BAUD, J.P.: X-ray studies of strontium incorporation into bone mineral in vivo. Proceedings of the Sixth European Symposium on Calcified Tissues, Lund 1968 (ed. J.F. DYMLING, G.C.H. BAUER). Calc.Tiss.Res. 2, suppl.6 (1968).
BAUD, C.A., KIMOTO, S., HASHIMOTO, H.: Étude de la distribution du calcium dans l'os haversien avec le radioanalyseur à microsonde électronique. Experientia 19, 524-525 (1963).
BAUD, C.A., LEE, H.S.: Étude par diffraction des rayons X de l'incorporation in vivo du carbonate dans la substance minérale osseuse. C.R.Acad.Sc.Paris, Série D, 268, 2956-2957 (1969).
BAUD, C.A., LEE, H.S.: X-ray diffraction study of carbonate incorporation in vivo into bone mineral substance. J.dent.Res. 49, 689 (1970).
BAUD, C.A. MOGHISSI-BUCHS, M.: Données radiocristallographiques sur les changements de cristallinité en rapport avec le taux de fluor incorporé in vivo dans la substance minérale osseuse. C.R.Acad.Sc. Paris 260, 1793-1794 (1965a).
BAUD, C.A. MOGHISSI-BUCHS, M.: Étude par diffraction des rayons X de la fixation in vivo du strontium dans la substance minérale osseuse. C.R.Acad.Sc.Paris 260, 5390-5391 (1965b).
BAUD, C.A., POUËZAT, J.A.: Données microradiographiques quantitatives sur le minéral osseux dans l'ostéoporose post-traumatique. Rhumatologie 25, 75-77 (1973).
BAUD, C.A., POUËZAT, J.A., VERY, J.M.: Ultrastructure and distribution of mineral salts in POA tissue. In: Current facts on Paraosteo-arthropathy (POA) (ed. A.B. ROSSIER et al.). Paraplegia 11, 70-71 (1973).
BAUD, C.A., SLATKINE, S.: Données radiocristallographiques sur l'incorporation du fluor dans la substance minérale osseuse in vivo. C.R.Acad.Sc.Paris 255, 1801-1802 (1962).
BAUD, J.P.: Photométrie automatique de microradiographies. Thesis, University of Geneva School of Physics 1969.
BERNSTEIN, D.S., COHEN, P.: Use of sodium fluoride in the treatment of osteoporosis. J.clin.Endocrinol.Metab. 27, 197-210 (1967).
BOHR, H.: Chemical analyses and microradiographic investigations on bone biopsies from cases of osteoporosis and osteomalacia as compared with normal. In: Bone and Tooth, Proceedings of the First European Symposium (ed. H.J.J. BLACKWOOD), pp. 405-409. Oxford: Pergamon 1964.
EANES, E.D., ZIPKIN, I., HARPER, R.A., POSNER, A.S.: Small-angle X-ray diffraction analysis of the effect of fluoride on human bone apatite. Arch.oral Biol. 10, 161-173 (1965).
GOLDSMITH, N.F., JOHNSTON, J.O., URY, H., VOSE, G., COLBERT, C.: Bone-mineral estimation in normal and osteoporotic women. J. Bone Joint Surg. 53-A, 83-100 (1971).
GRØN, P., McCANN, H.G., BERNSTEIN, D.: Effect of fluoride on human osteoporotic bone mineral. J. Bone Joint Surg. 48-A, 892-898 (1966).
HÖHLING, H.J., HALL, T.A., BOOTHROYD, B., COOKE, C.J., DUNCUMB, P., FITTON- JACKSON, S.: Untersuchungen der Vorstadien der Knochenbildung mit Hilfe der normalen und elektronenmikroskopischen Electron Probe X-ray Microanalysis. Naturwissenschaften 54, 142-143 (1967).
JOWSEY, J., GERSHON-COHEN, J.: Clinical and experimental osteoporosis. In: Bone and Tooth, Proceedings of the First European Symposium (ed. H.J.J. BLACKWOOD), pp. 35-48. Oxford: Pergamon 1964.
LANGER, B., DE SIEBENTHAL, J., BAUD, L., BAUD, C.A., MACH, R.S.: Effets de l'administration prolongée de diverses calcitonines dans l'ostéoporose humaine. Etude clinique, histologique et microradiographique. Ann. Méd. interne 122, 149-155 (1971).
LENART, G., BIDLO, G., PINTER, J.: Some basic problems in the examination of the calcium hydrogen phosphates of bone. Clin. Orthop. 83, 263-272 (1972).

LOPEZ, E., LEE, H.S., BAUD, C.A.: Étude histophysique de l'os d'un Téléostéen, Anguilla anguilla L., au cours d'une hypercalcémie provoquée par la maturation expérimentale, C.R.Acad.Sc.Paris, Série D, 270, 2015-2017 (1970).
LÖRCHER, K., NEWESELY, H.: Calcium carbonate (Calcite) as a separate phase besides calcium phosphate apatite in medullary bone of laying hens. Calc.Tiss.Res. 3, 358-362 (1969).
MAZESS, R.B.: Discussion on Dr. ROSE's paper. In: Osteoporosis(ed. U.S. BARZEL), p. 130. New York: Grune and Stratton 1970.
MUENZENBERG, K.J., GEBHARDT, M.: Brushite, octacalcium phosphate, and carbonate-containing apatite in bone. Clin.Orthop. 90, 271-273 (1973).
NILSSON, B.E.: Post-traumatic osteopenia. Acta orthop. scand. suppl. 41 (1966).
NILSSON, B.E., WESTLIN, N.E.: Osteoporosis following injury to the semilunar cartilage. Calc.Tiss.Res. 4, 185-187 (1969).
POSNER, A.S.: Relationship between diet and bone mineral ultrastructure. Fed.Proc. 26, 1717-1722 (1967).
POSNER, A.S., EANES, E.D., HARPER, R.A., ZIPKIN, I.: X-ray diffraction analysis of the effect of fluoride on human bone apatite. Arch. oral Biol. 8, 549-570 (1963).
RAMBAUT, P.C., DIETLEIN, L.F., VOGEL, J.M., SMITH, M.C.: Comparative study of two direct methods of bone mineral measurement. Aerosp. Med. 43, 646-650 (1972).
RUSSELL, J.E., TERMINE, J.D., AVIOLI, L.V.: Abnormal bone mineral maturation in the chronic uremic state. J.clin.Invest. 52, 2848-2852 (1973).
SORENSON, J.A., CAMERON, J.R.: A reliable in vivo measurement of bone-mineral content. J. Bone Joint Surg. 49-A, 481-497 (1967).
TERMINE, J.D., POSNER, A.S.: Amorphous/crystalline interrelationships in bone mineral. Calc.Tiss.Res. 1, 8-23 (1967).
THIEBAUD, M., ZENDER, R., COURVOISIER, B., BAUD, C.A., JACOT, C.: The action of fluoride on diffuse bone atrophies. In: Fluoride in Medicine (ed. T.L. VISCHER), pp. 136-142. Bern: Huber 1970.
VOSE, G.P.: Quantitative microradiography of osteoporotic compact bone. Clin.Orthop. 24, 206-212 (1962).
WOLLAST, R., BURNY, F.: Study of bone mineralization at the microscopic level using an electron probe microanalyser. Calc.Tiss.Res. 8, 73-82 (1971).
YOUNG, R.A., VAN DER LUGT, W., ELLIOTT, J.C.: Mechanism for fluorine inhibition of diffusion in hydroxyapatite. Nature 223, 729-730 (1969).

Quantitative Aspects of Calcium Metabolism and Homeostasis

F. BRONNER

I. Introduction

During the past twenty years kinetic analysis of calcium metabolism, i.e. the measurement of the major calcium fluxes in the whole animal with the aid of radiocalcium, has progressed to the point where there is general agreement on the usefulness and limitations of this approach. For example, true calcium absorption or the general rate of bone deposition can be estimated only by kinetic means, since other measurements, as intestinal flux in an Ussing chamber, or the count of resorption cavities in a section of bone, reflect specific local events and cannot readily be used to estimate tissue flux at the organ or whole animal level. Kinetic studies, on the other hand, do not point to the underlying molecular mechanisme, though they may serve to rule out one of several possible mechanisms. Moreover, a comparison of kinetic studies in man, rat and cow (BRONNER, 1973b) permits the conclusion that similar processes occur in all three species and that these processes are brought out by this type of analysis.

One of the principal uses to which kinetic analysis can be put is to develop and test models of regulation. For example, one might suppose that intestinal absorption (v_a) and skeletal deposition (v_{O+}) of calcium are related. A number of such studies have been done and the general conclusion is that v_{O+} is relatively independent of v_a (BRONNER and AUBERT, 1965; SCHWARTZ et al., 1965; PHANG et al., 1969). To test a variety of such relationships it is useful to have developed a model of regulation and to see whether experimental observations support the model. In what follows one such model of calcium homeostasis will be analyzed+.

II. A Cybernetic Model of Calcium Regulation

Three systems are involved in plasma calcium regulation: the gut, the skeleton, and the kidney. If plasma calcium is to remain at or near given value - typically 2.5 mM in mammals - the sum of inputs and outputs from the tree systems must necessarily equal zero. In the growing animal, calcium is added to the plasma by intestinal absorption and extracted from plasma by renal excretion and net deposition in the skeleton. When the skeletal calcium balance is negative, as in aging animals or in certain disease states, the skeleton adds calcium to the blood, but in the normal situation, the skeletal calcium balance varies from positive to near zero. Urinary calcium excretion (v_u) and the skeletal calcium balance (Δ) tend to vary directly with

+The model, originally proposed in 1965 (AUBERT and BRONNER, 1965), has been extensively tested (BRONNER and AUBERT, 1965; BRONNER et al., 1967; HURWITZ et al., 1969, SAMMON et al., 1969; 1970) and refined (BRONNER, 1973a, 1973b).

calcium absorption. For example, one can vary net calcium absorption (S_i) in growing rats at will, simply by raising or lowering the calcium content of an otherwise unchanged diet (BRONNER and AUBERT, 1965). V_u and Δ of these animals will vary similarly and in the same direction as S_i. It is therefore reasonable to classify the gut as a system whose output tends to disturb the plasma calcium and to classify the kidney and skeleton as systems whose outputs regulate or control plasma calcium (AUBERT and BRONNER, 1965). Fig. 1 is a schematic representation of this statement.

In Fig. 1, the constituent fluxes or signals are classified in the same way as the net fluxes resulting from them. Obviously alternative classifications are possible.

Classifying the gut as a disturbing system is not meant to imply that it is not subject to control. Biological systems characteristically contain highly sophisticated controls. However, the output of a regulated system may still function as a disturbing signal for another system and this is the intent of the classification used here. This classification would also be in error if there existed a feedback path

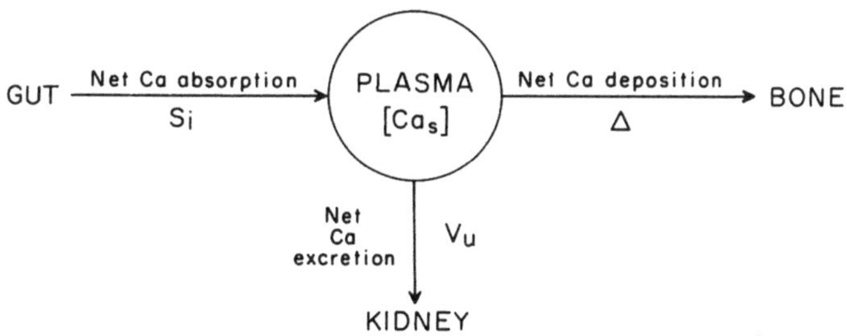

System	Net Signal	Constituent Signals	Regulatory Classification
PLASMA	$[Ca_s]$ Plasma Ca concentration		Controlled
BONE	Δ Ca balance	$= V_{o+}$ (Ca into bone) $-$ V_{o-} (Ca out of bone)	Controlling
KIDNEY	V_u Ca excretion	$= V_{fi}$ (Ca filtered) $-$ V_{re} (Ca reabsorbed) $+$ V_{se} (Ca secreted)	Controlling
GUT	S_i Net Ca absorption	$= V_a$ (Ca absorbed) $-$ V_{ndo} (fecal endogenous Ca)	Disturbing

[Note: By definition $S_i = V_a - V_{ndo} \equiv V_i$ (Ca intake) $-$ V_F (total fecal)]

Fig. 1. Regulatory classification of parameters of calcium metabolism (adapted from SAMMON et al., 1970, and reproduced by permission of the Am. J. Physiol.)

between gut and bone. This has often been suggested, but so far no report, with the possible exception of that by Boyle et al. (1971), has indicated an acutely functioning feedback path. Therefore no such feedback loop is included.

Fig. 2 depicts a control diagram embodying the definitions and classifications of Fig. 1. The diagram omits hormonal controls and is drawn according to the convention of linear control systems.

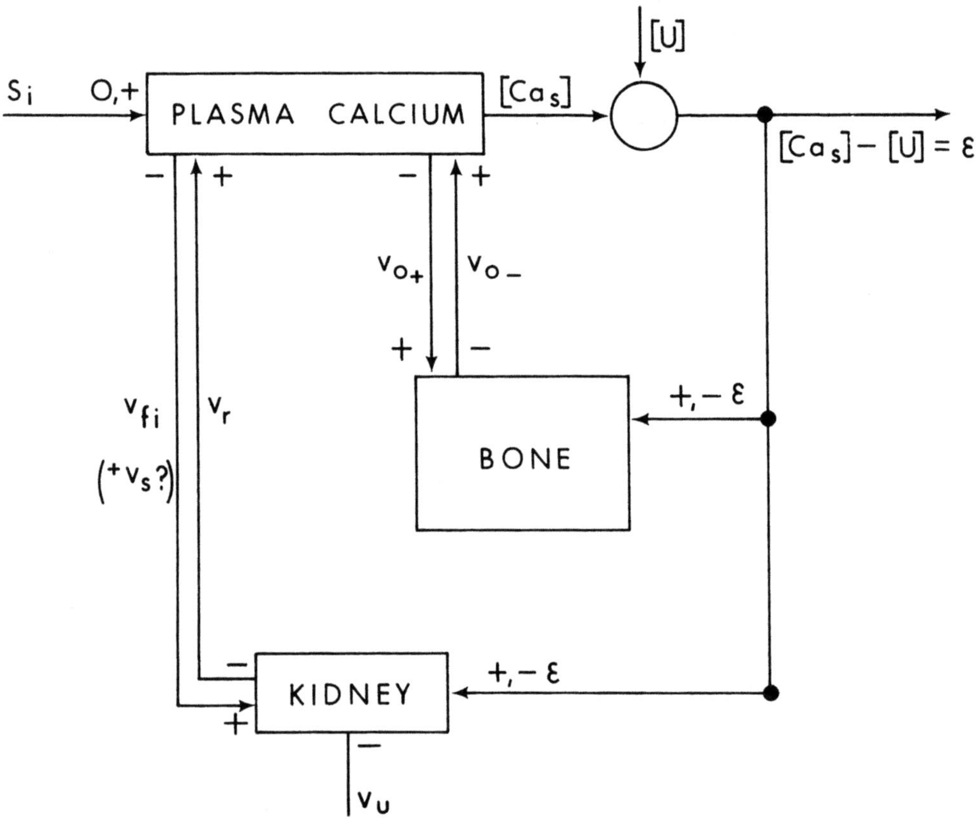

Fig. 2. Diagram of mammalian plasma calcium regulation, with hormonal controls omitted, as in thyroparathyroidectomized animals

S_i = net calcium absorption (i.e. calcium load from gut) mg Ca/day

v_{o+} = bone calcium deposition, mg/day

v_{o-} = bone calcium resorption, mg/day

v_{fi} = calcium filtered in kidney, mg/day

v_r = calcium reabsorbed in kidney, mg/day

v_s = calcium secreted in kidney, mg/day

$[Ca_s]$ = plasma calcium concentration, mg/100 ml

$[U]$ = plasma calcium reference value, mg/100 ml

ε = error, i.e. $[Ca_s]$ - $[U]$

When calcium enters the bloodstream via absorption from the intestine, it tends to raise the plasma calcium concentration $[Ca_s]$, drawn as a signal being compared in a comparator with a "set" or "reference" value, $[U]$. This is the value at which the body is attempting to hold the plasma calcium. The difference between actual plasma calcium and reference value is termed the "error signal".

The error signal actuates the two controlling systems, bone and kidney. When the error is positive, i.e. when the plasma calcium concentration, $[Ca_S]$, exceeds the reference value, bone responds largely by decreasing bone calcium resorption, v_{o-}, while bone calcium deposition goes up only a little (BRONNER and AUBERT, 1965; SAMMON et al., 1970). The result is an increase in the mass of bone calcium. At the same time the kidney responds by increasing urinary calcium excretion, v_u. The precise mechanism is not understood. It may involve diminished reabsorption as well as an increase in the amount filtered. When the error is negative, i.e. when $[Ca_S]$ drops below $[U]$, the processes are reversed. Bone responds by an increase in v_{o-}, with only a minor decrease in v_{o+}, while kidney diminishes v_u by increasing the amount reabsorbed. This may be accompanied by an obligatory decrease in the amount filtered. It is assumed that bone and kidney respond in direct proportion to the error, i.e. that they function as proportional controls.

The plus, minus and zero symbols in the diagram indicate direction of flow. For example, v_{o+} is positive with respect to bone and negative with respect to plasma calcium. The signs do not mean that a positive error signal necessarily evokes a positive flow in the controller. Indeed (see Eqs. 3-6) a positive error signal causes v_{o-} to decrease more than v_{o+} to increase.

Fig. 2 contains formulations that are as yet removed from what is known. Thus nothing is known of the mechanism(s) by which mass flows are converted to signals. Conceivably this could involve calcium-dependent conformational changes of specific membrane proteins, as calcium binding proteins, a number of which have been discovered in various tissues (intestine: WASSERMAN et al., 1968; kidney: PIAZOLO et al., 1971; brain: WOLFF and SIEGEL, 1972; muscle: HARTSHORNE and PYUN, 1971). Nor has a comparator been identified. However, one can imagine that the total physiological state of the organism affects turnover and conformational state of calcium-dependent processes in such a way that a number of specific molecular mechanisms, taken together, could function as a comparator.

The relative importance of kidney and bone in overcoming the disturbing signal varies in different species and is a function of age. For example, in young growing male rats, only two percent of the incoming disturbing signal (i.e. net absorption, S_i) is handled by the kidney (SAMMON et al., 1970), while in adult women the kidney handles at least two thirds and bone at best one third of the disturbing signal (HALL et al., 1969)+.

+This statement is only meaningful when the calcium balance is positive. When the balance is zero or negative, as is frequent in adult humans, bone acts to supply needed calcium. HALL et al., (1969) have shown that in adult women $\Delta = -114 + 0{,}77\ S_i$ (units: mg/day). In other words, unless net calcium absorption is 148 mg/day, the "average woman" would be in negative balance. Stated another way, net calcium input from the gut does not become a disturbing signal until it exceeds 148 mg/day.

III. Hormonal Controls

1. Parathyroid Hormone (PTH)

In the rat, parathyroidectomy (SAMMON et al., 1970) or removal of the calcitonin-producing cells (BRONNER et al., 1967; SAMMON et al., 1969) has no effect on the relative importance of bone and kidney, bone remaining the primary regulator of plasma calcium. However, parathyroid hormone has a profound effect on the efficiency of regulation. This is illustrated by the following equations (taken from SAMMON et. al., 1970):

(1) $$[Ca_s] = 6.02 + 0.021\ S_i$$

(2) $$[Ca_s] = 10.80 + 0.0021\ S_i$$

Eq. 1 describes plasma calcium (in mg Ca/100ml) as a function of net calcium absorption (S_i, mg Ca/day) in surgically parathyroidectomized (PTX) animals and Eq. 2 describes this relationship in normal rats. In the normal rat a calcium load of 100 mg/day from the intestine raises the steady state plasma calcium from 10.8 to 11.0 mg/100ml, or by less than 2 percent. In the ablated animal a comparable load would raise the steady state plasma calcium to 8 mg/100ml, or by 33 percent. Clearly the PTX animal is less able to regulate his plasma calcium around a steady value. This is also brought out by acute loading studies. When PTX animals received an intraperitoneal calcium (BRONNER et al., 1968) the acute plasma calcium error - i.e. the difference between plasma calcium in the absence of a load and at any moment following administration of the load - was greater than in normal animals. Moreover, it took appreciably longer for the plasma calcium to return to its preload value in the PTX than in the normal animals.

As one would expect, PTX animals respond to a negative error - as when the plasma calcium is depressed by EDTA infusions - less effectively than normal animals (COPP, 1957, 1960).

Thus parathyroid hormone contributes to more precise plasma calcium regulation and permits overcoming both positive and negative errors of plasma calcium more rapidly. Nevertheless, the fact that even in ablated animals the plasma calcium returned to the preload value relatively rapidly (BRONNER et al., 1968; SAMMON et al., 1970) is clear indication of controls capable of functioning in the absence of the parathyroid glands. Consequently, the parathyroids contribute to, but do not exert sole regulation of plasma calcium. This regulation must involve error sensing, parathyroid hormone release, and action on the target tissues, bone and kidney. It is these tissues that regulate plasma calcium.

In bone, parathyroid hormone modifies cellular processes leading to calcium deposition (v_{o+}) and resorption (v_{o-}) in such a way that there is higher turnover in the normal as compared to the PTX animal and there is a change in the relationship between v_{o+} and v_{o-}, as follows:

(3) $$v_{o+} = 67.2 + 0.11\ S_i$$

(4) $$v_{o-} = 67.5 - 0.87\ S_i$$

(5) $$v_{o+} = 36.5 + 0.29\ S_i$$

(6) $$v_{o-} = 36.9 - 0.69\ S_i$$

(units: mg Ca/day)

Eqs. 3 and 4 apply to normal and Eqs. 5 and 6 to PTX animals (SAMMON et al., 1970). According to these equations, the PTX animal relies less on bone calcium resorption (v_{o-}) than the normal animal to dispose of a calcium load from the gut. Moreover, the overall intensity of turnover in the PTX animal is much lower (37 vs. 67 mg/day).

Less is known about what happens in kidney. It is probable that PTH increases renal calcium reabsorption (TALMAGE and KRAINTZ, 1954; KLEEMAN et al., 1961), but no data are available to evaluate this quantitatively, as in Eqs. 3-6.

Fig. 3 is a regulatory diagram showing the addition of parathyroid function. In the model, a negative error is drawn as actuating the parathyroid glands to release hormone that acts on bone and kidney. In addition, as in Fig. 2, both positive and negative errors are shown acting directly on bone and kidney. Bone responds to negative errors by increasing v_{o-} more than V_{o+} and to positive errors by decreasing v_{o-} more than v_{o+}. This differential is accentuated by parathyroid hormone.

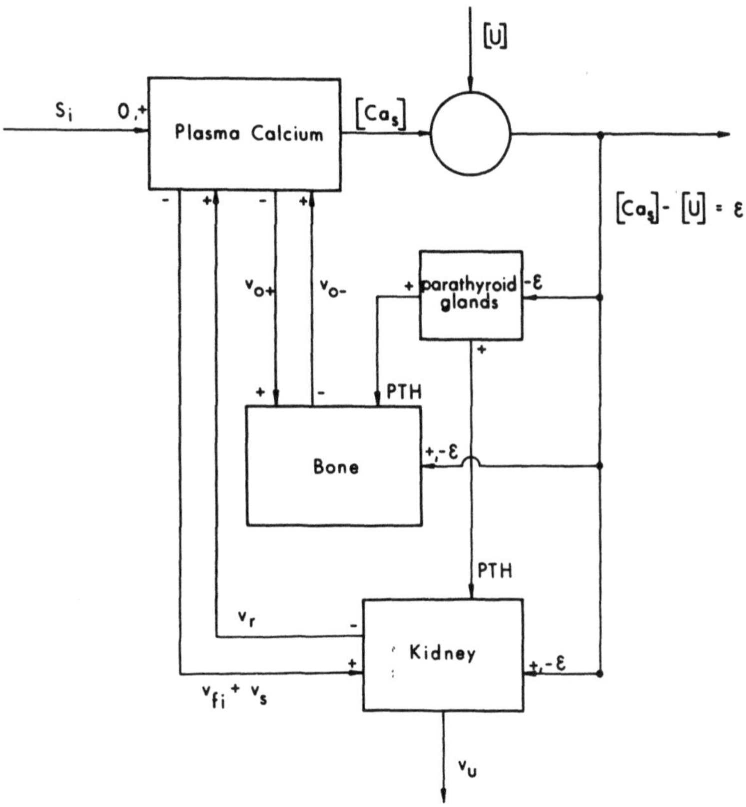

Fig. 3. Diagram of plasma calcium regulation including parathyroid hormone (PTH) action, but omitting other hormonal controls. For definition of other symbols, see fig. 2. (From BRONNER, 1973a. Reproduced by permission of Academic Press, Inc.)

Kidney responds to a negative error by decreasing v_u and to a positive error by increasing v_u. The precise relationship between v_r and v_{fi} ($+v_s$, if v_s exists) and the effect of parathyroid hormone thereon are not known, though v_r is al least partly under parathyroid control (AGUS et al., 1973). It is important to point out that whereas parathyroid hormone release occurs in proportion to a negative error signal only, the metabolic effects of the hormone on bone and kidey are such as to lead to more precise error control, whether the plasma calcium error is positive or negative.

As a result of the action of parathyroid hormone, steady-state error of plasma calcium is virtually absent and acute error correction is significantly improved. In man-made linear control systems, the addition of an integral to a proportional control eliminates steady-state error and minimizes the error-time curve without leading to overshoot (GOLDMAN, 1960). Consequently, it seems reasonable to propose that parathyroid hormone adds an integral control to the proportional control due to bone and kidney when the latter are not under parathyroid dominance (AUBERT and BRONNER, 1965).

2. Calcitonin (CT)

If calcitonin had a similar but opposite action to parathyroid hormone, one would expect that removal of the cells producing this hormone would have both acute and steady-state effects on plasma calcium homeostasis. For example, one would expect that animals without this hormone would control their steady-state calcium less efficiently than normal animals. One would have predicted also changes in intensity and relative importance of bone calcium deposition and resorption rates and/or renal calcium reabsorption and filtration rates.

Detailed studies (BRONNER et al., 1967, 1968; SAMMON et al., 1969) have shown that calcitonin has no obvious steady-state effect on calcium metabolism. Animals deprived of their endogenous supply of calcitonin, i.e. thyroidectomized, supplied with ℓ-thyroxine in the food and bearing functional parathyroid autografts, did as well as euthyroid controls with functional parathyroid autografts, as far as steady-state plasma calcium control is concerned.

Calcitonin removal had no significant effect on either intensity or relative regulatory importance of v_{o+} and v_{o-}, nor did it affect in any major way the calcium and phosphorus content of the long bones. Finally, urinary calcium excretion was unaffected by calcitonin removal and the relative regulatory importance of bone and kidney was unchanged, bone retaining its primary role in rats deprived of calcitonin.

However, calcitonin does play an important acute regulatory role. Thus when animals without endogenous calcitonin are subjected to intraperitoneal calcium loads (TALMAGE et al., 1965; BRONNER et al., 1967) it takes much longer for their plasma calcium to return to the base level than in normal animals. Moreover, measurements of plasma calcitonin levels have shown that above the normal reference calcium level calcitonin release varies positively and proportionately with plasma calcium (CARE et al., 1968).

In man-made systems the addition of a derivative control to a preexisting proportional control provides a mechanism for faster error correction without the danger of overshoot (GOLDMAN, 1960). The time-error curve of a derivative plus proportional control is similar to an integral plus proportional control. However, in a situation of integral plus proportional controls there is no steady-state error, while in a situation of derivative plus proportional controls a steady-state error exists.

On the basis of these considerations, BRONNER et al. (1968) have proposed that calcitonin is like a derivative control, i.e. a transient and not a steady-state control.

These considerations form the basis of the regulatory diagram shown in Fig. 4. Calcitonin is shown to be released in response to the derivative of a positive plasma calcium error. The action of calcitonin would lead to a temporary drop in v_{O-}, with no effect on v_{O+} (MILHAUD et al., 1965; O'RIORDAN and AURBACH, 1968). This would cause the plasma calcium to drop to its reference value, when calcitonin action would be minimal. Such a model predicts that hormone release would be slower as the difference between the actual plasma calcium value and its reference level becomes smaller. This has now been confirmed experimentally (WEST et al., 1973). (For an analysis of the relationship between plasma calcitonin and calcium levels, see BRONNER, 1973a).

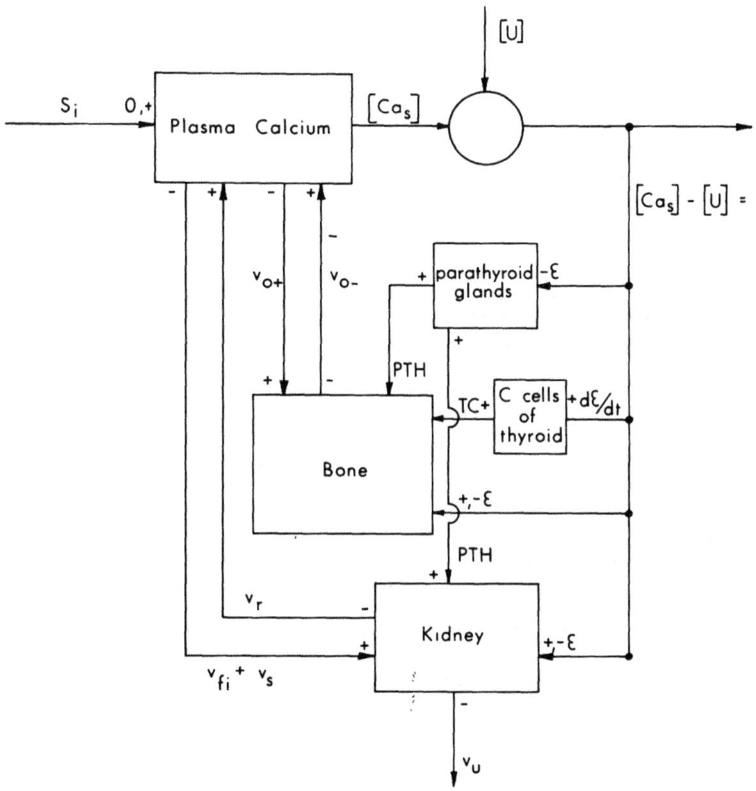

Fig. 4. Diagram of plasma calcium regulation, including parathyroid hormone (PTH) and calcitonin (CT) action. For definition of symbols, see Figs. 3 and 4. (From BRONNER, 1973a. Reproduced by permission of Academic Press, Inc.)

IV. Bone as Regulator

The preceding analysis leads to a new perspective on the role of bone. From what has been said above, it is clear that bone functions as the primary regulator of plasma calcium, whether or not the parathyroids are present. However, the mode of regulation depends on parathyroid function. In animals without parathyroids, regulation is less efficient and regulatory capacity is diminished. It is difficult to test the intermediate case, partially functioning parathyroids. This is partly so because of compensatory overgrowth and function and partly because plasma parathyroid hormone levels cannot yet be measured in rats. SAMMON et al. (1969) studied animals bearing parathyroid autografts and compared calcium metabolism and function with that of normal animals, *i.e.* animals whose parathyroids had been left intact. They commented on the fact that the animals with parathyroid autografts, in the quantitative terms of the analysis employed, were not quite "normal". In particular, regulation by bone seemed less efficient, though the overall capacity seemed unchanged, viz.

(7) $v_{o^-} = 67.5 - 0.89\ S_i$ Normal

(8) $v_{o^-} = 68.3 - 0.74\ S_i$ Parathyroid Autograft

(9) $v_{o^-} = 36.9 - 0.69\ S_i$ PTX

(units: mg Ca/day)

These equations show that in the absence of a load, when S_i=0, v_{o^-} was about the same in the normal rat and that bearing a parathyroid autograft. However, the latter tended to handle a somewhat smaller proportion of the incoming load (S_i) *via* diminution of bone calcium resorption. In the fully ablated animal, capacity was much lower since v_{o^-} was 36.9 mg/day when S_i was 0. Moreover, the PTX animals handled the smallest proportion of the incoming load (S_i) *via* bone calcium resorption, v_{o^-}.

As already pointed out, the absence of calcitonin has little effect on the steady-state regulatory function of bone (SAMMON et al., 1969), but has marked effect on the capacity for acute or transient regulation.

In the vitamin D-deficient animal (HURWITZ et al., 1969) the regulatory capacity and efficiency of bone are markedly impaired. Indeed the vitamin D-deficient rat behaves as if it were parathyroprivic, even though its parathyroids function near maximum capacity (AU and RAISZ, 1965). Whatever the mechanism, it is clear that vitamin D is required for normal bone structure and function. Without vitamin D, even in the presence of adequate amounts of minerals and parathyroid hormone, the regulatory function of bone is impaired.

If an animal is calcium-deficient, the regulatory capacity of bone is altered. For example, SHAH et al. (1968) have shown that animals raised on high calcium diets from weaning were able to maintain a near-normal plasma calcium for two weeks when placed on diets that contained only 0.01% Ca, but that similar animals raised from weaning on a low calcium diet (0.06%) were unable to do so, their plasma calcium dropping to near 6 mg%. Animals on the high calcium diets had high bone calcium levels and high values for v_{o^+}; animals on the low calcium diets had low bone calcium and low values of v_{o^+}. Severe phosphate deficiency also leads to marked changes in bone composition (BAYLINK et al., 1971) and its regulatory capacity (CUISINIER-GLEIZES et al., 1973). However detailed knowledge of the interrelationship between bone composition and regulatory capacity is still lacking.

What emerges then is a complicated set of relationships between bone structure and function. Any attempt at hierarchical ordering must necessarily be tentative. Nevertheless, one may classify the requirements of bone in order for it to be able to regulate plasma calcium in the following order:

least important	1. calcitonin
↓	2. parathyroid hormone
	3. vitamin D
most important	4. calcium, phosphorus and other constituents of bone

This is obviously a gross oversimplification. It neglects a host of cell constituents such as enzymes and also neglects other bone functions (support, storehouse of many elements,etc.), the expression of which is doubtless related to bone structure and turnover. Moreover, important feedback loops are bound to exist between these functions of bone and calcium regulation. Nevertheless, this attempt at classifying the role of the major factors in bone function and therefore in calcium homeostasis and metabolism may be of value in clarifying a complex field.

References

AGUS, Z.S., GARDNER, L.B., BECK, L.H., GOLDBERG, M.: Effects of parathyroid hormone on renal tubular reabsorption of calcium, sodium and phospate. Am. J. Physiol. 224, 1143-8 (1973).

AU, W.Y.W., RAISZ, L.G.: Effect of vitamin D and dietary calcium on parathyroid activity. Am. J. Physiol. 209, 637-642 (1965).

AUBERT, J.P., BRONNER, F.: A symbolic model for the regulation by bone metabolism of the blood calcium level in rats. Biophys. J. 5, 349-358 (1965).

BAYLINK, D., WERGEDAL, J., STAUFFER, M.: Formation, mineralization, and resorption of bone in hypophosphatemic rats. J. Clin. Invest. 50, 2519-2530 (1971).

BOYLE, I.T., GRAY, R.W., DeLUCA, H.F.: Regulation by calcium of in vivo synthesis of 1,25-dihydroxycholecalciferol and 2,25- dihydroxycholecalciferol. Proc. Nat. Acad. Sci. (USA) 68, 2131-2134 (1971).

BRONNER, F.: Calcium Homeostasis, In: Engineering Principles in Physiology (J.H.U. Brown and D. Gann,eds.), Vol. I, p. 227-248. New York and London: Academic Press 1973a.

BRONNER, F.: Kinetic and cybernetic analysis of calcium metabolism. In: Calcium and Phosphorus by J.T. Irving. New York and London: Academic Press 1973b.

BRONNER, F., AUBERT, J. P.: Bone metabolism and regulation of the blood calcium level in rats. Am. J. Physiol. 209, 887-890 (1965).

BRONNER, F., SAMMON, P.J., STACEY, R.E., SHAH, B.G.: Role of thyrocalcitonin in the regulation of the blood calcium level. Biochem. Med. 1, 261-279 (1967).

BRONNER, F., SAMMON, P.J., NICHOLS, C., STACEY, R.E., SHAH, B.G.: Thyrocalcitonin and plasma calcium homeostasis in the rat. In: Parathyroid Hormone and Thyrocalcitonin (Calcitonin), (R.V. Talmage and L.R. Bélanger, eds.) p. 353-369. Amsterdam: Excepta Medica Fndn 1968.

CARE, A.D., COOPER, C.W., DUNCAN, T., ORIMO, H.: The direct measurement of thyrocalcitonin secretion rate *in vivo*. In: Parathyroid Hormone and Thyrocalcitonin (Calcitonin). (R.V. Talmage and L.R. Bélanger, eds.) p. 417-427. Amsterdam: Excepta Medica Fndn. 1968.

COPP, D.H.: Calcium and phosphorus metabolism. Am. J. Med. 22, 275-285 (1957).

COPP, D.H.: Parathyroids and homeostasis of blood calcium. In: Bone As A Tissue (K. Rohdahl, J.T. Nicholson and E.M. Brown, eds.) p. 289-299. New York: McGraw Hill 1960.
CUISINIER-GLEIZES, P., DEBOVE, P., GEORGE, A., MATHIEU, H.: Phosphate deficiency and parathyroid function. Fed. Proc. 32, 270 Abs (1973).
GOLDMAN, S.: Cybernetic aspects of homeostasis. In: Mineral Metabolism - An Advanced Treatise (C.L. Comar and F. Bronner, eds.), Vol. IA, p. 61-101. New York and London: Academic Press 1960.
HALL, B.D., MACMILLAN, D.R., BRONNER, F.: Vitamin D-resistant rickets and high fecal endogenous calcium output. A report of two cases. Am. J. Clin. Nutr. 22, 448-457 (1969).
HARTSHORNE, F.J., PYUN, H.Y.: Calcium binding by the troponin complex and the purification and properties of troponin A. Biochem. Biophys. Acta 229, 698-711 (1971).
HURWITZ, S., STACEY, R.E., BRONNER, F.: Role of vitamin D in plasma calcium regulation. Am. J. Physiol. 216, 254-262 (1969).
KLEEMAN, C.R., BERNSTEIN, D., ROCKNEY, R., DOWLING, J.T., MAXWELL, M.H.: Studies on renal clearance of diffusible calcium and the role of the parathyroids in its regulation. In: The Parathyroids (R.O. Greep and R.V. Talmage, eds. p. 353-382. Springfield: Thomas 1961.
MILHAUD, G., PERAULT, A.M., MOUKHTAR, M.S.: Étude du mécanisme de l'action hypocalcémiante de la thyrocalcitonine. Compt. Rend. Acad. Sci. (Paris) 261, 813-816 (1965).
O'RIORDAN, J.L.H., AURBACH, G.D.: Mode of action of thyrocalcitonin. Endocrinology 82, 377-385 (1968).
PHANG, J.M., BERMAN, M., FINERMAN, G.A., NEER, R.M., ROSENBERG. L.E., HAHN, T.J.: Dietary perturbation of calcium metabolism in normal man: compartmental analysis. J. Clin. Invest. 48, 67-77 (1969).
PIAZOLO, P., SCHLEYER, M., FRANZ, H.E.: Isolation and purification of a calcium binding protein from human tissues. Hoppe Seyler Z. Physiol. Chem. 352, 1480-1486 (1971).
SAMMON, P.J., STACEY, R.E., BRONNER, F.: Further studies on the role of thyrocalcitonin in calcium homeostasis and metabolism. Biochem. Med. 3, 252-270 (1969).
SAMMON, P.J., STACEY, R.E., BRONNER, F.: Role of parathyroid hormone in calcium metabolism and homeostasis. Am. J. Physiol. 218, 479-485 (1970).
SCHWARTZ, E., PANARIELLO, V.A., SAELI, J.: Radioactive calcium kinetics during high calcium intake in osteoporosis. J. Clin. Invest. 44, 1547- 1560 (1965).
SHAH, B.G., SAMMON, P.J., BRONNER, F.: Effect of calcium intake levels on the response to exogenous rat thyrocalcitonin. In: Calcitonin. Proceedings of the Symposium on Thyrocalcitonin and the C Cells, p. 282-289. London: Heinemann 1968.
TALMAGE, R.V., KRAINTZ, F.W.: Progessive changes in renal phosphate and calcium excretion in rats following parathyroidectomy or parathyroid administration. Proc. Soc. Exptl. Biol. Med. 87, 263-267 (1954).
TALMAGE, R.V., NEUENSCHWANDER, J., KRAINTZ, F.W.: Evidence for the existence of thyrocalcitonin in the rat. Endocrinology. 76, 103-107 (1965).
WASSERMAN, R.H., CORRADINO, R.A., TAYLOR, A.N.: Vitamin D-dependent calcium binding protein: Purification and some properties. J. Biol. Chem. 243, 3978-3984 (1968).
WEST, T.E.T., O'RIORDAN, J.L.H., COPP, D.H., BAKER, R.F.L., CARE, A.D.: The effect of hypocalcemia on the secretion of calcitonin. J. Endocrin. 56, 463-470 (1973).
WOLFE, D.J., SIEGEL, F.L.: Purification of a calcium binding phosphoprotein from pig brain. J. Biol. Chem. 247, 4180-4185 (1972).

A Time-Varying Model for Calcium Metabolism Including Diurnal Variations*

A.M. PERAULT-STAUB, C. LEFEVRE, J.F. STAUB, P. BREZILLON AND G. MILHAUD

I. Introduction

Many methods have already been used for modeling calcium metabolism (MILHAUD et al., 1957, MARSHALL, 1969). If their characteristics were widely differing, they were in common based on both mathematical procedures and physiological data which require to be reconsidered to-day. Recently, three of the authors reported that, in the rat, plasma calcium fluctuates following a clear circadian rhythm (PERAULT-STAUB et al., 1973). Our present study includes detailed knowledge of plasma ^{45}Ca decay and shows that rhythmicity may be considered as an essential property of the processes involved in calcium metabolism. Moreover, during the modeling, we have attempted to include the new concepts of calcium homeostasis (TALMAGE, 1970; MATTHEWS et al., 1971).

II. Experimental Procedure

A compartimental analysis is used, defined by the following equations:

$$\overset{o}{Q}_i = \sum_{\substack{j=1 \\ j\neq i}}^{n} R_{ij} - \sum_{\substack{k=1 \\ k\neq i}}^{n} R_{ki} + d_{io}$$

$$\overset{o}{q}_i = \sum_{\substack{j=1 \\ j\neq i}}^{n} R_{ij}/Q_j \; x q_j - \sum_{\substack{k=1 \\ k\neq i}}^{n} R_{ki}/Q_i \; x q_i + e_{io}$$

The first set of equations describes the mass variation (Q_i) of the n compartments and the second set, the variation of the quantity of tracer (q_i) in each of them. The mass flow of material entering the compartment i is noted R_{ij}, the output mass flow, R_{ki}, and the irreversible entry rates of unlabeled and labeled material, d_{io} and e_{io}, respectively. These parameters, R_{ij} and R_{ij}/Q_j, are not necessarily constant and can vary with time (for nomenclature see BROWNELL et al., 1968).

+The authors are grateful to Dr. G. COUTRIS for his valuable suggestions during the preparation of this manuscript.

Blood samples were collected from groups of rats which have been given 100 µC of ^{45}Ca, as $CaCl_2$ i.v., at time O (45 days old Wistar C.F. rats, fed with a normal semi-synthetic diet during the 12 dark hours of each day). Plasma calcium and ^{45}Ca concentrations were estimated from 5 minutes to 8 days: except during the first experimental time, sampling was done every 4 hours from the 1st to the 5th and from the 7th to the 8th day. As shown in Figure 2A, a clear rhythmicity appears in the radioactive decay from the 12th hour on; the amplitude of the ^{45}Ca diurnal fluctuation depends on tracer-time. Obviously, such diurnal variations are synchronized with the external events (activity, food intake, light and dark) and considered as the consequence of daily changes in the controlling system.

III. Modeling Procedure

The analysis involves three steps:

A) Classification of calcium metabolism processes. The first step consists of breaking down these processes into the following subsystems:

- A "milieu extêrieur" (O) involving both calcium input through the digestive tract and calcium output through kidney and gut.
- A controlled system (1) directly exposed to (O). The kinetic characteristics of (1) are obviously the plasmatic ones, as plasma samples were used.
- A storage system (2), able to face any unexpected variation coming from (O).
- A controlling system (A & B), located between (1) and (2). It controls (1) through cellular mechanisms consistent with the new proposed concepts of calcium homeostasis : calcium pump (TALMAGE, 1970), calcium transport through biological membranes (NEUMAN and RAMP, 1971), calcium influx and efflux from mitochondria (BORLE, 1973; MATTHEWS et al., 1971) or other calcium storage pools. The data led us to assume that oscillations exist at the controlling system level itself. Thus, a time function needed to be introduced and consequently the model became a time-varying one. It had to be placed into the (A & B) system and for a first approach, a sine wave was used. Fig. I illustrates this first modeling step.

B) Analogous model able to describe changes of labeled calcium at the plasma level. Modeling was first built from the tracer curve (Fig. 2A). The basic design had to include the 4 previously described subsystems, tied together as shown in Fig. 1. The analysis was based on two postulates:

Postulate 1: At O time of the simulation, the injected radioactivity is abruptly diluted in the volume of (1). In no case can this volume be larger than that estimated from the plasma radioactivity in the rat at 5 minutes.

Postulate 2: The quantity of radioactivity which irreversibly exits from (1) between O and t time is considered as the difference between the injected radioactivity and the residual radioactivity in the rat, at t time.

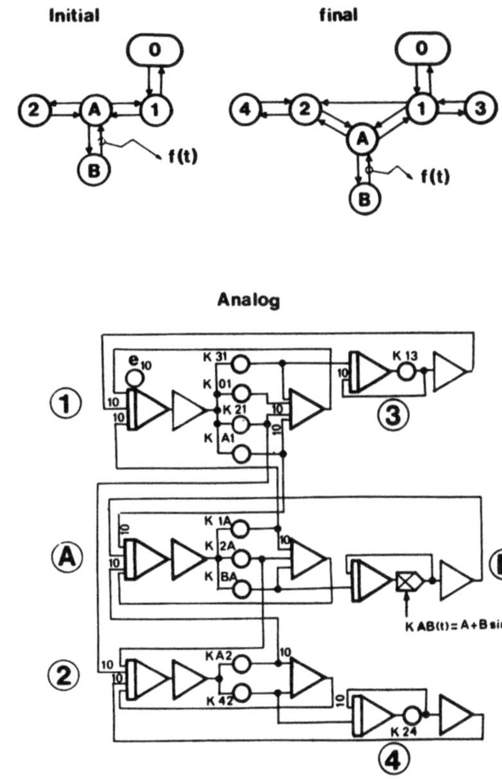

Fig. 1. (upper graph) Initial and final models including the subsystems, tied together as described: (O), "milieu extérieur"; (1), controlled system; (2), storage system; (A & B), controlling system; (3) and (4), additional systems. (lower graph) Simulation diagram for the final model equations. All operational blocks can be considered as analogous computing elements

The optimisation of the analogous model was performed on a computer EAI 580 using a quadratic function to minimize the instantaneous error. With this technique, we can extract the numerical values of parameters of a non-steady differential system of high order, direct resolution of which would be rather difficult and sometimes impossible.

The initial model (Fig. 1) reproduced the mean value of the real ^{45}Ca plasma curve from the 36th hour on to the 8th day. However, incorporation of changes in the model's design was necessary in order to take into account both the entity of the ^{45}Ca curve from 5 minutes to 8 days and the amplitude of the observed diurnal fluctuations. We had therefore to introduce:

- a new subsystem noted (3) rapidly interchanging with (1). As a consequence, the drop in the plasma ^{45}Ca was enhanced from 0 to 2 hours and, in contrast, diminished from 2 to 36 hours, as expected.

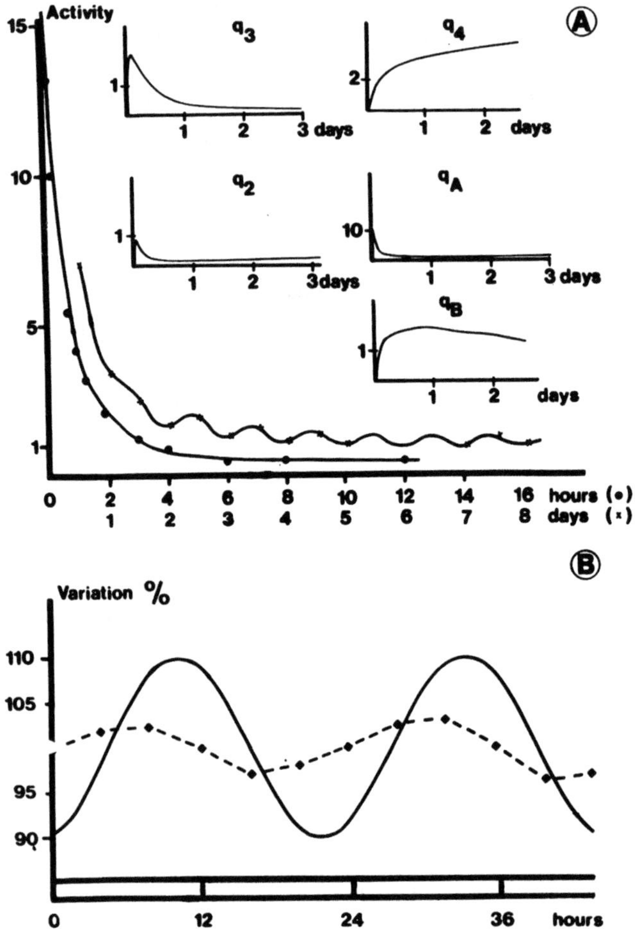

Fig. 2. (A) (large graph): 8 days pattern of changes in the level of radioactive calcium observed in plasma. Radioactive calcium (activity) is expressed as c.p.m. for 20 µl of plasma. •——• represents the activity decay from 5 minutes to 16 hours (1 → 10000 c.p.m.); ×——× represents the activity decay from 12 hours to 8 days (1 → 500 c.p.m.). (small graphs) Predicted activity (q) in the compartments (2), (3), (4), (A & B) from 0 to 3 days. (B) Diurnal variations of plasma calcium expressed as percent of the daily mean value. Observed (•---•); predicted (——) in the final scheme by simulation of a constant entry rate

- <u>a direct junction from (1) to (2)</u>: the consequence was a minimization of the irreversible radioactive exit to (0). In order to keep unchanged the characteristics of (2), we added a new subsystem (4), interchanging with (2) and able to store a big quantity of radioactivity. An apparent radioactive reflux back from (4) to (1) would appear only after a long period of time (several days).

The final scheme and its simulation diagram are presented in Fig. 1. Simulation of the corresponding system of equations led to a correct behavior of the system (1) from 5 minutes to 8 days. The diagrams for the simulation of the other subsystems are presented in Fig. 2A.

It should be noted that a sinusoid (linear oscillation) placed between A and B, does not permit a very good representation of the circadian rhythm observed at the level of the ^{45}Ca plasma curve. Further work is therefore required to determine the appropriate form of oscillation - probably non-linear - which is needed (OATLEY and GOODWIN, 1971).

C) First approach to an analogous model describing the calcium metabolism. Knowing the different ratios rate/mass, K_{ij}, which characterize calcium transfer between the subsystems and out of the system, K_{01}, it is possible to simulate the non-labeled calcium model of calcium metabolism. We introduce calcium from (O) to (1) to simulate calcium absorption, at a rate sufficient to maintain (1) at a constant mean level.

Fig. 2B illustrates the daily variation pattern, predicted in (1) when simulation of a constant calcium entry rate is applied. The total daily calcium mass entering the system is in equilibrium with the total irreversible exit. We are unable to reproduce adequately the real circadian fluctuations of plasma calcium, but amplitudes of the predicted fluctuations are too large (about 23% against 7% experimentally (PERAULT-STAUB et al., 1973) and the time position of the two extreme values are not the observed ones (PERAULT-STAUB et al., 1973). The incorrect behavior of this model may result from two facts:

1. A constant calcium entry rate is not correct. We suggest the existence of daily variations in the entry rate which should be relatively smaller during the light-unfed period than during the dark-fed period. Thus, for example, if the total mass of calcium entering during the dark-fed period is 3 times greater than that during the 12 following hours, amplitudes of the predicted fluctuations are reduced to about 7 percent, i.e. almost correct.

2. The diurnal changes in plasma result, among other things, from the conjunction of pattern of calcium entry and action of controlling system (A & B). As previously noted, the sine-wave in the (A & B) system is not adequate and it is difficult actually to predict the exact form of the entry pattern. However, it appears that the proposed model is compatible with a daily calcium entry different from the exit, that is including positive or negative calcium balance. This balance, placed at the output-input level is transferred at (4) level, the calcium mass of which is so big that this modification is entirely damped at the (1) level.

IV. Conclusion

The present study has to be considered as a first step toward the building of a comprehensive calcium metabolism scheme. It already provides additional insight into this metabolism. Indeed, the classical calcium metabolism schemes were based upon the stability of plasma calcium level and a steady decrease of plasma ^{45}Ca. The experimental data used here suggest that oscillations should be considered as a fundamental part of the control processes. A simple mathematical time-varying model is presented, which takes into account such daily events. The originality of this model consists in the incorporation of: (1) a determined pattern for daily calcium absorption and (2) a positive or negative calcium balance as part of the exchangeable calcium pool.

References

BORLE, A.B.: Fed. Proc. 32, no 9, 1944 (1973).

BROWNELL, G.L., BERMAN, M., ROBERTSON, J.S.: Int. J. of Appl. Radiations and Isotopes, 19, 249 (1968).

MARSHALL, J.H.: In: Mineral Metabolism. Calcium Physiology. Vol.III (Comar and Bronner, eds.), p. 1. New York: Academic Press 1969.

MATTHEWS, J.L.,MARTIN, J.H., ARSENIS, C., EISENSTEIN, R., KUETTNER, K.: In: Cellular Mechanisms for Calcium Transfer and Homeostasis (Nichols and Wasserman, eds.), p. 239. New York: Academic Press 1971.

MILHAUD, G., AUBERT, J.P.: In: Radioisotopes in Scientific Research, Intern. Conf. UNESCO, p. 120. London: Pergamon Press 1957.

NEUMAN, W.F., RAMP, W.K.: In: Cellular Mechanisms for Calcium Transfer and Homeostasis (Nichols and Wasserman, eds.), p. 197. New York: Academic Press 1971.

OATLEY, K., GOODWIN, B.C.: In: Biological Rhythms and Human Performance (Colquhoun, ed.), p. 1. New York: Academic Press 1971.

PERAULT-STAUB, A.M., STAUB, J.F., MILHAUD, G.: Proceedings of the 9th Europ. Symp. on Calcified Tissues (Vienna, Czitober and Eschberger, eds.) p. 393 (1973).

TALMAGE, R.V.: The Am. J. of Anatomy, 129, no 4, 467 (1970).

Quality Control and Correlations of Clinical Methods for Studying Metabolic Bone Disease*

H. G. HAAS, T. LAUFFENBURGER, M. A. DAMBACHER, J. GUNČAGA, C. LENTNER, A. J. OLAH AND R. K. SCHENK

Abstract

1. The extent to which clinical methods reflect bone metabolism remains largely uncertain. In the present study, parameters of bone turnover (plasma alkaline phosphatase, urinary hydroxyproline excretion, calcium balance and calcium kinetic results, and histomorphometric data of bone biopsies) have been standardized and compared. Correlations between these various parameters were calculated for two defined groups of patients.

2. An overall reliability (methodical precision + biological reproducibility) of ± 10 - 15% SD was recorded for the biochemical, the calcium-kinetic and the morphometric data and of ± 50 mg calcium per day for the balance investigation respectively. These figures apply to strictly controlled conditions on a metabolic ward.

3. For the correlations between the various metabolic and histologic parameters two groups of subjects have been used: patients with postmenopausal osteoporosis before and after NaF treatment and subjects with Paget's disease on and off the phosphonate EHDP. The correlation coefficients between the parameters of bone formation (extent of osteoid seams in the bone biopsy, calcium accretion rate V_{O+} and alkaline phosphatase) exceeded .86 with $p < .001$. An equally good correlation between the radiocalcium turnover data and the urinary hydroxyproline excretion was found ($r = .97$, $p < .001$). On the other hand, no correlation could be detected between the number of osteoclasts in the bone biopsies and the calcium mobilization rate V_{O-}, and only a fair correlation was seen between the latter index and the urinary hydroxyproline excretion.

   Thus, the bone formation and bone turnover of patients with metabolic bone disease may be assessed with precision and reasonable certainty, whereas the measurement of bone resorption remains an unsolved issue in man.

+The data of this investigation will be published as full paper elsewhere.

Chairmen: C.A. Baud and J.F. Dymling

Morphometric Microdensity Studies of Hard Tissue Sections, Utilising Optical Density Contour Maps and Associated Area Co-ordinated Computerised Data*

L. F. GORE, H. W. FERGUSON AND R. L. HARTLES

I. Introduction

Comparative studies of the mineralisation processes in control and experimental animals demanded the replacement of a single-point method of measuring microdensity by a more comprehensive and accurate quantitative method of evaluating the concentration and distribution of hydroxyapatite in their bones and teeth.

The methods used in the preparation of undecalcified sections and their subsequent microradiography are described by McQUEEN, (1972). In the present study, care has been taken to use only plano-parallel sections, of approximately 100 microns thickness, with an overall deviation of less than 2 microns. At least ten measurements were taken on each section, each measurement being a sum value obtained from the use of opposing gauge anvils.[+]

A more satisfactory control of the physical principles influencing the reliability of the contact microradiographs was achieved by x-raying the normal, and experimental tissues, together with a reference step wedge, on the same plate.

No embedding medium was used for the tissue, thereby avoiding the additional absorption due to the resin. The absorption equivalent of 100 μm thickness of these plastics is approcimately 3 μm of aluminium. Assuming that the lower the concentration of mineral in the tissue the more resin is proportionally present (BOOTHROYD, 1964), it follows therefore that the lowest levels of hydroxyapatite represent an artificially higher absorption value, and that the density contrast between these lower concentration levels is therefore reduced.

All of the microradiographs used were required to match a standard density gradient, derived from optical density readings of the aluminium step wedge.

II. Methods

Photomicrographs of all microradiographs were taken prior to microdensitometry. A polaroid microscope camera attachment was used for this purpose. The photographs were used to assist orientation of the specimen image on the microdensitometer table, and for subsequent comparative evaluation of density images. A standard Joyce Loebl double-beam recording microdensitometer was converted to a fully automatic 2-dimensional data acquisition system by the addition of stepping motors and potentiometer-linked punch-tape output.

[+]Feinpruf "millitron" measuring system

In order to obtain as representative a picture as possible of the comparative changes in the control and experimental cat femora, as much of the disc as possible, compatible with scanning time, was recorded on punch tape. The bone discs rotated manually after each rectangular area, (raster), was scanned, to provide four tapes taken from opposing sectors of each bone disc. In total, each raster provided a punch tape containing approximately 20.000 digitised optical density readings, taken in 5 micron steps, from an area approximately 1.000 microns by 5.000 microns.

Fig. 1 shows two bone discs, Step wedge and Raster Sectors. The time taken to scan each raster was about 25 minutes. Several readings were taken from the center of each step of the aluminium wedge. This procedure was made possible by manual control of the stepping motors and the use of the single point density 'READ' facility on the Auto-Densi-Dater. By avoiding the influence of optical artifacts, and physical irregularities, a reliable set of average step readings was thus obtained. The aluminium step/density equivalents, together with the section thickness in microns and the relevant disc-stored+ programs requests were submitted on a separate lead-in tape spliced to each spool of data tape.

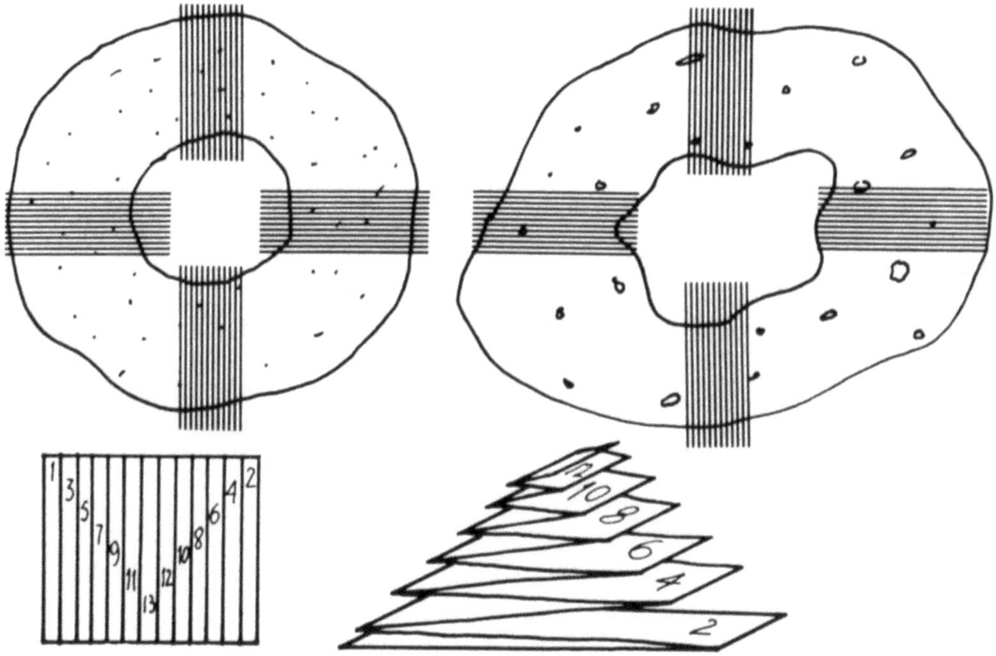

Fig. 1. Drawing of control and experimental bone discs showing raster areas. Construction of the accompanying aluminium step wedge is also shown

+The three computer programs controling data conversion are stored on magnetic discs for immediate access.

III. Data Conversion Methods

Stage 1. The transformation of the digital information by the computer was accomplished in three stages. The punch-tape information was transferred to a magnetic tape and a disc-stored program used to convert each digital value into its equivalent value of hydroxyapatite. A standard formula was used in which the variable factor of aluminum thickness was obtained from a reference 'curve' giving equivalent thickness of aluminum to optical density. The section thickness Factor 'T' was included as a tape constant. The resultant values of hydroxyapatite were obtained from the line Printer, i.e. in the same line and step formation, as dictated by the original scan pattern.

Stage 2. The appropriate data cards were prepared to retrieve the second disc program. These were submitted together with the types of variable data as noted below ..... X .. Y .. Area co-ordinates, number of lines and steps, step size, effective slit dimensions and the various density levels requiring mathematical analysis.

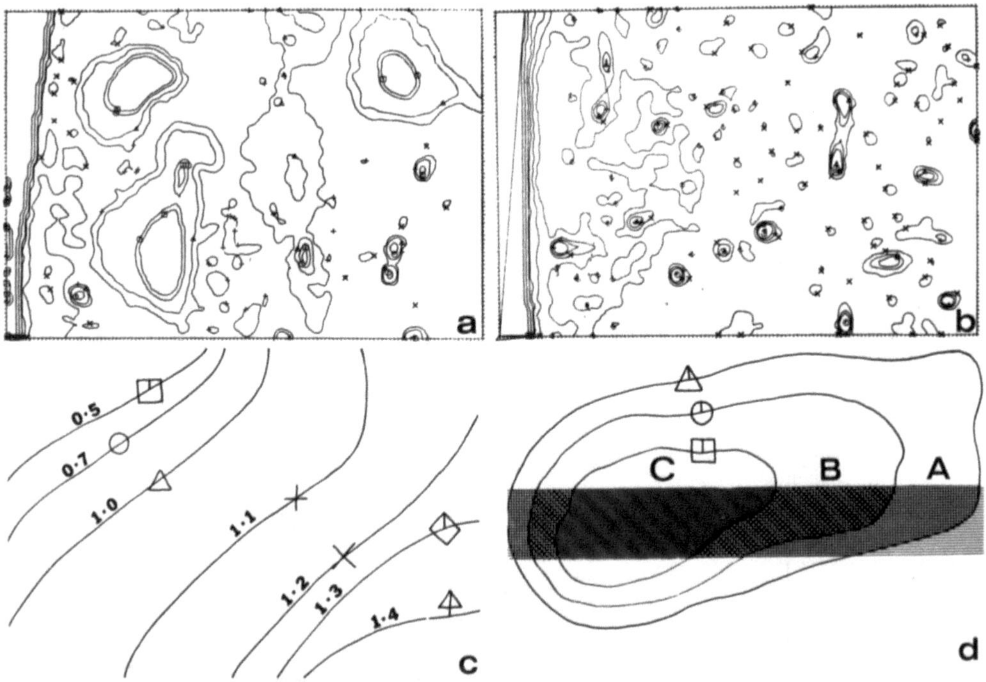

Fig. 2. (a) Graph plot of 'osteoporotic' bone (cat). (b) Graph plot of normal bone (cat). (c) Code symbols for isoplots from 0.5 to 1.4 AV Conc. g/ccm HA. (d)

Typical measurements from a selected area

	A	B	C
Area cm^2	0.345^{-4}	0.168^{-4}	0.450^{-5}
AV Conc. g/ccm	0.840	0.307	0
Total g.	0.145^{-6}	0.257^{-6}	0

This stage provided: (a) Total area occupied by each given density level of hydroxyapatite (expressed in sq. cm.). (b) Total hydroxyapatite present in the tissue scanned, which was also available as a differential analysis of various density levels as required (expressed in sq. cm.). (c) Average concentration of hydroxyapatite in each density level (expressed in grams/cubic cm). The X and Y co-ordinates submitted determined the area from which the above information was derived.

Stage 3. The third disc-stored program transferred the converted data to a graphic form. Contour lines were drawn between similar values of hydroxyapatite, thus creating a visual interpretation of the mathematical plateaux revealed by the computer.

Fig 2a shows a typical isoplot. A basic program has drawn a contour map of the whole area scanned depicting seven levels of density. (The code mark assigned to each density contour is shown in Fig. 2c. From this preliminary map (or the Isodensitrace mentioned later) specific area references may be resubmitted, e.g. to obtain data from one or more osteones, Fig. 2d, or, by excluding black border areas of emulsion, summate the total area of the black osteone cavities.

IV. Isodensitrace

We have retained the original pantographic capability of the microdensitometer to produce a colored two-dimensional enlarged map of the image scanned. This utilises the familiar drop-line technique, CRABB and MORTIMER (1967) and MILLER et al. (1964).

The following technique was used to provide co-ordinates, from which computer-stored information could be directly related to areas in the Isodensitrace. The Auto-Densi-Dater was programed to provide a single line of digital readings from the first and last lines of the raster. The image is subsequently returned to the first line of the raster in order to collect normal continuous sequential line data covering the whole of the raster.

Having completed this data collecting scan of the raster, the image was once again returned to the original starting point. Without altering the microradiograph's position on the scanning stage, the 'Y' axis stepping motor unlocked and the ratio-arm of the continuous 'Y' movement was locked in. A continuous scan was then commenced, with appropriate 'X' axis movement of the specimen and recording stages. It will be seen that 'X' co-ordinates can be obtained from either the first or last lines of the raster. Alignment of 'stepping data' with the continuous trace image(IDT) is achieved by using a print-out of the first and last line of the area scanned. A line drawn to connect the end-of-line characters of these two lines produces a 'Y' line reference, this consists of end-of-line characters which can be used by the computer as a reference point. The two lines of data words are accurately positioned by matching abrupt changes in digital value with obvious visual features. The spatial value assigned to each data word[+] is a product of the original stepping interval and the ratio arm used, e.g. a stepping interval of 5μ would be equivalent to 5000 when using a 100:1 ratio arm. Digital data may therefore , this instance, be related to the IDT with an accuracy of 0.5 mm.

[+]A 'Data Word' is the three digit number representing each density value reading.

V. Discussion

The techniques described provide a method of automatic collection, assessment and comparison of optical-density-related data from representative areas of bone formed under varying conditions. This method should prove of value to workers who are using microdensitometric techniques and who have access to computer units.

The preparation of IDT maps is of value in the initial stages of a project to recognise areas of interest. Subsequently their value is more limited.

Whichever microdensitometer instrument is used, we would, however, recommend the need for a permanent graphic record of the areas scanned, such as that provided by a Graph plotting computer terminal. This provides not only a visual assessment of mineral distribution, but also an essential safeguard against digital misalignment or image inversion.

VI. Summary

The study of diet-induced bone dyscrasias in cats demands the careful examination of micromorphic appearances, together with related quantitative evaluation. Only by automated microdensitometer techniques can sufficient data be obtained for reliable assumptions on bone dynamics.

A comparative evaluation technique is recommended together with an emphasis on visual representation using isopleths of various types. A double-beam recording microdensitometer has been modified to give a punch-tape output, and a program designed, using a KDF9 computer, to provide conversion of the recorded optical density values to hydroxyapatite concentration, together with various area-associated mathematical computations and concentrations levels depicted in graphical form.

References

BOOTHROYD, B.: The problem of demineralisation in thin sections of fully calcified bone. J. cell Biol. 20, 165-173 (1964).

CRABB, MORTIMER: Two-dimensional microdensitometry. Brit. dent. J. 122, 337-343 (1967).

McQUEEN, C.M., MONK, I.B., HORTON, P.W. and SMITH, D.A.: Preparation of undecalcified bone sections for auto- and microradiography. Ca. Tissue Research 10, 23-30 (1972).

MILLER, C.S., PARSONS, F.G., KOFSKY: Simplified two-dimensional microdensitometry. Nature 202, 1196-1200 (1964).

On Calcium Metabolism during Immobilization

H. Bohr

It is well known that decisive changes in metabolism take place during immobilization. Thus it was shown by DEITRICK, WHEDON and SHORR (1948) that excretion of calcium in the urine and feces increased, resulting in a negative calcium balance. This has been confirmed by among others, MACK (1965), ROSE (1966) and LYNCH et al. (1967), and a corresponding loss of calcium from the skeleton has been demonstrated through densitometric measurements (MACK, 1965).

In the present investigation the urinary excretion of calcium and the bone density were recorded for patients with Coxa Plana being treated with prolonged bed rest. The children were from 4 to 12 years old and healthy apart from the hip disease. Bed rest was carried out strictly and combined with extension of both legs for periods of 6 weeks with a 2-weeks' interval, usually over 12 to 18 months. It was followed by gradual mobilization on crutches with weight-bearing on the healthy leg for the first 3 months, then on both legs for another 3 months until the crutches were finally discarded. Food was of normal variety, including at least half a liter of milk daily, and the intake of calcium was estimated about 7 00 to 1000 mg/day. Urine was collected in periods of 6 days and calcium content determined by the method of KRAMER and TISDALL (1921). Measurements of bone density performed photometrically on X-ray photos, as described by among others, MACK, BROWN and TRAPP (1949), HENNY (1956), and HEUCK and SCHMIDT (1960). Radiographs of the hips, the heel bones and the wrists were taken on admission to hospital and at regular intervals thereafter. For calibration, ivory stepwedges were placed along the bone during the exposures. The photometric analysis was performed with a special apparatus constructed for this purpose, where the film is mechanically carried forward between the light source and a photocell, and the intensity of the transmitted light in a 2-mm wide zone is recorded with a potentiometer. The measurements on the femoral head were made in the frontal view from the joint space to the epiphyseal plate, corresponding to a sagittal section through the middle of the epiphyseal bone. As regards the heel bone, measurements were made on X-ray photos in the lateral view from the upper to the lower border 2 cm in front of the tuber calcanei, while the radial bone was measured on X-ray photos of the wrist 1 cm proximal to the distal epiphyseal plate. The density of the bones measured was expressed in arbitrary units related to the step wedge; since only comparisons between the same bone at intervals of time are made, expressions per unit bone volume were not appropriate

The results of the measurements of calcium output are shown in Fig. 1. Each figure presents the average value of about 30 cases S.D. about $\pm$ 5 percent. In a few cases the calcium output was determined before immobilization, where the values obtained corresponded to the calcium excretion in normal children of that age. It is seen that the calcium output rises during the first two weeks of immobilization to reach a maximum value of about 200 mg/day in week 4. During the following two months calcium output falls to about 140 mg/day, then gradually decreases at a slower rate. At the end of one year the output is still somewhat elevated and resumes normal values only when mobilization

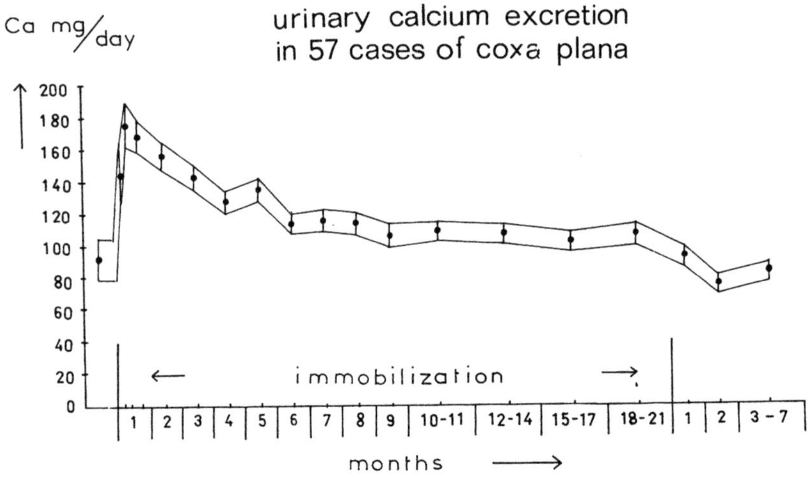

Fig.1. Urinary calcium output during immobilization and subsequent gradual mobilization

starts. Calcium concentration runs a similar course. The results of the densitometric measurements are given in Table 1 where the changes in the density of the normal femoral head, as well as the os calcis and the radial bone on both sides, are recorded as a percentage of the initial measurements. The standard deviation in the measurements on the hip is somewhat larger than that of the other bones, due to the greater amount of soft tissue which makes the measurements very sensitive to quite small displacements of the calibration wedges. It is seen that the density of the normal femoral head decreases by 21% during the first 3 months of immobilization, and by 29% after 6 months of immobilization. This reduction is retained until the end of immobilization, when the density rises, but does not reach normal values until 9 months later. The density of the os calcis has decreased by 17 and 23% respectively at 3 and 6 months of immobilization, and follows almost the same course as the femoral head. In the radial bone, however, only a 4% decrease was seen, probably owing to the lack of immobilization of the upper extremities. There was no significant difference between the values for the diseased and healthy side as regards the heel bone and the radial bone.

The decrease in density of the femoral head and the os calcis takes place mainly during the first 3 months of immobilization, corresponding to maximum urinary excretion of calcium. The increased excretion of calcium in the urine during that period amounts to about 8 g, and as the loss of calcium through the feces according to previous authors (DEITRICK, WHEDON, SHORR, 1948; MACK, 1965) is about twice the loss in the urine, the total loss of calcium should amount to about 25 g. During the same period the density of the os calcis and the normal femoral head decreases by about 20%. If this decrease in density were representative of the whole skeleton, it would correspond to a loss of about 50 g calcium. The discrepancy suggests that osteoporosis was most pronounced in the bones of the lower extremities, probably due to the more efficient immobilization of these. During the remaining period of immobilization there were only small changes in the density of the bones, indicating that an equilibrium is established between

Table 1. Densitometric measurements in cases of C-P (Density expressed as percentage difference from initial values with standard deviation of the mean in brackets)

Measurements	No cases	Months of immobilization				Months after e.l.		
		3	6	9	12	3	6	9
Femoral head	29	- 21%	- 29%	- 26%	- 29%	- 19%	- 16%	- 3%
normal side		(± 4)	(± 4)	(± 5)	(± 4)	(± 5)	(± 5)	(± 7)
Heel bone	17	- 17%	- 23%	- 18%	- 18%	- 15%	- 9%	+ 4%
right + left		(± 2)	(± 3)	(± 2)	(± 3)	(± 3)	(± 3)	(± 3)
Radial bone	31	- 4%	- 4%	- 4%	- 1%	+ 2%	+ 5%	+ 5%
right + left		(± 1)	(± 1)	(± 2)	(± 2)	(± 2)	(± 2)	(± 3)

bone formation and bone resorption. This is in agreement with the view that it is possible to distinguish between the early stage of immobilization with an increased metabolism of the bones, where resorption predominates, and a secondary more chronic stage with a decreased metabolism, where bone formation and bone resorption are equally reduced (HEANEY, 1964; MACH, 1971; EICHLER, 1970). Following mobilization, bone formation increases with gradual recovery of the normal bone density.

Abstract

In children with Coxa Plana the excretion of calcium in the urine and densitometric measurements of different bones were recorded during treatment with prolonged bed rest and gradual mobilization. It was shown that the calcium output increases during the first 2 months of immobilization to reach a maximum value of about 200 mg/day in the week 4. The calcium output gradually decreases to normal values when mobilization starts. The density of the os calcis and the normal femoral head decreased by about 20% in the first 3 months, but during the remaining period of immobilization there were only small changes, followed by gradual recovery of normal bone density during mobilization.

References

DEITRICK, J.E., WHEDON, G.D., SHORR, E.: Effects of immobilization upon various metabolic and physiological functions of normal men. Am. J. Med. 4, 1-36 (1948).

EICHLER, J.: Inaktivitätsosteoporose. Aktuelle Orthopaedie 3. Stuttgart: Georg Thieme 1970.

HEANEY, R.P.: Disuse Osteoporosis. Dynamic Studies of Metabolic Bone Disease. Oxford: Blackwell Scientific Publ. 1964.

HENNY, G.C.: Roentgenographic estimation of the mineral content of bone. Radiol. 54, 203-209 (1950).

HEUCK, F., SCHMIDT, E.: Die quantitative Bestimmung des Mineralgehaltes der Knochen aus dem Röntgenbild. Fortschr. Röntgenstr. 93, 523-554 (1960).

KRAMER, B., TISDALL, F. F.: J. Biol. Chem. 47, 475 (1921).

LYNCH, Th.N., JENSEN, R.L., STEVENS, P.M., JOHNSON, R.L., LAMB, L.E.: Metabolic effects of prolonged bed rest: Their modification by simulated altitude. Aerospace Med. 38, 10-20 (1967).

MACH, J.: Die Inaktivitätsosteoporose. Berlin: VEB Verlag Volk und Gesundheit 1971.

MACK, Pauline B.: Calcium loss studies during human bed rest. NASA, S.P. 64, 169-176 (1965).

MACK, Pauline B., BROWN, W.N. and TRAPP, H.D.: The quantitative evaluation of bone density. Am. J. Roentgenology 61, 808-825 (1949).

ROSE, A.G.: Immobilization osteoporosis. Brit. J. Surg. 53, 9, 769-774 (1966).

Limitations in Bone-Mass Measurements with ^{241}Am

G. E. RÖSINGH, G. HART, J. B. v. D. SCHOOT AND K. L. M. BON-NIJSSEN

In orthopedics there is a need for reliable parameters of sufficient precision to detect changes of bone-mass in vivo in parts of extremities. During investigations on femoral head-necrosis (RÖSINGH, et al., 1972) it was decided to undertake work in this direction. The investigations of SORENSON and CAMERON (1967) seemed a reliable starting point. However, ^{125}I shows insufficient penetration of γ-radiation (27.4 KeV) in heavy parts of extremities. It is mentioned in the literature that the precision of measurements with an ^{241}Am source approximates that of an ^{125}I source, whereas its harder radiation (59.6 KeV) gives a better penetration. Detailed information, however, is only given by NILSSON (1966) who used another method. A disadvantage of the high energy of ^{241}Am-radiation is the decrease in difference between mass absorption-coefficients of water and hydroxyapatite with a factor 2/3 compared with that of ^{125}I (OMNELL, 1957). Therefore it was feared that the precision of the measured parameter (the bone-mass per unit of bone-length) would be reduced considerably by this loss of contrast.

In view of our purpose and the facts mentioned above, an ^{241}Am source was used and the time in millisec. for a pre-set number of 10.000 counts was determined. The separate measurements were carried out with the source-detector unit at standstill. In between the measurements the source-detector unit was moved over a fixed distance of 2 mm in the direction of the desired cross-section of the object. In our set-up the influence of stochastic variation of source decay on the precision of measurements is constant and independent of the bone-mass. In general, measurements are carried out with the source-detector apparatus moving at constant speed across the object. In that way one has to count during pre-set time intervals, which implies a decrease of precision with increasing bone-mass.

In our method mathematical and statistical analysis of point-measurements is also facilitated, since the model can be defined better; measurements of each separate point can now be regarded correctly as the determin tion of one point of the curve reflecting the bone-mass per square-unit in the cross-section under consideration. The precision of each separate measurement is then given by the stochastic decay-variation. This situation offers the possibility of using a known curve-fitting technique. The Fourier-analysis seems the best method for the case under consideration. An application will be discussed later on.

The technical construction of the measuring design was as follows. The source consists of a ceramic sphere of 3 mm diameter on which 45 mCi ^{241}Am was adsorbed. A more powerful source would - by increasing self-absorption - yield only a relatively slight raise in output. The gamma-radiation was caught in a 5 cm thick NaI'TI crystal of an old linear scintigraph and amplified with 1100V. Apertures of steal and lead source and detector diaphragms were 3 mm and 4 mm respectively in front of the source, 5 mm and 6 mm in front of the detector. The 59.6 KeV radiation of ^{241}Am was measured in a 10 V wide window on the photo-peak using pulse height analysis.

To gain experience and to compare results with data from the literature, measurements in the fore-arm were carried out. Although for multipurpose use the distance between source and detector was adjustable in the holder, in this investigation it was constantly kept at 25 cm. The arm was submerged in a plastic tray with a water height of 6 cm in the field of investigation; it was fastened and its place recorded as accurately as possible. A series of about 50 measurements per investigation were performed and the total time per investigation was at least 10 min. Because of incomplete fixation of the fore-arm the testee had to hold his arm as motionless as possible during the investigation.

The bone-mass per unit length was computed from the equation: $M_B = k\Sigma_i \, LnT_i/To$; k is a constant dependent on mass absorption- coefficients and densities; T_i is the time measured in localisation $_i$ and To is the time measured outside the arm (10 measurements at each side). The unbiasedness of the estimator (M_B) of the bone-mass per unit length was examined with the Fourier-analysis. As can be anticipated it was found that the unbiasedness is dependent on the number of measurements in bone and the regularity of the curve expressing bone-mass per unit square as a function of localisation in the cross-section. The Fourier-analysis gives a mathematical relationship between the number of bone measurements and its curve regularity which has to be fulfilled for the unbiasedness of the estimator M_B. Experiments showed that the relationship generally could be fulfilled if at least 9 or 10 points of measurement were situated in the bone.

As far as the precision of the results is concerned it may be mentioned first that 10.000 counts-measurements through liquid media gave S.D. which did not deviate from the theoretical S.D. of 1%. Sources of error in the measuring equipment consequently could be neglected. By investigations on a dead male arm immersed in formalin, sources of error unrelated to the living patient were traced. Particularly, series of measurements could be performed in exactly the same cross-section without disturbance by movement of the testee or time-related changes in the bone. The precision for the bone-mass in 10 successive series of measurements in the same cross-section was equal to the value expected on grounds of the variation in source decay.

The influence of inadequate fixation on the measured bone-mass of the fore-arm in the watertray was investigated in the two directions of displacement. First the cadaverarm was displaced in the direction of source movement after each series of measurements in a sequence of 10 series of measurements. The precision remained concurrent with the source variation. Next the bone-mass in successive cross-sections of radius and ulna on distances of $\frac{1}{2}$ cm was determined. This revealed only small changes in bone-mass per cross-section near the wrist in the radius (Fig. 1). A cross-section at 15% of the ulna length above the wrist seemed a favorable area for investigations, the more so as only in this area 9 or 10 measurements were inside the bone, maximum bone-remodeling response in this metaphyseal area could be expected and a reasonable distance between radius and ulna was guaranteed. At last the influence of different distances between the measuring points on the precision was determined. Changing the distance from $\frac{1}{2}$ mm to 1 mm and to 2 mm did not influence precision; a significant decrease of precision was found for distances of 3 mm and 4 mm. Making allowance for these data other influences on precision were studied in the living subject. So 11 non-successive series of measurements within one week were performed on one of us. The S.D. of bone-mass per unit-length was now 2.7% which is higher than could be expected on grounds of source-variation alone. A Fourier-analysis demonstrated also differences related to the dispersion of bone-mass over the cross-section. With respect to this dispersion two groups of similarity existed of respectively 4 (1, 9, 10, 11) and 7 series of measurements. The S.D. of bone-mass within these groups was 1,1% and 1,3% which approximated the source variation. It was evident that even in most favorable circumstances, displacements of the arm and involuntary movements impaired precision. Also a marked decrease of transmission was found in some soft tissue areas. At the site of the membrana inter ossea an absorption of 4% of the total absorption in radius and ulna was always found, which cannot be caused by the diameter of the photon beam alone (Fig. 2). It may be supposed therefore that formation of fibrous tissue in bone marrow cavities will simulate an increase in bone-mass. A first successful attempt to measure changes in bone-mass by biological processes was made after reconstruction of a radius-pseudoarthrosis with a homogenous bone graft in a 26 year-old sailor. Three non-successive investigations were made at 5 months and $7\frac{1}{2}$ months after the operation. The disuse atrophy of the bone already present before operation decreased by 10% under the influence of increasing functional weight bearing, which means about 1% per week. It remained significantly below the normal value of the sound arm however.

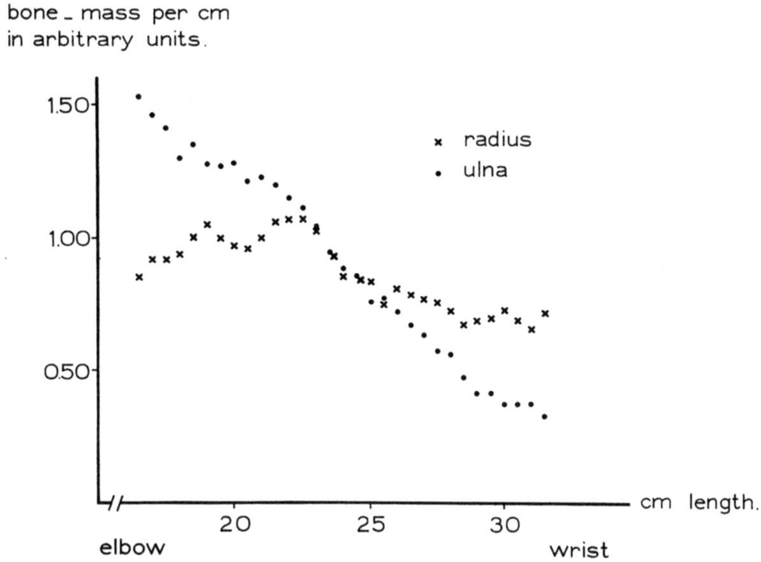

Fig. 1

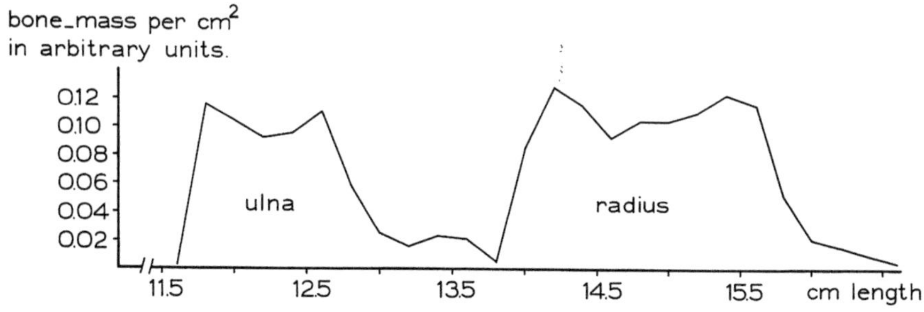

Fig. 2

Measurements of the fore-arm in kidney-transplanted patients were a total failure. As possible causes may be mentioned in the first place the difficulty these patients had in keeping their arm motionless during the investigation. Secondly, one has to consider that the low bone values may play a role. The stochastic source variation having an influence on bone values independent of the amount of bone-mass therefore gets more influence on the relative precision as the bone-mass decreases. Other projects concerning fore-arm measurements are still in progress.

The overall impression of this kind of investigation is that one is dealing with a method with a very high degree of precision. Variation in localising in repeated measurements of the same cross-section and stable positioning remain the most important sources of error. The most important problem, however, comes from the observation of decreased transmission of radiation through dense connective tissue related to values outside the arm. It implicates that tendons in upper parts of extremities as well as connective tissue formation in bone marrow cavities may lead to an apparent rise of bone-mass value.

References

NILSSON, B.R.: Post traumatic osteopenia. Acta Orthop. Scand. suppl. 91 (1966).

OMNELL, K.A.: Quantitative roentgenologic studies on changes in mineral content of bone *in vivo*. Acta Radiol. suppl. 148 (1957).

RÖSINGH, G.E., STEENDIJK, R., VAN DEN HOOFF, A.: Necrosis and deformation of the femoral head following fracture of the femoral neck. Arch. Chir. Neerl. XXIV - II, 165 (1972).

SORENSON, J.A., CAMERON, J.R.: A reliable *in vivo* measurement of bone-mineral content. J. Bone Jt Surg. 49 A, 481 (1967).

Intestinal Absorption and Retention of Calcium Measured by Whole-Body Counting*

S. P. NIELSEN, O. BÄRENHOLDT AND O. MUNCK

I. Abstract

A method for measurement of fractional intestinal absorption of an oral dose ^{47}Ca by whole-body counting is described. The intestinal absorption of Ca is calculated as the whole-body counting rate 6 days after the administration of radio-calcium, divided by the counting rate 6 days after a subsequent equal dose of ^{47}Ca given intravenously, minus the extrapolated value which would then have occured if no injection had taken place. The method is based on the application of a power function model. It supplies additional information about the fractional turnover of ^{47}Ca. The relationship between intestinal absorption and whole-body retention of Ca is outlined. Normal values for 21 adults are given.

Measurement of whole-body retention of an oral dose of ^{47}Ca (R) is of practical value in some clinical situations. R is a composite parameter, being dependent on intestinal Ca absorption, losses of Ca from the body, and probably Ca turnover in the skeleton. Furthermore, R is bound to be dependent on geometry factors and radiation absorption (e.g. degree of obesity), since the linear attenuation coefficient for ^{47}Ca is 6%/cm of tissue.

In the following we shall demonstrate that, by intravenous (i.v.) calibration, measurement of R can be combined with measurement of intestinal absorption of Ca (A). By this method the influence of geometry factors and radiation absorption is excluded and additional Ca-metabolic data are obtained.

II. Methods

Procedure: A low-background, iron-shielded three-detector linearly scanning whole-body counter was used (International Atomic Energy Agency, 1970). ^{47}Ca was counted in the energy interval 1.240-1.376 MeV. The background of the empty counter was .82 cps, and the whole-body background due to ^{40}K was .10-.32 cps more than the background of the empty counter in 21 normal adults.

^{47}Ca (Radiochemical Centre, Amersham, specific activity 500 μ Ci/mg) was given as an oral dose of 3 μ Ci, together with .5 g $CaCl_2$, $2H_2O$ (.136 g Ca) dissolved in 250 ml distilled water, on time zero. Whole-body counting was performed immediately hereafter and with one or two hours intervals until a maximum was reached. Thereafter countings were

+ This investigation was supported by grants from Ebba Celinders legat, Rigsforeningen til gigtens bekæmpelse and P.A. BRANDTS legat.

performed daily for 8 days or on day 5, 6, 7 and 8. Immediately after the counting on day 8, an i.v. injection of about one μCi ^{47}Ca from the same batch was given quantitatively, using a membrane needle. An oral load of $CaCl_2$, $2H_2O$ was given as described above, 30 minutes before the injection. Again, countings were performed daily hereafter, or on day 13, 14 and 15. The exact amounts of ^{47}Ca given orally and intravenously were measured by weight differences of syringes (= M_{oral} and M_{inject}, respectively). All countings were corrected for background and physical decay to time zero.

<u>Calculations and Definitions:</u> Whole-body retention of ^{47}Ca after an oral dose was measured as described by SJÖBERG (1970) by dividing the counting rate 7 days after the oral administration N_7 by the maximum value on the first day N_{max}.

We applied a power function model for our studies (MARSHALL, 1969). The disappearance curve for ^{47}Ca after passage of unabsorbed ^{47}Ca is rectilinear in a log-log plot (Fig. 1a) for more than 20 days (Fig. 1b). Therefore, extrapolation of the oral curve in the log-log plot is possible, and hence a "hundred per cent value" given by the i.v. dose can be obtained by subtracting the extrapolated value 6 days after the injection N'_{14}, from the then measured value N_{14} (Fig. 1 a). The counting rate 6 days after the oral administration N_6 divided by the thus obtained "hundred per cent value" is a measure of intestinal absorption, if $M_{oral} = M_{inject}$. Correction for differences in M_{oral} and M_{inject} is made according to the equation:

$$A = \frac{N_6}{N_{14} - N'_{14}} \cdot \frac{M_{inject}}{M_{oral}}$$

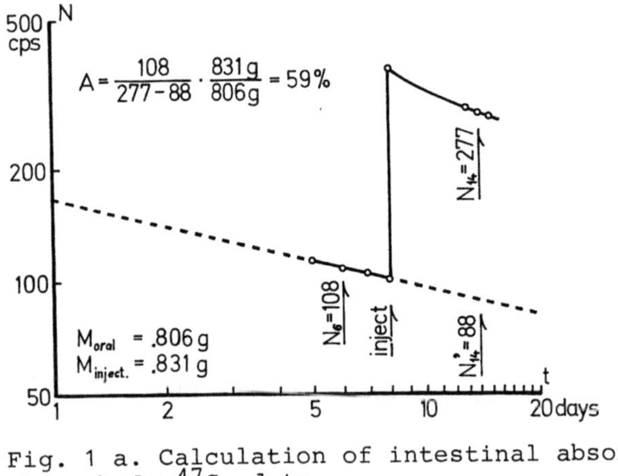

Fig. 1 a. Calculation of intestinal absorption of calcium A from whole-body ^{47}Ca data

'T' is the time at which the extrapolated disappearance curve in the log-log plot crosses the line denoting the maximum counting rate N_{inject} after injection (Fig. 1 b, below).

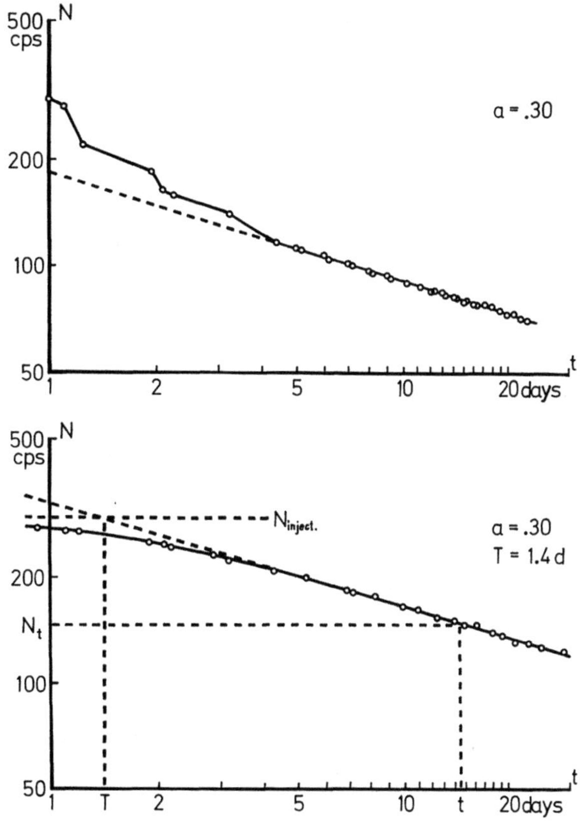

Fig. 1 b. Whole-body ^{47}Ca disappearance curves in a normal adult after oral ^{47}Ca administration (above) and after intravenous injection of ^{47}Ca three months later (below). Identical slopes

The shape of the whole-body curves implies that from about day 5 (Fig. 1 b, below) the fractional turnover of radio-calcium is inversely proportional to the time t after the administration i.e. $\frac{dM_t}{M_t dt} = \frac{a}{t}$, where M_t is the amount of radio-calcium in the body at the time t, if $t \gg T$. The interrelation between the four parameters in Table I, if $t \gg T$, is given by the following equation: $M_t = A\, M_{oral} (\frac{t}{T})^a$.

This can be seen from the following: We assume that $\frac{M_t}{M_{inject}} = \frac{N_t}{N_{inject}}$. Therefore $\log \frac{M_t}{M_{inject}} = \log \frac{N_t}{N_{inject}}$. From Fig. 1 b, below, it can be seen, if $t \gg T$, that $\log \frac{N_t}{N_{inject}} = a \log \frac{t}{T} = (\frac{t}{T})^a$, where $\log \frac{t}{T}$ represents a difference of the abscissa and $\log \frac{N_t}{N_{inject}}$ the corresponding

Table 1. ^{47}Ca whole-body data from normal adults. Six of the 21 subjects were re-investigated after 3 months (replicates).

	Retention after 7 days R %	Intestinal absorption A %	Slope a	T days	n
Mean	30	46	-.32	1.7	21
SD	8	16	.08	.5	
SD of replicates	4	5	.04	.4	6

difference of the ordinate. It follows from this that $\log \frac{M_t}{M_{inject}} = \log \left(\frac{t}{T}\right)^a$.

Therefore $M_t = M_{inject} \left(\frac{t}{T}\right)^a$.

The amount of radioactivity M_t retained at the time t after oral administration of M_{oral} is, accordingly, if $t \gg T$, after intestinal passage of unabsorbed ^{47}Ca:

$$M_t = A\, M_{oral} \left(\frac{t}{T}\right)^a.$$

III. Result and Comments

Normal Subjects: Twenty-one adults (volunteers from the laboratory staff, all in good health, 7 males and 14 females 20-77 years old) participated in the study. They were fasting for 12 hours before and 5 hours after the oral and intravenous ^{47}Ca administration. No dietary precautions were taken. The mean values and SD for R, A, a and T are shown in Table 1.

SJÖBERG (1970) in 27 normal Swedish adults, studied in an identical fashion, found nearly the same R-values (mean 36%, SD 8%) as we did. $\frac{M_7}{M_{oral}}$ is somewhat smaller than $R = \frac{N_7}{N_{max}}$ because N_{max} is extraordinarily reduced by radiation absorption in the tissues due to the location of a large fraction of the radioactivity in the gastrointestinal tract.

In our subjects $\frac{R}{M_7 / M_{oral}} = 1.06$ (SD .02) on an average. This value is significantly different from 1.00 ($P < 0.01$). Hence measurement of R without correction is misleading. This will especially be true in obese persons.

Our mean value for a in normal adults is nearly identical with the mean value (.30) found by MARSHALL (1969). The mean A is bound to be higher than the mean R, because losses of radio-calcium will influence R and not A. Contrary to A, R is therefore likely to be dependent on the function of the kidneys.

Clearly, the slope a is dependent on bone Ca turnover and losses of Ca from the body (by the intestinal and renal route, and by sweat). T is possibly dependent on the diffusion constants in bone. It is used in this context as an operational parameter only. Its physiological meaning and its importance as a clinical parameter remain to be established.

The method described can easily be applied in clinical investigative work, although its value is limited by the requirement that the patient must be in a steady state situation regarding Ca metabolism for two weeks.

References

Directory of whole-body radioactivity monitors. International Atomic Energy Agency, Vienna, 1970 Edition (DK 3.1).

SJÖBERG, H.E.: Retention of orally administered ^{47}Calcium in man under normal and diseased conditions studies with a whole-body counter technique. Acta Med. Scand. Suppl. 509 (1970).

MARSHALL, J.H.: Measurements and Models of Skeletal Metabolism. In: Mineral Metabolism vol. III, pp. 2-122. C.L. Comar and F. Bronner, Eds. New York, London: Academic Press 1969.

II. Pharmacology and Metabolism of Vitamin D

Chairmen: S. Balsan and R. H. Wasserman

Studies on the Metabolism, Mode of Action and Pharmacology of Vitamin D and Related Analogs*

A.W. Norman, W.H. Okamura, M.N. Mitra and R.L. Johnson

In recent years intensive efforts by a number of laboratories have been directed toward the elucidation of the many parameters involved in the regulation of calcium and phosphorus metabolism. Three of the most important of these biological regulators are calciferol (vitamin D), parathyroid hormone, and calcitonin. With each of these regulators, outstanding developments and advances have been made in recent years. It is the purpose of this article to outline some of the developments that have occured, specifically with regard to our understanding of the mechanism of action of vitamin D in terms of its intervention in calcium homeostasis.

Perhaps the most striking advance has been the recognition that the mode of action of calciferol, which is chemically classified as a steroid[++] is highly similar to that of many other classical steroid hormones, such as estrogen, aldosterone, hydrocortisone, or testosterone (NORMAN, 1968). NORMAN and HENRY (1974a,b) have elaborated on the thesis that 1,25-dihydroxycholecalciferol is a steroid hormone produced by the kidney in response to various physiological signals. With this concept in mind, it is possible to identify new relationships between calciferol (the hormone), its secretory organ (the kidney), its target tissues (the intestine, skeleton, and perhaps kidney), and various disease states related to vitamin D.

+ This work was supported in part by U.S. Public Health Service grants AM-09012 and AM-14,750. A.W. NORMAN is a recipient of a PHS Career Development Award 1-KD-AM-13,654. We would like to acknowledge the invaluable assistance and discussions with Professors J.W. COBURN, S. MASSRY, A. BRICKMAN, D. HARTENBOWER, R. FRIEDLER, and Drs. H. HENRY, H.C. TSAI, A.M. SPIELVOGEL, and R.G. WONG and Mr.R.J. MIDGETT in preparation of this article. Also the technical assistance of Mr.T. CONKLIN in the preparation of some analogs of 1,25-$(OH)_2$-D_3 is gratefully acknowledged.

++Calciferol is used as a generic term for vitamin D, without distinction as to the structure of the side chain. Ergocalciferol and cholecalciferol, respectively, are vitamins D_2 and D_3. Ergocalciferol is the usual form of vitamin D employed clinically and used to fortify foods, while cholecalciferol is the natural form present in mammals. According to the International Union of Pure and Applied Chemistry Commission on the nomenclature of biological chemistry, cholecalciferol (vitamin D_3) may be defined as a steroid. The chemical name is 9,10-secocholesta-5,7,10 (19)-trien-8-ol. One International Unit (IU) of cholecalciferol (vitamin D_3) or ergocalciferol (vitamin D_2) has been defined to be 0.025 μg (65.0 pmoles) (League of Nations, 1935). No official definitions of units have been formulated for 25-hydroxycholecalciferol, 1,25-dihydroxycholecalciferol, or for any of the other calciferol metabolites. In this review 1.0 unit of 25-hydroxycholecalciferol and 1,25-dihydroxycholecalciferol is defined to be 65.0 pmoles which is approximately equivalent to 0.025 μg. Some of these problems are discussed by NORMAN (1972).

It is now firmly established that cholecalciferol undergoes an obligatory two-step metabolism prior to the production of the biologically active species, 1,25-dihydroxycholecalciferol. First calciferol which is either produced photochemically in the skin from 7-dehydrocholesterol or absorbed from the diet is transported to the liver where it becomes hydroxylated at the end of its side chain at carbon-25. This enzyme is believed to involve cytochrome P-450 and to be localized in the microsomal fraction of the cell (BHATTACHARYYA and DELUCA, 1973).

After hydroxylation by the liver, 25-hydroxycholecalciferol is transported to the kidney, where it undergoes a second hydroxylation, at carbon number 1, to produce 1,25-dihydroxycholecalciferol. A variety of data obtained in many laboratories support the contention that the kidney is the only site of the production of this steroid. When high specific activity radioactive 25-hydroxycholecalciferol is administered to vitamin D-deficient nephrectomized rats, no 1,25-dihydroxycholecal - ciferol is detected in any tissue (FRASER and KODICEK, 1970; WONG, 1972). Further, WONG et al. (1972) have clearly shown that only 1.25-dihydroxycholecalciferol and not calciferol or 25-hydroxycholecalciferol can stimulate intestinal calcium transport or enhance bone calcium mobilization in vitamin D-deficient, nephrectomized rats. In contrast to the liver hydroxylation system, the kidney hydroxylation system is localized exclusively in the mitochondrial fraction of the kidney cortex (MIDGETT et al., 1973). The enzymatic hydroxylation is inhibited by carbon monoxide, thus suggesting that it too is a cytochrome P-450 mediated reaction. Further, HENRY and NORMAN, 1974, have carried out an extensive phylogenic study of the distribution of this enzyme activity in 23 different species. To date we have unequivocally detected the presence of 25-hydroxycholecalciferol-1-hydroxylase activity in various species of mammalia, including primates, and in aves, reptilia, amphibia, and osteichthyes. Results from experiments such as these thoroughly substantiate the generalized distribution of the enzyme required for the production of 1,25-dihydroxycholecalciferol as being in the kidney. Clearly, the production of the steroid hormone, 1,25-dihydroxycholecalciferol, by the kidney has a universal distribution, at least in vertebrates.

Recent data obtained in several laboratories suggest the intriguing possibility that the activity of the 25-OH-D_3-1-hydroxylase present in the renal tissue is subject to regulation, so that the enzyme activity is inversely proportional to serum calcium levels. A variety of conflicting reports have purported to show that the activity is regulated by either dietary calcium (OMDAHL et al., 1972), serum calcium (BOYLE et al., 1971), circulating levels of parathyroid hormone (GARABEDIAN et al., 1972; GALANTE et al., 1972a; RASMUSSEN et al., 1972), circulating levels of calcitonin (GALANTE et al., 1972b) or the concentration of inorganic phosphate in renal tissue (TANAKA and DELUCA, 1973). At the present, it is not possible to state definitively whether any or all of these observations are applicable under all physiological circumstances. The fact that the 25-OH-D_3-1-hydroxylase is localized in the mitochondrial fraction of the cell imposes certain severe restrictions on the biochemical mechanisms by which this enzymatic activity might be increased. An elevation of enzyme activity may result from (a) an allosteric activation or inactivation of pre-existing enzyme molecules, (b) increased or decreased biosynthesis of enzyme molecules, or (c) increased or decreased biodegradation of enzyme molecules. A particularly challenging problem is to delineate the nature of the signal transmitted from the external membrane of the kidney cell to its mitochondria. Several logical candidates include the cellular concentrations of ionized calcium and inorganic phosphate. In view of the possible involvement of the peptide hormones, one should also not neglect a possible role of cyclic AMP. The external membrane of the renal tubular cell could respond to a variety of signals: these might include either of the peptide hormones, calcitonin or parathyroid hormone, or changes in the concentrations of serum calcium or phosphate, or some unknown and as yet unidentified factor. Thus, it seems appropriate to study this complex maze of possible interrelationships at several levels of organization, including the whole animal, slices of kidney tissue, isolated renal tubules, and isolated renal cortical mitochondria. It remains to the future to ascertain the detailed nature of these relationships in terms of the maintenance of calcium homeostasis. It also should be mentioned that in addition to 25-hydroxycholecalciferol and 1,25-dihydroxycholecalciferol, three other metabolites of vitamin D have also been characterized; these include 24, 25-dihydroxycholecalciferol (HOLICK et al., 1972a) and 25,26-dihydroxycholecalciferol (SUDA et al., 1970) and 1,24-25-trihydroxycholecalciferol (DeLUCA, 1973). However, the function, if any, of these latter three metabolites is not known at the present time.

After production of the 1,25-dihydroxycholecalciferol by the kidney, the steroid hormone is carried to the target tissues by specific serum binding proteins. Estimates of the plasma levels of calciferol (BELSEY et al., 1971) and 25-hydroxycalciferol (HADDAD and CHYU, 1971) have been reported as have studies of the turnover of cholecalciferol and its metabolites (SMITH and GOODMAN, 1971). From such studies NORMAN and HENRY, (1974a), have calculated that while the circulating plasma concentrations of cholecalciferol and 25-OH-cholecalciferol are 65.000 pg/ml (6.5×10^{-8} M) the steady state level of $1,25\text{-}(OH)_2$-cholecalciferol is at least 3ooo x lower, probably in the range of 10-100 pg/ml ($2.4\text{-}24.0 \times 10^{-11}$ M).

This emphasizes the exceedingly low circulating concentrations of $1,25\text{-}(OH)_2$-cholecalciferol and highlights its impressive biological activity. Indeed it is essential that any receptor for $1,25\text{-}(OH)_2$-cholecalciferol in the target intestine or bone have a very high affinity for the steroid, since it is circulating in such low concentrations. In my laboratory we have focused on its interaction in the target intestine where it stimulates or mediates an increased absorption of calcium. We have shown that within 10-12 hours there is a maximal stimulation of intestinal calcium transport (MYRTLE and NORMAN, 1970). Also associated with the increase in intestinal transport is the production or appearance of a calcium binding protein (CaBP) (SPIELVOGEL, 1973; TAYLOR and WASSERMAN, 1970).

Our laboratory has proposed (TSAI, WONG and NORMAN, 1972; TSAI and NORMAN, 1973a) that the mode of action of 1,25-dihydroxycholecalciferol is very much akin to that of other classical steroid hormones. That is, it produces its biological response by association with a nucleus of the target cell and stimulates a "read-out" of genetic information which is essential to the development of some of the characteristic physiological response. Fig. 1 provides a summary of data we have obtained in recent years which supports this contention. The top series of panels reviews the structure of the steroids involved in the two-step metabolism of calciferol to 1,25-dihydroxycholecalciferol. The second set of panels shows the subcellular localization of radioactivity which results in the target intestinal mucosa after intracardial, intraperitoneal, or oral doses of radioactive cholecalciferol, or 1,25-dihydroxycholecalciferol. The maximum localization of radioactivity in the intestine occurs in 16 hours, 12 hours, or 4 hours, respectively after administration of physiological doses of these steroids (65-3000 pmoles). In each instance, the largest proportion of radioactivity is associated with the nuclear fraction of the cell. The

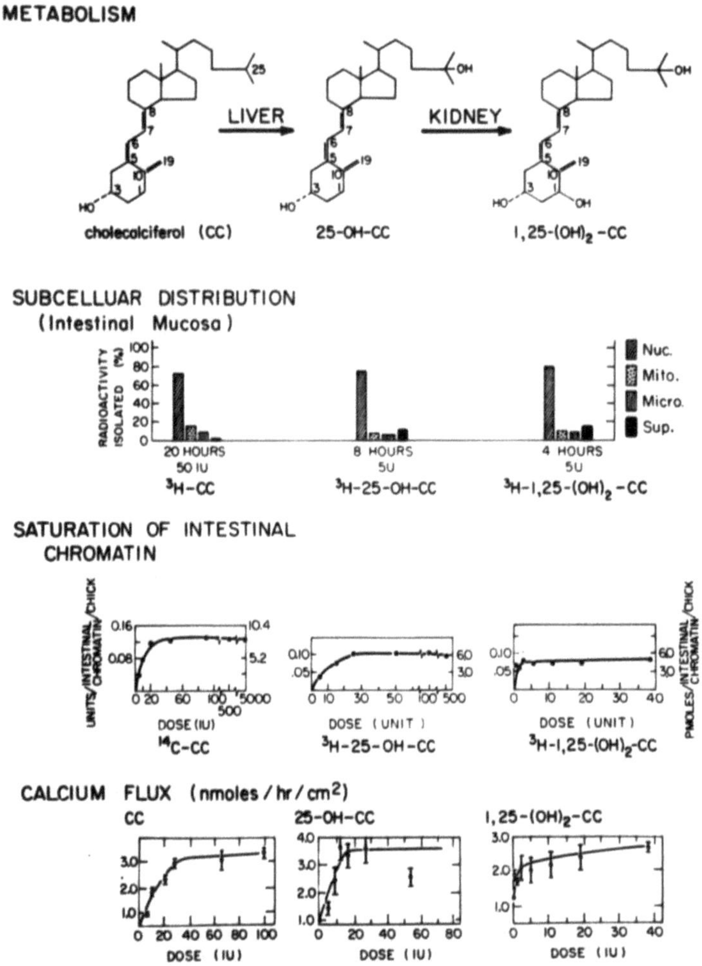

Fig. 1. Summary of calciferol metabolism (top line), subcellular localization of calciferol + metabolites (middle line), saturation of intestinal chromatin receptor with 1,25-$(OH)_2$-D_3 (third line) and stimulation of intestinal calcium transport (bottom line) after doses of calciferol (left panels), 25-OH-calciferol (middle panels), or 1,25-$(OH)_2$-calciferol (right panels)

time required for maximal nuclear localization is of the expected order, in view of the required metabolic transformations of cholecalciferol and its 25-hydroxy derivative. As shown in the third row of panels, as we administered increasing doses of these three steroids, there is only a finite amount of radioactivity which subsequently localizes in the intestinal chromatin binding sites. In each instance, we have examined the chemical form of this radioactivity which is associated with the nucleus and its chromatin fraction. In fact it always is 1,25-dihydroxycholecalciferol, irrespective of which steroid was administered to the animal. The binding capacity of the intestinal chromatin receptor became saturated after a dose of 1950, 1400, or 26 picomoles of cholecalciferol, 25-hydroxycholecalciferol or 1,25-dihydroxycholecalciferol respectively. As shown in the bottom set of panels, exactly the same maximum dose of steroids which resulted in a saturation of the intestinal chromatin receptor, also resulted in a maximum stimulation of intestinal calcium transport. Further, the lag time in hours required to saturate the chromatin receptor sites was in good agreement with the lag time necessary to produce the maximum calcium transport. That is to say, the length was longest in both instances with cholecalciferol, was intermediate in length with 25-hydroxycholecalciferol, and was shortest with the steroid 1,25-dihydroxycholecalciferol. Thus, we are attempting to establish the validity of the concept that there is a very strong cause-and-effect relationship between the appearance of 1,25-dihydroxycholecalciferol in the target intestinal mucosa and a subsequent development of the biological response, increased intestinal calcium transport.

In other experiments not shown here, TSAI and NORMAN (1973b) have conclusively shown that prior to the association of 1,25-dihydroxycholecalciferol with its binding protein in the chromatin, that it is obligatorily associated with a binding protein present in the cytosol ofthe target intestine. This cytosol-binding protein apparently functions to transfer the steriod from the outer cell membrane to the nucleus. After arrival of the steroid in the nucleus, there ensues an as yet undefined series of steps wherein there is initiation or stimulation of the synthesis of RNA and protein. TSAI and NORMAN (1973b) have recently demonstrated that physiological doses of 2-6 U (50-150 ng) of 1,25-dihydroxycholecalciferol will maximally stimulate a synthesis of RNA by 4-6 hours in the intestine. This agrees nicely with the maximal localization of the steroid in the intestinal chromatin at 4 hours and a maximal stimulation of intestinal calcium transport by 8-9 hours.

One point that is obvious from the complex set of interrelationships and interactions discussed above is that there are a multiplicity of possible sites and steps where the calciferol endocrine system may be disrupted. Thus, it is not surprising that there is such an array of disease states which are known to be related in some fashion to vitamin D. Fig. 2 presents a summary of many of these. Several possible means of discussing these disease states include an analysis from the point of view of: (a) their abnormal responsiveness to vitamin D therapy as evidenced by a hypersensitivity, antagonism or resistance to the steroid, or (b) a systems analysis point of view wherein a disease may be related to a particular problem either in production of the biologically active form of vitamin D at certain organs or to the elicitation of the biological response at other target organs. Given the primary thesis of this article, that 1,25-dihydroxycholecalciferol is the biologically active form of vitamin D, it is obvious why the kidney plays such an important and integral role in health and diseases which are known to be related to vitamin D. Any disease state which directly impairs the production of 1,25-dihydroxycholecalciferol by the kidney or which interferes with any of the multiplicity of steps involved in calcium homeostasis leading to an abnormal feedback signal to the kidney, will result in an altered production of the hormone. Thus, the kidney can be directly implicated or implicated in a secondary manner in a wide variety of metabolic diseases. In either case, if a deficiency in the biologically active form of calciferol ensues, a large spectrum of bone diseases may develop, such as rickets, osteomalacia, osteopenia, or fibrogenesis imperfecta ossium. However, these same end conditions may occur as a consequence of a wide variety of primary reasons, and from the point of view of the clinician, it is often a difficult and challenging task to analyze and predict what the nature of the primary disorder may be.

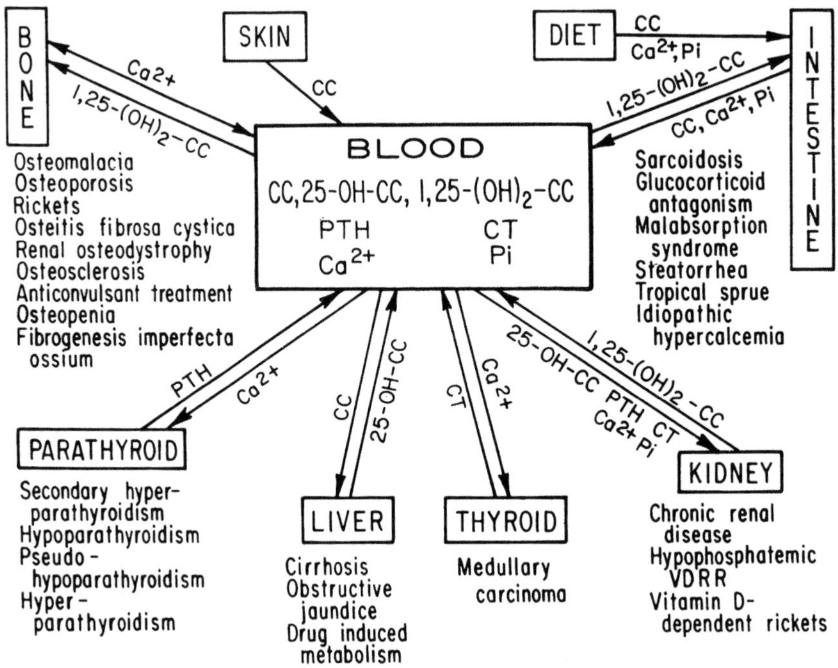

Fig. 2. Disease states in man related to calciferol

For a number of years, clinical findings have suggested similarities between patients with vitamin D deficiency and those with chronic renal failure (STANBURY and LUMB, 1962). Thus, impaired growth and deformities of long bones were described in children with chronic renal failure (FOLLIS, 1950). Moreover, such patients failed to respond to treatment with quantities of calciferol that were effective in those children with nutritional rickets. LIU and CHU (1941) originally showed that dihydrotachysterol (DHT) was effective in patients with renal disease while cholecalciferol was not. With the chemical characterization of 1,25-dihydroxycholecalciferol and the realization that the unique aspect of this steroid molecule was the presence of a hydroxyl at carbon number 1, it was readily apparent that DHT has certain structural similarities to the 1,25-dihydroxycholecalciferol (NORMAN, 1971). The conformation of the A ring of this steroid is inverted in relationship to calciferol resulting in the hydroxyl at carbon-3 occupying a "pseudo carbon number 1 position". As can be seen in Fig. 3, the hydroxyl at carbon number 3 of DHT has the same stereochemical position as the one hydroxyl of 1,25-dihydroxycholecalciferol. It was not long before the ingenuity of the chemist produced the compounds, 5,6-trans-cholecalciferol, 25-hydroxy-5,6-trans-cholecalciferol and 1-hydroxy-cholecalciferol. These should all be considered to be chemical analogs of the natural steroid 1,25-dihydroxycholecalciferol. The 5,6-trans isomers can be produced readily by employing simple chemical conditions that mediate the cis-trans isomerization of the 5,6 double bond. Chemical synthesis of 1-hydroxy-cholecalciferol represents a remarkable feat, inasmuch as it is very difficult to introduce via chemical means a hydroxyl group at carbon number 1. However, as shown in Table 1, six laboratories to date have accomplished this feat. Preliminary results obtained with all three of these analogs

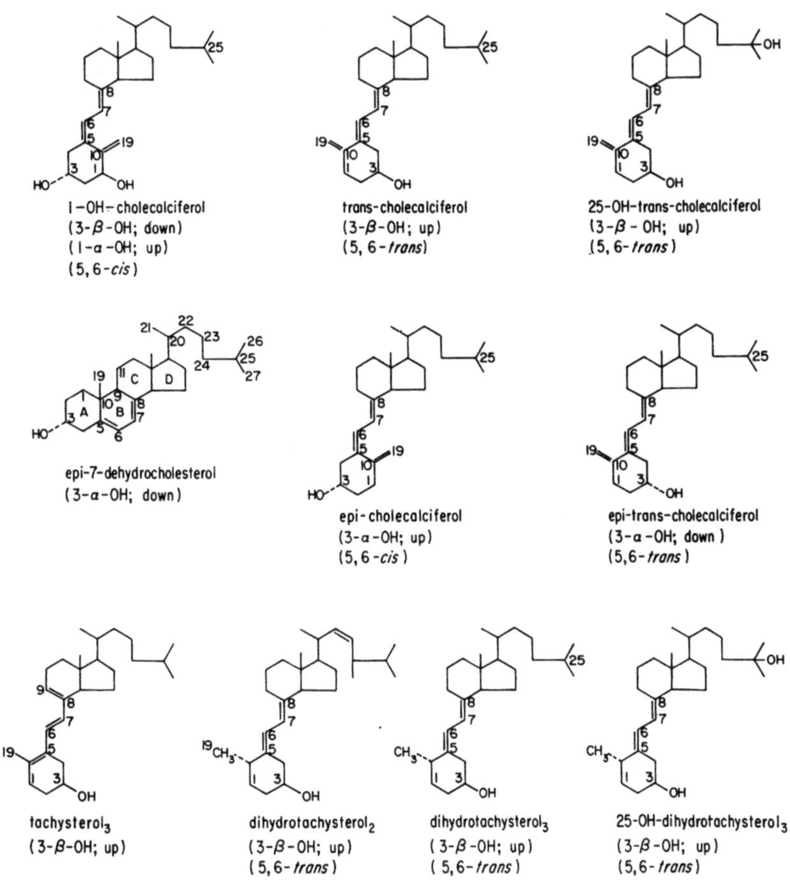

Fig. 3. Compounds related to vitamin D. None of the structures in this figure are naturally occuring steroids; all except the epi-series (middle line) have been chemically synthesized. Tachysterol$_2$ is an irradiation product of 7-dehydrocholesterol and is without any biological activity. The dihydrotachysterol steroids are produced by chemical reduction of the parent tachysterol; the designation, dihydrotachysterol$_2$ and dihydrotachysterol$_3$, indicates that the side chain is that of ergosterol or 7-dehydrocholesterol, respectively. Epi-7-dehydrocholesterol is identical in structure to 7-dehydrocholesterol except that the configuration of the hydroxyl at position 3 is reversed

suggest that the "pseudo-1" character of these compounds confers upon them the ability to produce biological effects only previously obtainable by administration of calciferol or 1,25-dihydroxycholecalciferol (NORMAN, OKAMURA and MITRA, manuscript in preparation; HOLICK et al., 1972b,c, 1973). In particular, the biological activity of these compounds has been examined in nephrectomized animals. Under these circumstances the lack of renal tissue should normally prevent any biological response from cholecalciferol or 25-hydroxycholecalciferol. Only steroids which do not require 1-hydroxylation and bypass this obligatory renal enzymatic step, would be capable of producing a biological response. Thus 1-hydroxycholecalciferol, 5,6-trans-cholecalci-

Table 1. Chemical synthesis of calciferol metabolites and analogs

	NETHERLANDS- Duphar	MADISON - DeLuca	CAMBRIDGE - Bell, Kodicek	TOKYO - Ikekawa	BOSTON - Barton	PARIS - Roussel	KALAMAZOO - Upjohn	RIVERSIDE - Norman, Okamura	BASEL - Hoffman LaRoche
Metabolites of Calciferol:									
25-OH-D_3	X	X	X	X		X	X	X	
24,25-$(OH)_2$- D_3		X	X	X					X
25,26-$(OH)_2$-D_3			X	X					
1,25-$(OH)_2$-D_3		X		X	X?				X
25-keto-27-nor-D_3			X						
Analogs of Calciferol:									
1-OH-D_3		X		X	X		X	X	X
5,6-trans-D_3		X	X			X		X	X
5,6-trans-25-OH-D_3		X	X			X			X
27-nor-25-OH-D_3		X	X					X	
26,27-dinor-25-OH-D_3		X							
23,24,25,26,27-pentanor-25-OH-D_3								X	
25-OH-DHT_3			X						

ferol, and the 25-hydroxy-5,6-trans-cholecalciferol all have been shown to produce a stimulation of intestinal calcium transport and bone calcium mobilization in the anephric rat. However, in general, the data are too preliminary to permit a precise determination of the relative effects on intestinal calcium transport vs. bone calcium mobilization or an estimation of their potential usefulness to the clinician. This clearly is an area where much further experimentation and evaluation will be carried out in the near future. Results from such

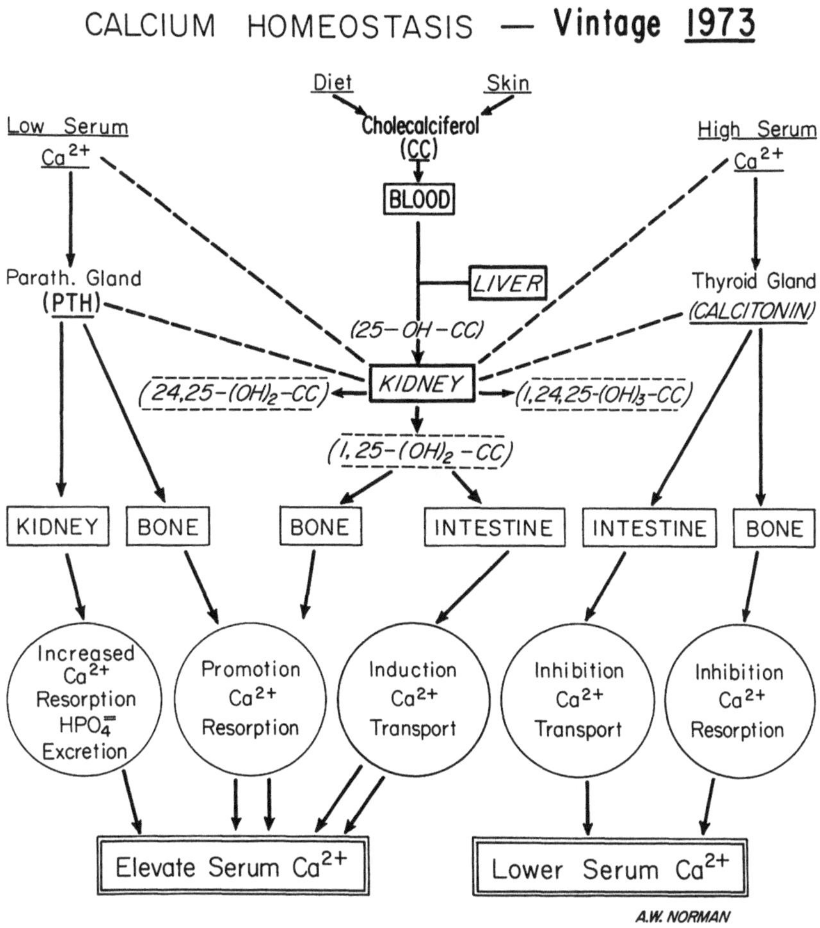

Fig. 4. Central role of the kidney as an endocrine organ in the regulation of plasma calcium concentration

studies hold the expectation that the clinician will be provided with a battery of steroids, both natural and analogs, which will permit him to manipulate calcium metabolism in a beneficial manner. Also shown in Table 1 is a synopsis of the other calciferol metabolites and analogs which have been chemically synthesized and which are potentially available for such studies.

At present, widespread clinical experience with the use of calciferol and its analogs is limited to cholecalciferol, ergocalciferol, and dihydrotachysterol$_2$ (DHT). COBURN, NORMAN, BRICKMAN and coworkers have pioneered in the clinical evaluation of 1,25-dihydroxycholecalciferol. Their findings are summarized in Table 2, and are discussed in detail in the indicated publications. Their results have emphasized the striking biological activity of this steroid in both normal man and a variety of disease states. Clearly much further work will be required to determine its precise clinical efficaciousness.

Table 2. Summary of clinical trials with 1,25-dihydroxycholecalciferol[a]

Condition	Trials	Duration of treatment (days)	Response[b]	Literature Citation
Normal	20	7	IC, UC, SC	BRICKMAN et al., 1974
Renal osteodystrophy	15	7-14	IV, UC, SC	BRICKMAN, COBURN, NORMAN, 1972
Renal osteodystrophy	11	60-90	in progress	in preparation
VDRR	4	7-12	IC, UC	BRICKMAN et al., 1973
Hypoparathyroidism	2	7-10	UC, SC	
Pseudohypoparathyroidism	3	7-10	IC, SC	COBURN et al., 1974

[a] These studies have been carried out by Professors J.W. COBURN and A.S. BRICKMAN, Dept. Medicine, UCLA, Los Angeles, in collaboration with the authors.

[b] IC = Intestinal $^{47}Ca^{2+}$ absorption; UC = Urinary Ca^{2+} excretion; SC = Serum Ca^{2+}.

Fig. 4 summarizes the relationships which have been discussed above which relate the metabolism of vitamin D to its functioning in the intestine and bone for the maintenance of a normal calcium homeostasis. There are now at least three primary regulators of calcium homeostasis: the peptide hormones, parathyroid hormone and calcitonin, and the sterol, 1,25-dihydroxycholecalciferol. Together these agents carry out in integrated fashion the control and regulation of serum calcium via their independent and interdependent actions on calcium and phosphorus metabolism at the intestinal, renal, and skeletal levels. Probably by far the most significant addition to our understanding of calcium homeostasis as shown in Figure 3 is the central role that the kidney plays in this system. On the one hand, it is the producer of the steroid hormone 1,25-dihydroxycholecalciferol which has actions at other distal organs (the intestine and the skeleton) and on the other hand, it is a target organ itself for the actions of parathyroid hormone, calcitonin, and possibly a calciferol metabolite. In view of the complex nature of all of these relationships shown in Fig. 4, and their relatively recent discovery, the relationships presented in Fig. 4 should only be considered to be tentative at this time.

It is apparent that in recent years many striking relationships have come into focus concerning the metabolism and functioning of vitamin D. Principal among these has been the highlighting of the central role of the kidney in the production of the biologically active form of the steroid 1,25-dihydroxycholecalciferol. Three immensely important areas which are under intensive investigation in many laboratories concern (1.) the role of the kidney as an endocrine organ producing the biologically active form of vitamin D, (2.) the regulation of the endocrine organ and its integration in the process of calcium homeostasis, and (3.) attempts to synthesize chemically new vitamin D analogs which will aid in understanding its mode of action and which may have potential therapeutic value. Clearly, much further work needs to be done in all three areas before a clear understanding of the metabolism, mode of action, and pharmacology will be at hand.

References

BELSEY, R., DELUCA, H.F., POTTS, J.T.Jr.: Competitive binding assay for vitamin D and 25-OH-vitamin D. J.Clin. Endo. Met. 33, 554-557 (1971).

BHATTACHARYYA, M., DELUCA, H.F.: The regulation of rat liver calciferol-25-hydroxylase. J. Biol. Chem 248, 2969-2973 (1973).

BOYLE, I.T., GRAY, R.W., DELUCA, H.F.: Regulation by calcium of in vivo synthesis of 1,25-dihydroxycholecalciferol and 21,25-dihydroxycholecalciferol. Proc. Nat. Acad. Sci. 68, 2131-2134 (1971).

BRICKMAN, A.S., COBURN, J.W., KUROKAWA, K., BETHUNE, J.E., HARRISON, H.E., NORMAN, A.W.: Actions of 1,25-dihydroxycholecalciferol in patients with hypophosphatemic vitamin D-resistant rickets. New England J. Medicine 289, 495-498 (1973).

BRICKMAN, A.S., COBURN, J.W., MASSRY, S.G., NORMAN, A.W.: Studies on the effects of 1,25-dihydroxy vitamin D_3 in normal man and in patients with advanced renal failure. Annals Internal Medicine (1974).

BRICKMAN, A.S., COBURN, J.W., NORMAN, A.W.: Action of 1,25-dihydroxycholecalciferol, a potent, kidney-produced metabolite of vitamin D_3, in uremic man. N.Eng. J. Med., 287, 891-895 (1972).

COBURN, J.W., BRICKMAN, A.S., KUROKAWA, K., MASSRY, S.G., BETHUNE, J. E., HARRISON, H.E., NORMAN, A.W.: Action of 1,25-dihydroxycholecalciferol in normal man and patients with hypophosphatemic resistant rickets, pseudohypoparathyroidism and uremia (S. Taylor Ed.). Endocrinology 1973. London: W. Heineman 1974.

DELUCA, H.F.: Recent advances in our understanding of functional vitamin D metabolism and its regulation. Endocrin. Abstract, p. 3 1973).

FOLLIS, R.H.: Renal rickets and osteitis fibrosa in children and adolescents. Bull. Johns Hopkins. 87, 593-615 (1950).

FRASER, D.R., KODICEK, E.: Unique biosynthesis by kidney of a biologically active vitamin D metabolite. Nature 228, 764-766 (1970).

GALANTE, L., MacAULEY, S., COLSTON, K., MacINTYRE, I.: Effect of parathyroid extract on vitamin D metabolism. Lancet i, 985-988 (1972a).

GALANTE, L., COLSTON, K.W., MacAULEY, S.J., MacINTYRE, I.: Effect of calcitonin in vitamin D metabolism. Nature, 238, 271-272 (1972b).

GARABEDIAN, M., HOLICK, M., DELUCA, H.F., BOYLE, I.T.: Control of 25-hydroxycholecalciferol metabolism by parathyroid glands. Proc. Nat. Acad. Sci. (US) 69, 1673-1676 (1972).

HADDAD, J.G., JR., CHYU, K.J.: Competitive protein-binding radio-assay for 25-hydroxycholecalciferol. J. Clin. Endocrin. Metab. 33, 992-995 (1971).

HENRY, H., NORMAN, A.W.: Phylogenetic distribution of the renal 25-hydroxycholecalciferol-1-hydroxylase. Nature (1974).

HOLICK, M.F., SCHNOES, H.K., DELUCA, H.F., GRAY, R.W., BOYLE, I.T., SUDA, T.: Isolation and identification of 24,25-dihydroxycholecalciferol, a metabolite of vitamin D_3 made in the kidney. Biochemistry 11, 4251-4255 (1972).

HOLICK, M.F., GARABEDIAN, M., DELUCA, H.F.: 5,6-trans-25-hydroxycholecalciferol: Vitamin D analog effective on intestine of anephric rats. Science 176, 1247-1248 (1972b).

HOLICK, M.F., GARABEDIAN, M., DELUCA, H.F.: 5,6-trans-isomers of cholecalciferol 1 and 25-hydroxycholecalciferol. Substitutes for 1,25-dihydroxycholecalciferol in anephric animals. Biochemistry 11, 2715-2719 (1972c).

HOLICK, M.F., SEMMLER, E.J., SCHNOES, H.K., DELUCA, H.F.: 1α-hydroxy derivative of vitamin D_3: A highly potent analog of 1-α-,25-dihydroxy vitamin D_3. Science 180, 190-191 (1973).

LEAGUE OF NATIONS: Quarterly Bulletin of the Health Organization: Memorandum on the International Standard for vitamin D and its application. Vol. IV, no. 3, 540-542 (1935).

LIU, S.H., CHU, H.T.: Studies of calcium and phosphorus metabolism with special reference to the pathogenesis and effect of dihydrotachysterol (AT10) and iron. Medicine 22, 103-161 (1941).

MIDGETT, R.J., SPIELVOGEL, A.M., COBURN, J.W., NORMAN, A.W.: Studies of calciferol metabolism VII. The renal production of the biologically active form of vitamin D, 1,25-dihydroxycholecalciferol: species, tissue and subcellular distribution. J. Clin. Endocrinol. 36, 1153-1161 (1973).

MYRTLE, J.F., NORMAN, A.W.: Vitamin D: A cholecalciferol metabolite highly active in promoting intestinal calcium transport. Science 171, 79-82 (1970).

NORMAN, A.W.: The mode of action of vitamin D. Biol. Reviews 43, 97-137 (1968).

NORMAN, A.W.: Evidence for a new kidney produced hormone, 1,25-dihydroxycholecalciferol, the proposed biologically active form of vitamin D. Amer. J. Clin. Nutrition 24, 1346-1351 (1971).

NORMAN, A.W.: Problems relating to the definition of an international unit for vitamin D and its metabolites. J. Nutrition 102, 1243-1246 (1972).

NORMAN, A.W., HENRY, H.: The role of the kidney and vitamin D in health and disease. Clin. Orthopedic and Related Research (1974a).

NORMAN, A.W., HENRY, H.: 1,25-Dihydroxycholecalciferol-A hormonally active form of vitamin D_3. Rec. Prog. Hor. Res. (1974b) In press.

NORMAN, A.W., MIRCHEFF, A.K., ADAMS, T.H., SPIELVOGEL, A.: Studies on the mechanism of action of calciferol III. Vitamin D-mediated increase of intestinal brush border alkaline phosphatase. Biochem. Biophys. Acta 215, 348-359 (1970).

OMDAHL, J.L., GRAY, R.W., BOYLE, I.T., KNUTSON, J., DELUCA, H.F.: Regulation of metabolism of 25-hydroxycholecalciferol by kidney tissue in vitro by dietary calcium. Nature 237, 63-65 (1972).

RASMUSSEN, H., WONG, H., BIKLE, D., GOODMAN, D.B.P.: Hormonal control of 25-hydroxycholecalciferol to 1,25-dihydroxycholecalciferol. J. Clin. Invest. 51, 2502 (1972).

SPIELVOGEL, A.M.: Calciferol (vitamin D) mediated intestinal calcium transport and the mechanism of action of the polyene antibiotic filipin. Ph. D. Dissertation, University of California, Riverside (1973).

STANBURY, S.W., LUMB, G.A.: Metabolic studies of renal osteodystrophy. Medicine (Baltimore) 41, 1-31 (1962).

STANBURY, S.W., LUMB, G.A., MAWER, E.B.: Osteodystrophy developing spontaneously in the course of chronic renal failure. Arch Intern. Med. 124, 274-291 (1969).

SUDA, T., DELUCA, H.F., SCHNOES, H.K., TANAKA, Y., HOLICK, M.F.: 25-26-Dihydroxycholecalciferol, a metabolite of vitamin D_3 with intestinal calcium transport activity. Biochemistry 9, 4776-4780 (1970).

TANAKA, Y., DELUCA, H.F.: The control of 25-hydroxyvitamin D metabolism by inorganic phosphorus. Arch. Biochem. Biophys. 154, 566-574 (1973).

TAYLOR, A.N., WASSERMAN, R.H.: Immunofluorescent localization of vitamin D-dependent calcium-binding protein. J. Histochem. and Cytochem. 18, 107-114 (1970).

TSAI, H.C., WONG, R.G., NORMAN, A.W.: Studies on calciferol metabolism IV. Subcellular localization of 1,25-dihydroxy vitamin D_3 in intestinal mucosa and correlation with increased calcium transport. J. Biol. Chem. 247, 5511-5519 (1972).

TSAI, H.C., NORMAN, A.W.: Studies on calciferol metabolism VIII. Evidence for the existence of a cytoplasmic receptor for 1,25-dihydroxy vitamin D_3 in the intestinal mucosa. J. Biol. Chem. 248, 5967-5975 (1973).

TSAI, H.C., NORMAN, A.W.: Studies on the mode of action of calciferol VI:Effect of 1,25-dihydroxy vitamin D_3 on RNA synthesis in the intestinal mucosa. Biochem. Biophys. Res. Comm. 54, 622-627 (1973b).

WONG, R.G.: The role of calciferol (vitamin D) and its biologically active metabolite 1,25-dihydroxycholecalciferol, in the metabolism of calcium. Ph.D. Dissertation, University of California (1972).

WONG, R.G., NORMAN, A.W., REDDY, G.R., COBURN, J.W.: Biologic effects of 1,25-dihydroxycholecalciferol (a highly active vitamin D metabolite) in acutely uremic rats. J. Clin. Invest. 51, 1287-1291 (1972).

Vitamin D-Dependent Calcium Binding Protein of Rat Renal Cortex*

C. L. HERMSDORF, T. FREUND AND F. BRONNER

I. Abstract

A vitamin D-dependent calcium binding protein (CaBP) has been partially purified from rat renal cortex. Its molecular weight is about 28.000 daltons, more than twice that of the vitamin D-dependent rat intestinal CaBP. The apparent pK of the partially purified renal CaBP is 5.0 and its binding capacity 22 nanomoles Ca bound/mg protein. No renal CaBP was found in kidneys from rachitic rats or in the medulla of normal animals. In contrast with intestinal CaBP, renal CaBP activity does not appear to vary markedly with calcium intake, plasma calcium concentration or urinary calcium excretion.

II. Introduction

The discovery of a vitamin D-dependent intestinal calcium binding protein (WASSERMAN et al., 1968; WASSERMAN and TAYLOR, 1968; DRESCHER and DELUCA, 1971; HITCHMAN and HARRISON, 1972; ALPERS et al., 1972) led to the search for such a protein in kidney. TAYLOR and WASSERMAN reported in 1967 that kidneys from rachitic chicks contained no CaBP and that vitamin D administration to such animals led to the appearance of the protein. Recently, the same workers (TAYLOR and WASSERMAN, 1972) expanded on their findings and have reported that small amounts of CaBP persist in frankly rachitic chicks. This persistence of CaBP was explained to have resulted from the relatively slow turnover of renal cells. SANDS and KESSLER (1971) identified a calcium binding component of dog kidney cortex in the 16.000g supernate and related its binding capacity to renal calcium excretion. PIAZOLO et al. (1971) isolated a highly purified CaBP from human kidney and reported that human renal cortex contains four times more CaBP than medulla. Its molecular weight was reported to be in the range of chicken intestinal CaBP (25-28.000, WASSERMAN et al., 1968). HURWICH et al. (1973) have reported the existence of a renal calcium binding factor in the rat and, on the basis of the Chelex test on heated kidney supernates, claim higher activity in the renal medulla than cortex. A vitamin D-dependency of the renal CaBP has been established only for the chick (TAYLOR and WASSERMAN, 1967; TAYLOR and WASSERMAN, 1972).

We now report the isolation of vitamin D-dependent CaBP from the cortical region of rat kidneys.

+We thank KUKIL and BURNS for technical assistance. Supported by NIH grants AM 14251 and DE 02593.
A complete report has appeared in Biochim. Biophys. Acta (in press).

III. Materials and Methods

Weanling male rats, purchased from Sprague-Dawley, Madison, Wisconsin, were divided into two groups and fed high calcium, low phosphorus semi-synthetic diets (modified from HURWITZ et al., 1969) containing either no (BR) or 2.200i.u. vitamin D_2/kg feed (BRD). By analysis the diets contained 1.4% Ca and 0.2%P. Approximately one month later, when the animals were phosphorus-deficient (HURWITZ et al., 1969) and when they were presumed vitamin D-deficient, they were killed, the kidneys rapidly removed, decapsulated and weighed.

When kidneys were wanted from animals on a low calcium diet, (I, 0.06% Ca, 0.2% P), male Sprague-Dawley rats weighing typically 120g were placed on the diet from 10-14 days and the kidneys obtained as above.

In some instances the kidneys were separated into three regions: the outer 2 mm cortical region, the inner medullary region, and the boundary region between cortex and medulla.

The kidneys, or kidney regions, were homogenized in iced Tris buffer (0.013M Tris-HCl, 0.12M NaCl, 3mM KCl, pH 7.4) with the aid of a Potter-Elvehjem Teflon homogenizer. The homogenate was centrifuged at 100.000g for 1 hr. The supernate was decanted, lyophilized, then redissolved and further fractionated by elution on Sephadex G-100 with 0.02M ammonium acetate and 1mM mercaptoethanol as elution buffer. Each fraction was assayed for calcium binding activity by a micromodification of the competitive Chelex binding assay (WASSERMAN et al., 1968; WASSERMAN and TAYLOR, 1968; PIAZOLO et al., 1971).

To determine the apparent calcium binding constant (K_d) and the binding capacity (n) the assay procedure was modified by varying the added calcium concentration (FREUND and BRONNER).

Molecular weight was determined on a calibrated Sephadex G-100 column. Protein was determined spectrophotometrically (LAYNE, 1957) and calcium by atomic absorption spectrophotometry (HURWITZ et al., 1969).

IV. Results

The absorbance and calcium binding profiles shown in Fig. 1 were obtained from kidney supernates that had been lyophilized, redissolved, and chromatographed on Sephadex G-100. In the case of material derived from the control animals (BRD group), there were three peaks of binding activity, here designated as peak A, peak B, and peak C. Since chromatography on Sephadex G-100 of Tris buffer alone yields binding activity in the peak C region, peak C binding represents an effect of ions on the assay. In the case of material derived from the rachitic animals (BR group) there was no peak B. Hence peak B material is vitamin D-dependent. Peak A, the void volume peak, was not abolished in rachitic animals.

To determine the molecular weight of peak B material, a larger supply was obtained from animals on a low calcium diet (I). Chromatography of these kidney supernates on Sephadex G-100 yielded absorbance and calcium binding profiles similar to those from the BRD animals. When this peak B material was chromatographed on a calibrated Sephadex G-100 column, its MW was estimated to be 28.000 daltons.

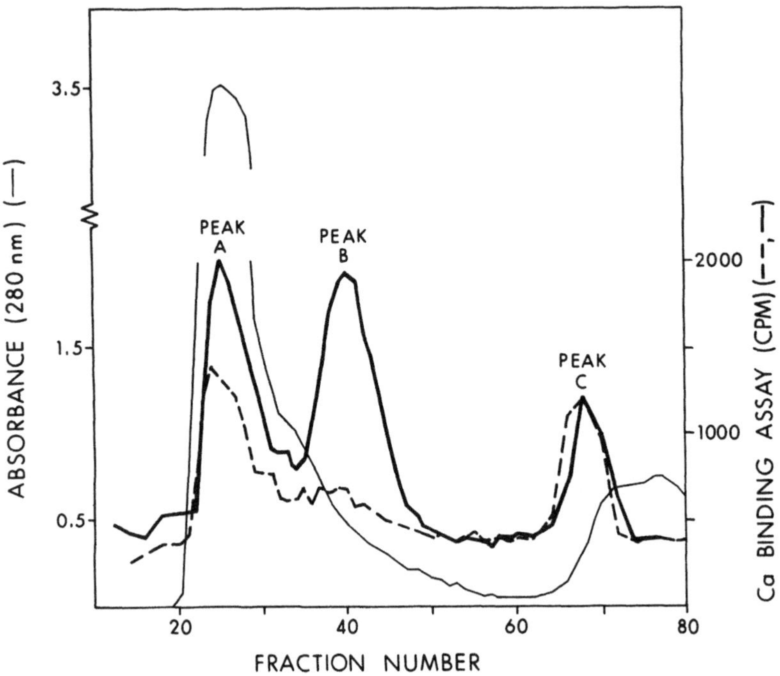

Fig. 1. Sephadex G-100 Elution Profiles of Kidney Supernates from Rachitic (BR) and Control (BRD) Animals. Approximately equal weights of lyophilized kidney supernates were dissolved and chromatographed. Absorbance was monitored and calcium binding assayed in each fraction. Maximum dextran absorbance was found in fraction 25, maximum peak B binding activity in fraction 40, and maximum binding activity due to ionic calcium in fraction 68. Note the presence of binding activity in the material from the control animals (heavy solid line) and its absence in material from the rachitic animals (dashed line)

When peak B material was incubated with trypsin and rechromatographed on Sephadex G-100, the eluate showed no binding activity. This can be taken as an indication of the protein nature of peak B material.

To characterize further the peak B material it was rechromatographed on Sephadex G-100 and binding activity of the eluate determined as a function of calcium concentration. A typical curve of specific binding activity ($\underline{n}$, nanomoles Ca bound per mg total protein) as function of the free (i.e. unbound) calcium ion concentration is shown in Fig. 2. Material at that stage of purification yielded an apparent pK of 5.0 and a value of $\underline{n}$ of ~ 22.

When kidneys were separated into cortical and medullary regions and then fractionated, peak B material was found in the cortical, but not in the medullary region. Tissue isolated from a kidney region intermediate between cortex and medulla yielded a small amount of peak B material.

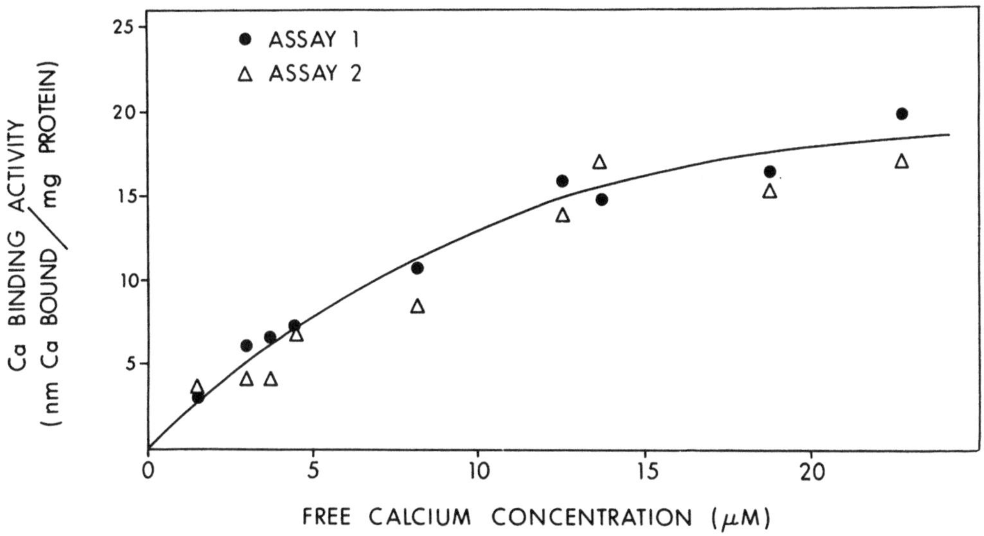

Fig. 2. Calcium Binding Activity of Peak B Material as a Function of Free Calcium Concentration. Kidneys were obtained from animals on a low calcium diet (I), the supernate chromatographed on Sephadex G-100, the resulting peak B material chromatographed on Sephadex G-100 and the new peak B material lyophilized and taken up in Tris buffer to a concentration of 0.8 mg protein/ml. From the data displayed, one can calculate an apparent pK of 5.0 and a binding capacity of 22 nanomoles Ca bound/mg protein

V. Discussion

It is apparent from the above results that rats have a vitamin D-dependent calcium binding protein located primarily in the cortical region of kidney.

This conclusion would not have been arrived at by examination of the kidney supernate for calcium binding. When such samples were analyzed by the Chelex assay, we were unable to differentiate between material from rachitic and control animals or material from the cortical and medullary regions. This may be the explanation why KRAWITT and KUNIN (1971) found only moderate depression of the intestinal binding activity in material from rachitic as compared to control rats, and why HURWICH et al. (1973) reported more calcium binding activity in medulla than cortex.

Since rat renal CaBP is vitamin D-dependent and occurs in the cortex, the region where calcium reabsorption and therefore active calcium transport are thought to occur (WALSER, 1969, p. 261), it seems reasonable to postulate that renal CaBP is involved in active calcium transport, perhaps in the recognition step. This postulate would be strengthened if CaBP could be shown to be calcium-dependent.

As yet we have been unable to demonstrate differences in renal CaBP activity similar to those found in intestinal CaBP (FREUND et al., 1973a), i.e. associated with differences in calcium intake, or associated with differences in plasma calcium levels or urinary calcium excretion.

Renal CaBP seems to be more than twice as large as the intestinal CaBP. When both renal and intestinal preparations were chromatographed on the same Sephadex G-50 column, the renal material had an elution ratio of 1.3, whereas the intestinal material whose molecular weight on a calibrated Sephadex G-50 column appeared to be approximately 12.000 daltons, had an elution ratio of 1.9 (FREUND et al., 1973b). It will be of interest to learn how these proteins differ yet have some functional similarities.

References

ALPERS, D.H., LEE, S.W., AVIOLI, L.V.: Identification of two calcium-binding proteins in human intestine. Gastroenterology 62, 559-563 (1972).

DRESCHER, D., DELUCA, H.F.: Vitamin D-stimulated calcium binding protein from rat intestinal mucosa-purification and some properties. Biochemistry 10, 2302-2307 (1971).

FREUND, T.S., BRONNER, F.: Regulation of intestinal calcium binding protein by calcium intake in the rat. Am.J.Physiol. (in press).

FREUND, T.S., BURNS, M., BRONNER, F.: Calcium binding protein (CaBP), vitamin D and Ca intake in rats. In: Calcified Tissue: Proc. 9th European Symposium on Calcified Tissues. (H. Czitober and J. Eschberger, Eds.) Vienna: Facta Publication 1973a, 263-269.

FREUND, T.S., HERMSDORF, C., CUISINIER-GLEIZES, P., BRONNER, F.: Calcium binding proteins (CaBP) of rat intestine and kidney. Fed. Proc. 32, 424Abs (1973b).

HERMSDORF, C.L., BRONNER, F.: Vitamin D-dependent calcium binding protein from rat kidney. Biochim. Biophys. Acta (in press).

HITCHMAN, A.J.W., HARRISON. J.: Calcium binding proteins in the duodenal mucosa of the chick, rat, pig and human. Can. J. Biochem. 50, 758-765 (1972).

HURWICH, B.J., GELMONT, D., MENCZEL, J., LENDER, M., MOR, E.: Calcium binding protein (CaBP) in normal and uremic human and rat kidneys. In: Calcified Tissue: Proc. 9th European Symposium on Calcified Tissues. (H. Czitober and J. Eschberger, Eds.) Vienna: Facta Publication 1973, 105-108.

HURWITZ, S., STACEY, R.E., BRONNER, F.: Role of vitamin D in plasma calcium regulation. Amer. J. Physiol. 216, 254-262 (1969).

KRAWITT, E.L., KUNIN, A.S.: Intestinal calcium-binding protein and rickets. Proc. Soc. Exp. Biol. Med. 136, 530-533 (1971).

LAYNE, E.: Spectrophotometric and turbidimetric methods for measuring protein. In: Methods of Enzymology (S.P. Colowick and N.O. Kaplan, Eds.) p. 447-454. New York: Academic Press 1957.

PIAZOLO, P., SCHLEYER, M., FRANZ, H.E.: Isolation and purification of a calcium binding protein from human tissues, Hoppe Seyler's Z. Physiol. Chem. 352, 1480-1486 (1971).

SANDS, H., KESSLER, R.H.: A calcium binding component of dog kidney cortex and its relationship to calcium transport. Proc. Soc. Exp. Biol. Med. 137, 1267-1273 (1971).

TAYLOR, A.N., WASSERMAN, R.H.: Vitamin D_3-induced calcium-binding protein: partial purification, electrophoretic visualization, and tissue distribution. Arch. Biochem. Biophys. 119, 536-540 (1967).

TAYLOR, A.N., WASSERMAN, R.H.: Vitamin D-induced calcium-binding protein: comparative aspects in kidney and intestine. Amer. J. Physiol. 223, 110-114. (1972).

WALSER, M.: Renal excretion of alkaline earths. In: Mineral Metabolism - An Advanced Treatise (C.L. Comar and F. Bronner, Eds.). Vol. III, pp. 260-261. Calcium Physiology. New York and London: Academic Press 1969.
WASSERMAN, R.H., CORRADINO, R.A., TAYLOR, A.N.: Vitamin D-dependent calcium binding protein: purification and some properties. J. Biol. Chem. 243, 3978-3986. (1968).
WASSERMAN, R.H., TAYLOR, A.N.: Vitamin D-dependent calcium binding protein: response to some physiological and nutritional variables. J. Biol. Chem. 243, 3987-3992 (1968).

The Effect of Disodium Ethane-1-Hydroxy-1, 1-Diphosphonate on the Metabolism of Vitamin D

E. B. MAWER, C. M. TAYLOR AND L. F. HILL

Administration of ethane hydroxy diphosphonate (EHDP) to animals inhibits mineralization of bone, probably by a direct action on bone (KING et al., 1971). Treatment with vitamin D (HILL et al., 1973) or 1,25-dihydroxycholecalciferol (BONJOUR et al., 1973) does not correct the defective mineralization.

EHDP-treated rats and chicks have been used to investigate the regulation of renal biosynthesis of 1,25-dihydroxycholecalciferol. Day-old cockerels or weanling rats were fed a vitamin D-free diet which contained 0.6% calcium and 0.1% phosphorus for 5 or 10 days. A solution of EHDP was given by subcutaneous injection in a dose of 40 mg/kg body wt/day.

To study vitamin D metabolism in vivo 250 ng (1,2-$^3H_2$, 4-^{14}C) cholecalciferol was injected intravenously into vitamin D-deficient rats. Control and EHDP-treated rats were killed from 1 to 7 days after injection of labeled cholecalciferol and tissue extracts chromatographed on sephadex columns to separate the cholecalciferol metabolites. Duodenal calcium transport was measured in vitro in the same animals using an everted gut sac technique. 1,25-$(OH)_2$CC was found in the intestinal mucosa of control rats from 1 to 7 days after injection of the parent vitamin. In EHDP-treated rats 1,25-$(OH)_2$CC was found after 1 day, but subsequently the concentration declined and this metabolite could not be measured in the mucosa after 7 days. Duodenal calcium transport paralleled the changes in mucosal 1,25$(OH)_2$CC so that whereas calcium transport remained high in control rats after injection of cholecalciferol, EHDP-treated rats showed only a short-lived response. Thus D-treated rats receiving EHDP had impaired calcium transport ; a further injection of 2.5ug of cholecalciferol did not stimulate transport in these animals but when 50 ng of 1,25-$(OH)_2$CC was injected intravenously, calcium transport increased.

Direct measurement of the effect of EHDP on renal 1-hydroxylase activity was carried out in vitro using chick kidney homogenates. Vitamin D-deficient chicks received 625 ng cholecalciferol orally and birds were killed from 1-9 days later. Chick kidney homogenates were incubated for 2h with 25-hydroxy (26,27 - $^3H_2$) cholecalciferol and extracts chromatographed as before. Kidneys obtained from EHDP-treated chicks converted 25-hydroxycholecalciferol to 24,25-dihydroxycholecalciferol (24,25-$(OH)_2$CC) and not to 1,25-$(OH)_2$CC. Control chick kidneys continued to synthesize 1,25-$(OH)_2$CC. Addition of EHDP to chick kidney homogenates in vitro had no effect on 1-hydroxylase activity. That the inhibitory effect of EHDP on the synthesis of 1,25-$(OH)_2$CC depends on the diet was shown in a further experiment. When chicks were fed a low calcium, low phosphorus diet(0.1% calcium, 0.012% phosphorus), treatment with EHDP and cholecalciferol did not prevent production of 1,25-$(OH)_2$CC by the kidney homogenates (Table).

The inhibitory action of EHDP on the 1-hydroxylase does not therefore appear to be a direct effect on the kidney as it depends both on the diet of the animals and prior treatment with vitamin D. The effect of EHDP in preventing mineral deposition in the skeleton means that the

Table 1. Effect of EDHP and diet on the *in vitro* production of 1,25-dihydroxycholecalciferol and 24,25-dihydroxycholecalciferol by chick kidney (p moles/g kidney tissue)

Diet	0.6% calcium 0.1% phosphorus		0.1% calcium 0.012% phosphorus	
	1,25	24,25	1,25	24,25
Control	0.46	0	0.48	0
EHDP	0	0.25	0.36	0

high absorption of calcium characteristic of young growing animals is inappropriate. The fall in 1,25-$(OH)_2$CC production and consequent drop in calcium absorption may be adaptation to this reduced requirement for calcium. It has been suggested that the renal 1-hydroxylase is regulated through changes in serum calcium or phosphorus, or the intracellular concentrations of these ions. In these experiments, there were no differences in the serum calcium of EHDP treated chicks on normal or low calcium diets despite the different metabolism of 25-(OH)CC in kidney homogenates. Similarly there were no consistent differences in serum phosphorus. Preliminary measurement of serum parathyroid hormone in the chick does not suggest that the fall in 1,25-$(OH)_2$CC is caused by reduced secretion of parathyroid hormone.

It is difficult to make direct comparisons between these animal experiments and adult man treated with much smaller doses of EHDP. In vitamin D-deficient patients not receiving EHDP, 1,25-$(OH)_2$CC was demonstrated consistently in the serum after injection of 1,2-$^3H_2$, 4-^{14}C cholecalciferol, whereas 24,25-$(OH)_2$CC could not be detected (MAWER et al., 1971). In three patients receiving EHDP with a similar state of vitamin D nutrition, 1,25-$(OH)_2$CC was detected though in smaller amounts, and in two of these patients 24,25-$(OH)_2$CC was also found. The significance of these effects of EHDP on vitamin D metabolism in man remains uncertain.

References

BONJOUR, J.P., DELUCA, H.F., BAXTER, L.A., FLEISCH, H., TRECHSEL, U.: Experientia 29, 743 (1973).

HILL, L.F., LUMB, G.A., MAWER, E.B., STANBURY, S.W.: Clinical Science 44, 335-347 (1971).

KING, W.R., FRANCIS, M.P., MICHAEL, W.R.: Clinical Orthopedics 78, 251-270 (1971).

MAWER, E.B., BACKHOUSE, J., LUMB, G.A., STANBURY, S.W.: Nature: New Biology 232, 188-189 (1971).

The Effect of Estrogen on Calcium Binding Protein Activity in Odentogenic Epithelium of the Rat

G. EILON AND J. MENCZEL

The bone, kidney and intestine are target organs of a certain group of steroids, vitamin D and its metabolites. One of the many manifestations of these steroids is the induction of CaBP (Calcium Binding Protein) in the intestine (TAYLOR and WASSERMAN, 1972). Relevance of this protein to calcium transport in the intestine was assessed (WASSERMAN and TAYLOR, 1966).

Another group of steroids, the estrogens, exerts some major effects on bone structure and metabolism, affecting inhibition of linear growth as well as the acceleration of development and condensation (BAKER and LEEK, 1946; LINDQUIST et al., 1960). Intestinal transport and absorption of calcium is also influenced by estrogens.

The rat incisor develops from an elliptical sheath called odentogenic epithelium, this epithelial tissue determines the size and outline of the future incisor.

An attempt was made to determine the presence of CaBP in the odentogenic epithelium and to study the effect of estrogen on CaBP activity.

I. Material and Methods

For the experiments described, male, Sabra H.U. strain rats weighing between 100-130 gr. were used. The incisors extracted and odentogenic epithelium (dental bulb) taken out. Homogenization was performed in a 0.1M tris-HC1 buffer pH 7.4 and centrifuged at 15.000 g for 20 min. Supernatants were assayed for CaBP relative specific activity by the chellex method described by WASSERMAN and TAYLOR (1966).

In vitro experiments were performed in 2.5ml Eagle + biocarbonate medium (Welcome) areated with 5% CO_2 + 95% O_2.

II. Results

A CaBP was detected in odentogenic epithelium. Heating for 10 minutes at 60°C did not produce a significant change in specific activity of CaBP, even a slight increase in activity was observed. Heating at 96°C for the same period of time and incubation with 50µg pronase for 30 minutes at 37°C inactivated the CaBP.

A dose of 1000u of vitamin D_3 was injected intramuscularly to rachitic animals fed on rachitogenic diet for 6 weeks. Odentogenic epithelium was extracted at different time intervals after vitamin D_3 administration and CaBP relative specific activity determined. A rise of 412±2.7%

in calcium binding activity was detected 14 - 16 hours after vitamin D_3 administration.

10µg of 17β estradiol were injected i.p. to normal and rachitic animals. A rise of 98±0.9% in calcium binding activity in the odentogenic epithelium was detected in normal animals 2 - 3 hours after the injection. In rachitic animals, a rise of 79±1.1% in calcium binding activity was detected 3 - 4 hours after the injection.

To understand the effect of estradiol on protein synthesis, <u>in vitro</u> experiments with actinomycin D were performed. Rat odentogenic epithelium was incubated in an Eagle medium in the presence of 0.05µg/ml of estradiol and 1µg/ml of actinomycin D. The calcium binding activity was examined every 30 minutes.

The result of these experiments are summarized in Fig. 1. Unexpectedly, actinomycin D did not suppress the CaBP activity; in the medium where estradiol alone was added the peak of CaBP activity was found after

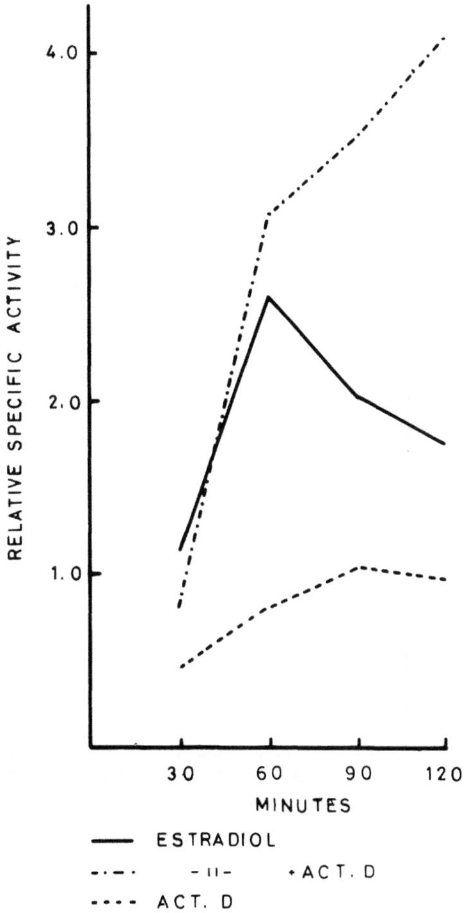

Fig. 1. Effect of 0.05µg/ml 17βestradiol and 1µg/ml actinomycin D on CaBP in odentogenic epithelium tissue in Eagle medium. Expressed in relative specific activity

60 minutes. A higher CaBP activity was detected in the medium where estradiol and actinomycin D were added as compared to the activity obtained when estradiol was given alone.

The higher CaBP activity obtained when estradiol and actinomycin D were given together could be explained by a "superinduction effect" probably due to suppression of a repressor protein by actinomycin D (GARRAN, 1964).

The effect of chloramphenicol on CaBP synthesis was also examined. 50µg/ml of chloramphenicol was used. Chloramphenicol suppressed the CaBP activity induced by estradiol and when given alone. Estradiol alone induced the expected level of CaBP activity. The results of these experiments are summarized in Fig. 2.

The suppression produced by chloramphenicol is a further indication that the calcium binding activity is due to a protein. The difference found between the effects of actinomycin D and chloramphenicol is probably due to the effect of chloramphenicol on the protein synthesis on the mitochondrial level.

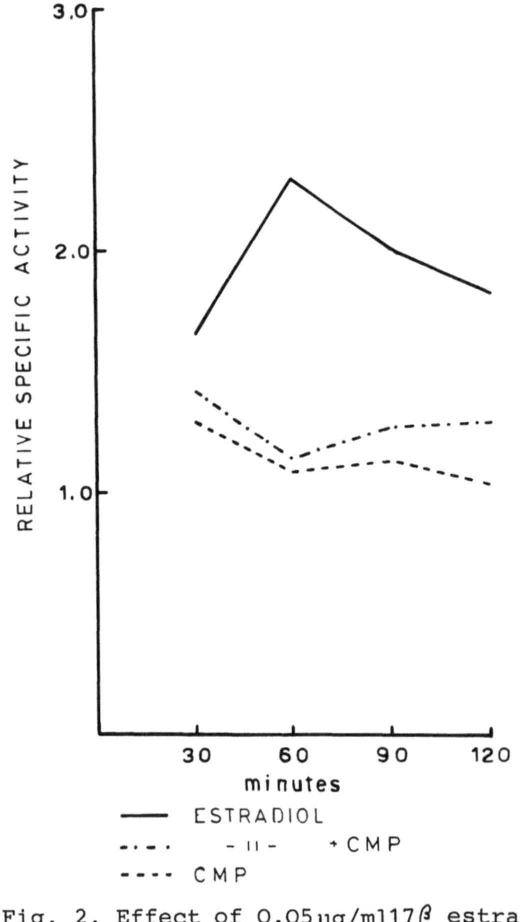

Fig. 2. Effect of 0.05µg/ml17β estradiol and 50ug/ml chloramphenicol (CMP) on CaBP in odentogenic epithelium tissue in Eagle medium. Expressed in relative specific activity

Odentogenic epithelium has an important role in the eruption and proliferation of the evergrowing incisor of the rat. These properties as well as the finding of a CaBP suggest that the odentogenic epithelium could be used in experiments involving bone metabolism.

The effect of estrogen on the CaBP activity is further evidence of the important role this hormone plays in calcium metabolism.

III. Summary

A vitanim D-dependent protein capable of binding calcium was found in the odentogenic epithelium extracted from rat incisors. Estrogen is able to induce a similar protein. The effect of estrogen is on protein synthesis on the mitochondrial level.

References

BAKER, B.L., LEEK, J.H.: The relationship of the parathyroid glands to the action of estrogen on bone. The American Journal of Physiology 147, 523 (1946).

GARREN, L.D., HOWELL, R.R., TOMKINS, G.M.: A paradoxical effect of actinomycin D: The mechanism of regulation of enzyme synthesis by hydrocortisone. Proceedings of National Academy of Science 52, 1121 (1964).

LINDQUIST, B., BUDY, A.M., MCLEAN, F.C., HOWARD, J.L.: Skeletal metabolism in enzyme-treated rats studied by means of Ca^{45}. Endocrinology 66, 100 (1960).

TAYLOR, A.N., WASSERMAN, R.H.: Vitamin D_3-induced calcium binding protein: Comparative aspects in kidney and intestine. American Journal of Physiology 223, 110 (1972).

WASSERMAN, R.H., TAYLOR, A.N.: Vitamin D_3-induced calcium binding protein in chick intestinal mucosa. Science 152, 791 (1966).

Chairmen: D. HIOCO and A.W. NORMAN

Effect of Vitamin D on the Bone in Anticonvulsant Osteomalacia

J. B. EASTWOOD, P. J. BORDIER AND H. E. DE WARDENER

The association between epilepsy and osteomalacia is well established (KRUSE, 1968; DENT et al., 1970) and is due to the administration of anti-epileptic drugs which influence the hydroxylation of cholecalciferol by induction of hepatic microsomal enzymes (HAHN et al., 1972). It is unlikely, however, to be only mechanism responsible for the development of anticonvulsant osteomalacia. This communication describes the study of a single case of anticonvulsant osteomalacia in which the biochemical changes and histological response of the bone to a physiological dose of vitamin D_3 were measured. In particular osteoid mineralisation as well as bone resorption and formation were quantitated.

Case Report: S.V. 18 years. Female. Ugandan Asian. In England for 4 years. Epilepsy was first diagnosed at the age of 7 and treatment with phenytoin and primidone was started. Full investigation including myelography revealed no cause. E.E.G. showed a focus of abnormal activity in the right temporal region. At the age of 16, she fractured the spine of the 7th cervical vertebra during an epileptic fit. One year later she was admitted to hospital with a fracture of the neck of the right femur following a fit. She was still taking primidone, 250 mgs. b.d. and phenytoin 100 mg. t.d.s. It was noted, in particular, that she had never taken phenobaritone.

The plasma calcium was 8.5mg/100ml, plasma phosphorus 2.1mg/100ml and alkaline phosphatase 38 KA units. A Looser's zone was noted in the neck of the left femur and the phalanges showed sub-periosteal erosions. Her daily diet was estimated to have contained 94 i.u. of vitamin D, 840mg of calcium and 1160mg of phosphorus. Investigations of the intestinal tract were found to be normal as follows; fecal fat 2.1g/day, xylose absorption 28% in 6 hours, radiological examination of the small bowel normal, jejunal biopsy normal. Renal function was normal; blood urea 16mg/100ml, creatinine 0.6mg/100ml. There was no glycosuria or aminoaciduria. Iliac bone biopsy showed evidence of osteomalacia in that there was excessive osteoid tissue in combination with a diminution in calcification front in that osteoid lamella lying closest to calcified bone. There was also evidence of folate deficiency as shown by macrocytic anemia, a megaloblastic marrow and response to a small dose of oral folate. This is a well recognised complication of anticonvulsant medication (GIRDWOOD and LENMAN, 1956). Later she had an attack of acute porphyria, the pattern of excretion of porphyrins being consistent with acute intermittent porphyria. It is possible that this also may have been related to her anticonvulsant medication.

A single dose of 1mg of vitamin D_3 was given intravenously; at the time the plasma calcium was 7.5mg/100ml, the plasma phosphorus 2.1mg /100ml and the alkaline phosphatase 38 KA units. The biochemical changes that occured are shown in Fig. 1. In the 2 weeks after the dose there was a rise in plasma calcium and a small rise in plasma phosphorus. The urinary calcium and phosphorus showed no change. There was a sharp rise in alkaline phosphatase and then a gradual fall. Two weeks after the 1mg i.v. dose, a small oral dose, 0.05mg/day, of vitamin D_3 was started. During the next 6 weeks the plasma calcium continued to rise; there was by then a fall in urinary calcium, an obvious rise of plasma phosphorus and a tendency for the urinary phosphorus to fall. There was also a continued fall in serum alkaline phosphatase. The rise in serum calcium illustrates the sensitivity of this patient to a moderate dose of vitamin D_3. The rise in serum phosphorus together with the fall in urinary phosphorus is similar to the change seen in nutritional rickets using a similar dose of vitamin D. It was probably due to a progressive fall in the secretion of parathyroid hormone since the serum immunoassayable parathyroid hormone level before treatment was 1.09ng/ml, and 8 weeks after the start of treatment was 0.12ng/ml (normal range $<$ 0.9ng/ml) (Dr. J.L.H. O'RIORDAN, Middlesex Hospital, London).

Bone biopsies were taken at the times shown in Fig. 1. An oral dose of a tetracycline was given 2 days before each biopsy as a marker for the calcification front. These through-and-through iliac bone biopsies were embedded in plastic and series of thin (6µ) undecalcified sections were prepared. The sections were stained with toluidine blue at pH 2.8. The histological changes were quantitated (Table 1). In the biopsy taken before treatment there was a marked excess of osteoid tissue. The volume of osteoid as a percentage of the cancellous bone tissue was 28% (normal, 20-29 yrs. of age, 7%±2) and the surface of trabecular bone, covered by osteoid, as a percentage of the total trabecular surface was 65% (normal, 20-29 yrs. of age, 20%±6). 31% of that osteoid lamella lying closest to calcified bone had a calcification front (normal, 20-29 yrs. of age, 81%±10). Two weeks after the single intravenous dose of 1mg of vitamin D_3 the extent of calcification front (88%) had returned to normal. In other words there had been a restoration of the normal process of mineralisation. At that time, however, neither the volume nor the extent of osteoid tissue had changed. In contrast, six weeks later the amount of osteoid had changed; the surface of trabecular bone covered by osteoid, though still abnormal, had fallen considerably and the volume of osteoid had fallen to within the normal range. The amount of calcification front was 79%.

Two measurements of bone remodeling were also made; first the active formation surface, that is the surface of osteoid covered by active osteoblasts (expressed as a percentage of the total trabecular surface) and secondly, the active resorption surface, that is those scalloped osteoclastic erosions actually undergoing resorption, as shown by the presence of osteoclasts. Active resorption surface was expressed as a percentage of the 'available' (total minus osteoid) surface i.e. calcified surface. The active formation surface was greater than normal before the single intravenous dose of vitamin D was given. In the biopsy taken 2 weeks later it had risen still further. This is in line with the transient rise in serum alkaline phosphatase. Eight weeks later the active formation surface had fallen to within the normal range (Table 1). The active resorption surface was abnormally high before treatment. Two weeks later it had risen even higher; after 8 weeks it was returning towards normal. This rise in osteoclastic activity is in contrast to the presumed fall in serum parathyroid hormone. It probably reflects the known permissive effect of vitamin D on the action of parathyroid hormone on the bone.

Table 1. Some aspects of quanitative bone histology in the patient

Biopsy	Total dose of vitamin D_3 given	Osteoid and Mineralisation			Bone remodeling	
		Volume of osteoid(as % of cancellous bone tissue)	Extent of osteoid(as % of total trabecular surface	Calcification front[a]	Active resorption surface;(as % of 'available' surface)	Active formation surface(as % of total cancellous surface)
Before treatment	-	28	65	31	3.4	11.0
2 weeks after 1mg i.v. vitamin D_3	1mg	33	71	88	12.0	14.8
8 weeks after start of vitamin D_3	3mg	6.5	40	79	3.3	8.5
Normal $\pm$ I.S.D.	20 - 29 yrs	7 ± 2	20 ± 6	81 ± 10	0.8 ± 0.4	7 ± 2

[a] Calcification front in that osteoid lamella lying closest to calcified bone, expressed as a percentage of the total surface of osteoid.

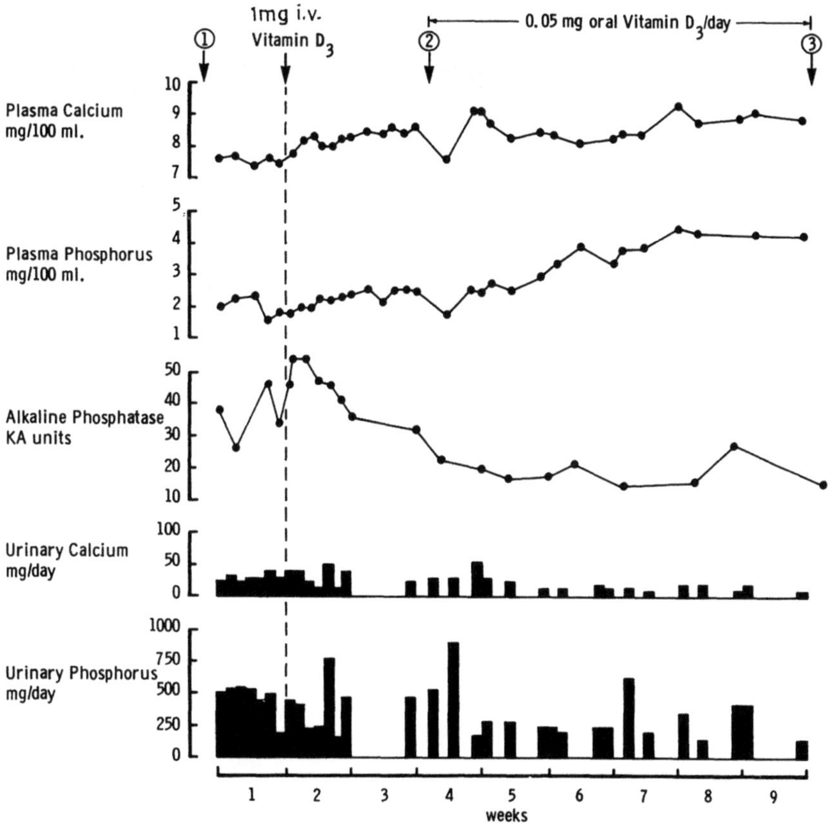

Fig. 1. Biochemical changes before and during the administration of vitamin D_3. Circled numbers 1, 2 and 3 refer to iliac bone biopsies

Summary

1. The biochemical changes and the histological response of the bone to the administration of a physiological amount of vitamin D_3 were measured in a patient with osteomalacia secondary to the administration of antiepileptic drugs.
2. 2 weeks after 1mg i.v. vitamin D_3 the extent of calcification front had returned to normal.
3. The biochemical and histological response was similar to that of nutritional osteomalacia, indicating that there is no impairment in the actions of vitamin D.
4. The early transient rise of serum alkaline phosphatase in conjunction with the rise of osteoblasts seen at 2 weeks suggests that vitamin D stimulated bone formation.
5. The increase in active resorption 2 weeks after the dose of vitamin D supports the concept that vitamin D has a permissive effect on the action of parathyroid hormone on osteoclastic bone resorption.

References

DENT, C.E., RICHENS, A., ROWE, D.J.F., STAMP, T.C.B.: Osteomalacia with long term anticonvulsant therapy in epilepsy. Brit. Med. J. 4, 69-72 (1970).

GIRDWOOD, R.H., LENMAN, J.A.R.: Megaloblastic anemia during primidone therapy. Brit. Med. J. 1, 146-7 (1956).

HAHN, T.J., BIRGE, S.J., SCHARP, C.R., AVIOLI, L.V.: Phenobarbital-induced alterations in vitamin D metabolism. J. Clin. Invest. 51, 741-748 (1972).

KRUSE, R.: Osteopathien bei antiepileptischer Langzeittherapie (Vorläufige Mitteilung). Mschr. Kinderheilk. 116, 378-381 (1968).

The Effect of Vitamin D_3 on Lanthanum Absorption: Suggestive Evidence for a Shunt Path*

R. H. WASSERMAN, A. N. TAYLOR and L. LIPPIELLO

Deciphering the path that calcium takes during the absorptive process will be of significant importance in understanding the calcium absorption mechanism and vitamin D action. The usual model (cf. Calc. Tissue Research 2, 301, 1968) depicts Ca entering the epithelial cell across the brush border, moving through the cytoplasma as the free ion or in association with a complexer, and then being extruded from the cell across the basal membrane by a Ca pump or by a Na-Ca exchange mechanism. Energy is required either directly via ATP input or indirectly via the maintenance of a sodium-gradient. This intracellular path model undoubtedly pertains to that calcium which is actively transported. However, there is sufficient evidence suggesting that Ca is also absorbed by a diffusional process, a process that is vitamin D-dependent, as is the active transport mechanism (cf. WASSERMAN, 1974). The question arises as to whether those calcium ions that are diffusionally transmitted use the same route as those calcium ions being actively transported. Certain kinetic data, such as the enhanced bidirectional calcium permeability of the intestine by vitamin D (WASSERMAN and KALLFELZ, 1962; MARTIN and DELUCA, 1969) and thermodynamic calculations (WASSERMAN and TAYLOR, 1969; WASSERMAN, 1974) suggest a possible non-cytoplasmic (or paracellular) path for calcium translocation.

Epithelial cells of cellular membranes, such as the intestine, are in close contact apically via the tight junction (zona occludens). The "dogma" for several years was that the tight junction represents an impermeable barrier to the passage of all substances. This sealing function presumably forces all absorbed materials to move through the absorptive enterocytes. However, recent studies have begun to challenge this concept. Observations from different quarters indicate that the intestine (MACHEN et al., 1972), gall bladder (FRÖMTER and DIAMOND, 1972; MACHEN et al., 1972) and kidney (WHITTEMBURY and RAWLINS, 1971) are "leaky" membranes and that their tight junctions are considerably more permeable than previously supposed.

A key probe in discerning the existence of these shunt paths has been lanthanum and other rare earths. Since lanthanum (as the trivalent cation) enters cells only sparingly or not at all, the demonstration that La^{+3} can move across an epithelial membrane would be suggestive of the presence of a shunt path. Electron microscopic visualization of La^{+3} within the tight junction and within the intercellular space would constitute necessary supportive information. MACHEN et al., (1972) were able to satisfy both criteria as evidence for a paracellular pathway in gall bladder.

+This work was supported by NIH Grant Am-04652 and U.S. A.E.C. Contract AT(11-1)-3167. The technical assistance of FRANCIS DAVIS, MARIE BRINDAK FULLWER, and KAREN NI is gratefully acknowledged.

I. Methods

Studies were undertaken on lanthanum chloride metabolism in rachitic and vitamin D_3-repleted chicks (4 weeks of age; vitamin D_3 dose = 500 I.U. 48 hrs. before experiment). The duodenal absorption of ^{140}La (as $LaCl_3$) and ^{47}Ca (as $CaCl_2$) was assessed by the in situ ligated segment technique (WASSERMAN, 1962). The dosing solution contained 150 mM NaCl, 2 mM PIPES buffer mixture (pH 6.8) and varying concentrations of $LaCl_3$ and/or $CaCl_2$, depending upon the experiment. The level of ^{140}La was usually 5 µCi/ml, and that of ^{47}Ca was usually about 0.1 µCi/ml. Percent absorption was calculated by taking the difference between the administered dose (100%) and the percent activity remaining after the absorption period. In some experiments, the uptake of ^{140}La or ^{47}Ca by the mucosal tissue was measured. Counting was done with a NaI well-type crystal counter and single channel analyzer to exclude the ^{47}Sc contribution to ^{47}Ca, or a small animal whole body counter (Tobor) in assessing the absorption of ^{140}La.

II. Results

As shown in Fig. 1, the duodenal absorption of ^{140}La was significantly increased by vitamin D_3 as compared to the non-treated controls when stable La^{+3} was 0.1 mM or 1.0 mM. At 10 mM La^{+3} the vitamin D_3 value was still greater than the rachitic value, but the difference was not statistically different ($p < .2$).

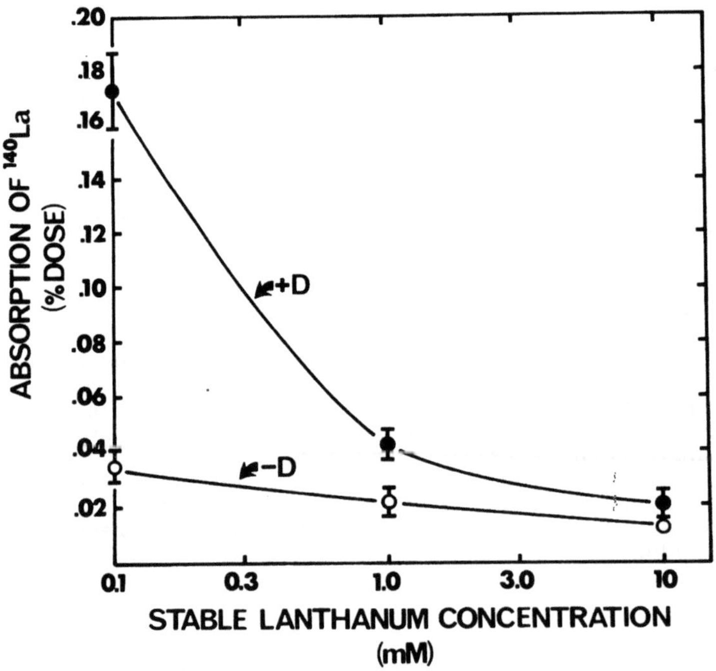

Fig. 1. Effect of vitamin D on the duodenal absorption of lanthanum in the rachitic chick. Values are the mean ± S.E. of 6 chicks per group

Table 1. Effect of Calcium and Vitamin D_3 on the Duodenal Absorption of ^{140}La in the Rachitic Chick[a]

Treatment	Ca^{+2} in dosing solution (mM)	^{140}La Absorption[b] (% dose)
Rachitic	-	0.039 ± 0.009
Rachitic	10	0.037 ± 0.004
Rachitic + D_3	-	0.144 ± 0.016[c]
Rachitic + D_3	10	0.084 ± 0.012[d]

[a] Chicks raised on a rachitogenic diet for 4 weeks. At 48 hrs. before experiment, vitamin D_3 treated groups received 500 I.U. D_3 by intramuscular injection (from WASSERMAN et al., 1974).

[b] Absorption time = 30 min. Absorption determined by the in situ ligated loop procedure (WASSERMAN, 1962) under ether anesthesia. Dosing solution contained 150 mM NaC1, 0.1 mM $LaC1_3$, 2 mM PIPES buffer, pH 6.8, plus 5 μc ^{140}La/ml. Values represent mean ± standard error of the mean; chicks/group.

[c] Significantly greater than all other groups at $p < .025$.

[d] Significantly greater than rachitic groups at $p < .01$.

The next series of experiments were undertaken to assess the effect of stable Ca^{+2} on ^{140}La absorption and the effect of stable La^{+3} on ^{47}Ca absorption. As shown in Table 1, Ca^{+2} (10 mM) significantly depressed ^{140}La absorption (0.1 mM) by vitamin D_3-replete chicks ($p < .025$). However, there was no effect of the calcium on ^{140}La absorption in the rachitic chick. It thus appeared that the presence of vitamin D_3 is required for Ca^{+2} to exert a depressing effect on the translocation of radiolanthanum. In another experiment (data not shown), it was demonstrated that lanthanum inhibited ^{47}Ca absorption but only in the vitamin D-replete chicks. No differences were noted in the rachitic animal, indicating again that the interaction between La^{+3} und Ca^{+2} requires the presence of vitamin D.

The data presented above suggest that La^{+3} is absorbed by a vitamin D-dependent process. The observation that La^{+3} depressed Ca^{+2} absorption and that Ca^{+2} depressed La^{+3} absorption further indicates a common component or site in the translocation of the rare-earth and calcium. La^{+3}, in another experiment, was also noted to yield a greater inhibitory effect on calcium absorption when the latter was at concentrations at which the diffusional mode is prominent.

III. Discussion

The interpretation of these and other related results (to be published elsewhere) is highly dependent upon the path that lanthanum takes during the course of intestinal absorption. Available information suggests that lanthanum only minimally penetrates into cells and this has been used to indicate the presence of a paracellular path across epithelial membranes (MACHEN et al., 1972). Using these considerations as a basis for the present experiments, it was observed that vitamin D enhanced La^{+3} (and Ca^{+2}) absorption by chick duodenum and that calcium depressed La^{+3} absorption, but only in the vitamin D-replete chick and not in the rachitic animal. Further La^{+3} inhibited Ca^{+2} absorption but, again, only in the vitamin D-replete chick. A hypothesis, therefore, that seems reasonable is that vitamin D affects ion movement via an extracellular path. However, it should be noted that the amount of lanthanum absorbed in the presence or absence of vitamin D is quite small, and it is conceivable that this small amount of lanthanum could be transferred either by a pinocytotic process or possibly by directly entering the cells across the microvillus membrane.

Electron micrographs of lanthanum localization in the present study showed that lanthanum can apparently penetrate through the tight junctions of chick intestine. This has yet to be quantitated in terms of the number of penetrated tight junctions and the effect of vitamin D on the degree of penetration; this information would be quite relevant in assessing the validity of the paracellular path hypothesis. Until these and other data become available, it is only justifiable to state that, at this time, the evidence now available on the behavior of lanthanum in the intestinal system is consonant with the presence of a paracellular path. It should be emphasized again that the hypothetical extracellular path would seem to pertain only to that calcium (and lanthanum) which moves across the intestine by the diffusional process and undoubtedly does not pertain to that calcium which is actively transported.

References

FRÖMTER, E., DIAMOND, J.: Nature 235, 9 (1972).
MACHEN, R.E., ERLIJ, D., WOODING, F.B.P.: J. Cell Biol. 54, 302 (1972).
MARTIN, D.L., DELUCA, H.F.: Arch. Biochem. Biophys. 134, 139 (1969).
WASSERMAN, R.H.: Handbook of Physiology (G. Aurbach, ed.), in preparation (1974).
WASSERMAN, R.H.: J. Nutrition 77, 69 (1962).
WASSERMAN, R.H., KALLFELZ, F.A.: Am. J. Physiol. 203, 221 (1962).
WASSERMAN, R.H., TAYLOR, A.N.: In: Mineral Metabolism, (C.L. Comar and F. Bronner, eds.), Vol. 3, 321-4/3. New York: Academic Press 1969.
WASSERMAN, R.H., TAYLOR, A.N., FULLMER, C.S.: Biochenical Symposium on Vitamin D (in press, 1974).
WHITTEMBURY, G., RAWLINS, F.A.: Plüger's Arch. 330, 302 (1971).

The Effects of Cyclic AMP, Hormones and Ions on the Conversion of 25-Hydroxycholecalciferol (25-HCC) to 1,25-Dihydroxycholecalciferol (1,25-DHCC) in Isolated Chick Kidney Tubules*

S.J. MacAuley, R.G. Larkins, A. Rapoport, T.J. Martin and I. MacIntyre

I. Introduction

25-HCC metabolism has been studied in isolated renal tubules obtained from digestion of chick kidneys with collagenase and hyaluronidase.

This intact cell preparation has advantages compared with a broken cell system: it allows the study of peptide hormones and of manipulated variations in the extracellular medium surrounding the cells.

The investigation examines effects of the calcium regulating hormones, cyclic nucleotides, certain ions and 1,25-DHCC conversion by the kidney.

II. Methods

1. Preparation

The method was based on that on BURG and ORLOFF (1962) and described by us previously LARKINS et al., (in press).

Histological examination of the preparation under the electron microscope showed the cells to be structurally intact with no apparent subcellular contamination.

$^{14}CO_2$ production from ^{14}C-U-glucose, (SHAIN and BARNEA, 1971) was linear over 4 hours indicating the cell preparation was viable and biochemically active.

2. Assessment of 25-HCC Metabolism

Routinely a 30-minute period of preincubation in the presence of the agent being tested was used prior to addition of 6-12ng of tritiated 25-HCC (26-(27)-methyl-^{3}H-25-HCC)-(sp. act. 13.1 or 6.1 Ci/mmol, Radiochemical Centre, Amersham). After a 30-minute incubation term at 37°C vitamin D metabolites were extracted, separated on Sephadex LH20 and counted as previously described by GALANTE et al. (1972).

+ This work was supported by the Medical Research Council, the Wellcome Trust, the Winston Churchill Memorial Trust (R.G.L.) and the Association of Commonwealth Universities - Senior Medical Fellowship (A.R.). The 1,25-DHCC was a generous gift from Dr. M.M. Pechet. The cAMP assay was kindly performed by Dr. B.R. Tulloch.

Re-chromatography of the Sephadex LH20 peak corresponding to 1,25-DHCC on a Dupont 830 high-speed liquid chromatograph showed that it was homogeneous and ran in the same position as a synthetic 1,25-DHCC reference standard.

Assessment of 25-HCC metabolism was quantitated by expressing the tritiated 1,25-DHCC peak as a percentage of the recovered radioactivity.

To compensate for peptide degradation the hormones were added to the preparation two times in equal amounts - the first just prior to preincubation and the second at the same time as 25-HCC administration. Cyclic nucleotides, when being tested, were also added in this way.

3. Adenylate Cyclase Assay

Crude particulate membrane preparations of chick kidney were tested for adenylate cyclase activity by the method of RAMACHANDRAN (1971) and WHITE and ZENSER (1971).

III. Results

1. Effect of 1,25-DHCC on 25-HCC Metabolism

3ng/ml of synthetic 1,25-DHCC (donated by Dr. M.M. Pechet, Research Institute for Medicine and Chemistry, Massachusetts, USA) incubated with the tubules for 3 hours prior to addition of tritiated 25-HCC significantly inhibited the formation of 1,25-DHCC (Table 1).

2. Effect of Extracellular Ions on 25-HCC Metabolism

Increasing the added calcium concentration in the presence of physiological phosphate ion concentration did not affect conversion. Similarly increasing the extracellular phosphate to 6 mM in the presence of physiological calcium ion concentration was also ineffective (Table 1). However, calcium depletion of the cells by washing in a calcium-free medium or in 0.1 mM EGTA caused significant inhibition of conversion (Table 1).

Strontium 5 mM increased conversion when the period of preincubation in the presence of strontium was short, i.e. from 15 - 30 minutes. This stimulation of conversion was abolished when preincubation in strontium was of longer duration (150 minutes).

3. Effect of Cyclic Nucleotides on 25-HCC Metabolism

Table 1 shows that 10^{-3}M cyclic AMP (cAMP) and 10^{-3}M and 10^{-4}M dibutyryl cyclic AMP (dbc AMP) significantly stimulated 1,25-DHCC production from 25-HCC. In subsequent experiments, 10^{-4}M dbc AMP was used as a positive control.

Table 1. Effect of extracellular ions and 1,25-DHCC on conversion of tritiated 25-HCC to 1,25-DHCC

Agent and Concentration	Preincubation Interval (mins)	No. of Replicates	% Recovered ^3H Eluting as 1,25-DHCC Mean $\pm$ SEM	Significance
A. Effect of 1,25-DHCC				
Control (0.6mM Ca	180	6	7.44 ± 0.59	P = 0.001
1,25-DHCC 3ng/ml	180	6	4.45 ± 0.35	
B. Effect of calcium and Phosphate				
Ca^{+2} 0 phosphate 0	30	3	24.91 ± 0.60	n.s.
Ca^{+2} 1.2mM phosphate 0	30	3	24.91 ± 3.88	n.s.
Ca^{+2} 4.8mM phosphate 0	30	3	27.40 ± 1.66	n.s.
Ca^{+2} 0 phosphate 2mM	30	3	25.99 ± 2.61	n.s.
Ca^{+2} 1.2mM phosph. 2mM	30	3	26.76 ± 0.61	n.s.
Ca^{+2} 4.8mM phosph. 2mM	30	3	25.08 ± 2.78	n.s.
Ca^{+2} 1.2mM phosph. 6mM	30	3	25.03 ± 0.63	n.s.
Ca^{+2} 0mM phosphate 6mM	30	3	22.87 ± 1.92	n.s.
C. Effect of Calcium Depletion				
Ca^{+2} 1.2mM	30	5	35.83 ± 2.86	
EGTA 0.1mM + Ca^{+2} 1.2mM	30	6	31.86 ± 2.08	n.s.
EGTA 0.1mM + Ca^{+2} 0	30	6	13.78 ± 0.96	P 0.001
D. Effect of Strontium				
Control (1.2mM Ca^{+2})	15	6	25.61 ± 0.65	P 0.01
Strontium Gluconate 5mM + (1.2mM Ca^{+2})	15	6	30.12 ± 1.09	
Control (1.2mM Ca^{+2})	180	6	11.06 ± 0.93	n.s.
Strontium Gluconate 5mM + (1.2mM Ca^{+2})	180	6	11.46 ± 1.47	

Table 2. Effect of cyclic nucleotides, bPTH and SCT on conversion of tritiated 25-HCC to 1,25-DHCC

Agent and Concentration		No. of Replicates	% Recovered ^3H as 1,25-DHCC Mean ± SEM	Significance
A. Effect of dbcAMP				
Expt. 1	Control	5	23.19 ± 1.73	
	dbcAMP 10^{-3}	5	33.17 ± 2.48	P 0.02
	cAMP 10^{-3}	5	32.49 ± 2./5	P 0.01
Expt. 2	Control	5	32.83 ± 2.25	
	dbcAMP	5	40.88 ± 1.87	P=0.025
B. Effect of bPTH				
Expt. 1	Control	5	41.69 ± 2.00	
	bPTH 10ng/ml	5	33.77 ± 2.48	P 0.05
	bPTH 500ng/ml	5	34.18 ± 1.52	P 0.02
Expt. 2	Control	5	13.54 ± 1.16	
	bPTH 10ng/ml	5	13.71 ± 0.58	n.s.
	bPTH 500ng/ml	5	14.20 ± 0.86	n.s.
	dbcAMP 10^{-4}M	5	20.15 ± 2.12	P 0.05
C. Effect of bPTH (Calcium Deplete Medium				
	Control + 0.1mM EGTA	6	13.78 ± 0.96	
	bPTH 50ng/ml + 0.1mM EGTA	6	18.78 ± 1.21	P 0.01
	bPTH 500ng/ml + 0.1mM EGTA	6	17.57 ± 1.66	P=0.08
	dbcAMP 10^{-4} 0.1mM EGTA	6	24.25 ± 2.21	P=0.001
D. Effect of SCT				
	Control	5	32.83 ± 2.25	
	SCT 10ng/ml	5	34.50 ± 1.86	n.s.
	SCT 500ng/ml	5	40.56 ± 1.58	P 0.025
	dbcAMP 10^{-4}M	5	40.88 ± 1.87	P=0.025

4. Effect of Bovine Parathyroid Hormone (bPTH)

a) With Calcium in the Medium

Table 2 shows that 10ng/ml and 500ng/ml of highly pruified bPTH (National Institute for Medical Research, Mill Hill, London))having a potency of 2500 u/mg produced slight but significant depression of 1,25-DHCC formation when control conversion was high but had no effect when control conversion was low in spite of the stimulation seen with the positive control 10^{-4}M dbcAMP. No stimulation by bPTH was ever seen when calcium was in the medium even when periods of incubation were varied or when normo-calcemic or vitamin D-replete chicks were used as described previously by LARKINS et al., (in press).

b) In a Calcium-Deplete Medium

Table 2 shows the effect of washing the cells in 0.1 mM EGTA and exposing them to 50 or 500ng/ml bPTH. In the presence of calcium, EGTA had no effect. In the absence of added calcium EGTA caused a depression of conversion. In this situation of calcium depletion bPTH caused a significant elevation of conversion over appropriate controls containing EGTA.

5. Effect of Salmon Calcitonin (SCT)

Table 2 shows that at the relatively high concentration of 500ng/ml, synthetic SCT caused a slight but significant stimulation of 1,25-DHCC production. SCT, inactivated by performic acid oxidation, had no effect in raising conversion.

6. Adenylate Cyclase Activity

bPTH caused a dose-related rise in adenylate cyclase activity and cellular cAMP levels in the chick tubules. Similar concentrations of SCT had no effect (LARKINS et al., in press).

IV. Discussion

The results described appear paradoxical. bPTH elevates cellular cAMP levels and cAmp stimulates conversion yet bPTH either has no effect or causes depression of the hydroxylation.

In addition SCT does not appear to stimulate adenylate cyclase activity but it does stimulate 1,25-DHCC formation. It is suggested that apart from any regulating effect cAMP may have on the conversion other factors may be equally or more important in determining 25-HCC-1-hydroxylase activity.

In the presence of calcium the stimulant effect of PTH-induced cAMP is balanced or exceeded by a depressant effect produced by Ca^{++} influx into the cell in response to bPTH. Such an influx of calcium with PTH has been shown to occur in monkey kidney cells by BORLE (1968) and calcium has been shown to inhibit 25-HCC-1-hydroxylase activity (FRASER and KODICEK, 1973; COLSTON et al., 1973).

In the absence of a calcium influx, i.e. in presence of EGTA, the stimulant effect of PTH induced cAMP elevation is allowed to proceed, resulting in the observed elevation in 1,25-DHCC.

The stimulant effect of high concentration of SCT may be due to a lowering of intracellular calcium (RASMUSSEN, 1972) as there was no apparent cAMP elevation.

An explanation for the elevated conversion seen after short preincubation periods in strontium may be that this divalent cation is competitively blocking calcium uptake into the cell, thereby lowering intracellular calcium levels and leading to stimulated conversion.

In conclusion, cAMP has an acute effect in vitro on the activity of the renal 25-HCC-1-hydroxylase enzyme, but variations in intracellular cAMP levels alone could not explain the results observed. The results suggest that intracellular calcium may be one factor acutely affecting the activity of this enzyme. In addition, 1,25-DHCC at low concentrations emerges as a potential regulator of its own formation.

References

BORLE, A.B.: Effect of purified parathyroid hormone on the calcium metabolism of monkey kidney cells. Endocrinology 83, 1316-1322 (1968).

BURG, M.B., ORLOFF, J.: Oxygen consumption and active transport in separated renal tubules. American Journal of Physiology 203, 327-330 (1962).

COLSTON, K.W., EVANS, I.M.A., GALANTE, L., MACINTYRE, I., MOS, D.W.: Regulation of vitamin D metabolism: factors influencing the rate of formation of 1,25-dihydroxycholecalciferol by kidney homogenates. Biochemical Journal 134, 817-820 (1973).

FRASER, D.R., KODICEK, E.: Regulation of 25-hydroxycholecalciferol-1-hydroxylase activity in kidney by parathyroid hormone. Nature: New Biology 241, 163-166 (1973).

GALANTE, L., MACAULEY, S.J., COLSTON, K.W., MACINTYRE, I.: Effect of parathyroid extract on vitamin D metabolism. Lancet I, 985-988 (1972).

LARKINS, R.G., MACAULEY, S.J., RAPOPORT, A., MARTIN , T.J., TULLOCH, B.R., BYFIELD, P.G.H., MATTHEWS, E.W., MACINTYRE, I.: Effects of nucleotides, hormones, ions and 1,25-dihydroxycholecalciferol on 25-hydroxycholecalciferol metabolism in isolated chick renal tubules. Clinical Science and Molecular Medicine (in press).

RAMACHANDRAN, J.: A new simple method for separation of adenosine 3':5' - cyclic monophosphate from other nucleotides and its use in the assay of adenyl cyclase. Analytical Biochemistry 43, 227-239 (1971).

RASMUSSEN, H.: The cellular basis of mammalian calcium homeostasis. Clinic in Endocrinology and Metabolism 1, 3-20 (1972).

SHAIN, S.A., BARNEA, A.: Some characteristics of the estradiol binding protein of the mature, intact rat uterus. Endocrinology 89, 1270-1279 (1971).

WHITE, A.A., ZENSER, T.V.: Separation of cyclic 3':5' nucleoside monophosphate from other nucleotides on aluminum oxide columns. Analytical Biochemistry 41, 372-396 (1971).

Extrarenal Metabolism of High Doses of 25-(OH)-D_3: a New Active Metabolite

M. GARABEDIAN, H. PAVLOVITCH AND S. BALSAN

Data available at present on the metabolites of vitamin D_3 seem to indicate (FRASER and KODICEK, 1970; OMDAHL et al., 1972; HOLICK et al., 1972; WONG et al., 1972) that 25-hydroxycholecalciferol (25-(OH)-D_3) has to be hydroxylated in the kidney in order to promote its biological activity. Yet, in a recent study (PAVLOVITCH et al., 1973) we have seen that 25-(OH)-D_3 has a clear-cut calcium mobilizing effect in anephric rats, at doses equal to or higher than 125 nmoles. In order to analyze this action of 25-(OH)-D_3 further we investigated a possible extrarenal conversion of 25-(OH)-D_3 into more polar active metabolites. This investigation led to the revelation of a new metabolite termed X_1, which possesses biological activity.

Weanling albino Wistar rats were fed a normal Ca (0.47%) normal P (0.30%) diet for 5 days, then a low Ca (0.02%) normal P diet for three days. On the 8th day half the animals were bilaterally nephrectomized, the others sham-operated. Immediately post-surgery they were given I.V. 0.25 or 625 nmoles of ^3H(26,27)-25-(OH)-D_3 in 95% ethanol. All animals were killed 24 hours later. Chloroform extract of serum, liver, bone, and small intestine were chromatographed on Sephadex LH 20.

I. Results

In serum and all tissues studied, three major peaks were observed on the chromatographic profile: Peak IV, Peak Va (HOLICK and DELUCA, 1971) and a more polar Peak X_1. Peak IV cochromatographs with crystalline 25-(OH)-D_3. Periodate cleavage studies of the Peak Va in all cases revealed it to be 80-85% sensitive to periodate cleavage as has been described for 24,25-dihydroxycholecalciferol (HOLICK et al., 1972), thus presumably the Peak Va for the most part described in these experiments represents 24,25-dihydroxycholecalciferol as well in intact as in anephric animals. The chromatographic behavior of Peak X_1 does not correspond to any of the more polar metabolites of vitamin D_3 so far described (PONCHON adn DELUCA, 1969; HAHN et al., 1972; FROLIK and DELUCA, 1972; HOLICK et al., 1973). Peak X_1 is also periodate sensitive, however, periodate cleavage experiments suggest that it is not a 24-hydroxylated form of the vitamin. Thus it seems unlikely that the Peak X_1 metabolite is identical with the 1,24,25-trihydroxyvitamin D_3 recently described by HOLICK et al. (1973).

When a small dose of 25-(OH)-D_3 is used, nephrectomy almost abolishes the conversion of 25-(OH)-D_3 into Va and X_1; in contrast, when a 625nmol dose is utilized, an appreciable proportion of total radioactivity appears in Peaks Va and X_1, and this is similar in intact and nephrectomized rats. The calcium mobilizing activity of Va and X_1 was investigated in rats fed a low Ca D-deficient diet. The results demonstrate that 62.5 pmol of Peak X_1 were as active as 625 pmol of Peak Va, in the elevation of plasma calcium concentration 24 hours after an intravenous injection. In fact, 312 pmol of Peak X_1 produced an equivalent response to the same dose of 1,25-dihydroxycholecalciferol, revealing its very potent activity in the mobilisation of calcium from body stores. Nephrectomy eliminates this activity.

II. Summary

The present data demonstrate the following:
1) conversion of 25-(OH)-D_3 into more polar metabolites occurs in the absence of kidney tissue when large enough doses of this metabolite are used.
2) the major derivatives found under such conditions, Va and X_1, are biologically active in D-deficient animals; X_1 has not previously been reported.
3) neither Va nor X_1 can be responsible for the calcium mobilizing action of large doses of 25-(OH)-D_3 since these Peaks are inactive in nephrectomized rats.

References

FRASER, D.R., KODICEK, E.: Nature. 228, 764-766 (1970).
FROLIK, C.A., DELUCA, H.D.: J. Clin. Invest. 51, 2900-2906 (1972).
HAHN, T.J., BIRGE, S.J., SCHARP, C.R., AVIOLI, L.: J. Clin. Invest. 51, 741-748 (1972).
HOLICK, M.F., DELUCA, H.F.: J. Lip. Res. 12, 460-465 (1971).
HOLICK, M.F., GARABEDIAN, M., DELUCA, H.F.: Science. 176, 1146-1147 (1972).
HOLICK, M.F., KLEINER-BOSSALLER, A., SCHNOES, H.K., KASTEN, P.M., BOYLE, I.T., DELUCA, H.F.: J. Biol. Chem. 248, 6691-6696 (1973).
HOLICK, M.F., SCHNOES, H.K., DELUCA, H.F., GRAY, R.W., BOYLE, I.T., SUDA, T.: Biochemistry, 11, 4251-4255 (1972).
OMDAHL, J.L., GRAY, R.W., BOYLE, I.T., KNUTSON, J., DELUCA, H.F.: Nature New Biol. 237, 63-64 (1972).
PAVLOVITCH, H., GARABEDIAN, M., BALSAN, S.: J. Clin. Invest. 52, 2656-2659 (1973).
PONCHON, G., DELUCA, H.F.: J. Nutr. 99, 157-167 (1969).
WONG, R.G., NORMAN, A.W., REDDY, C.R., COBURN, J.W.: J. Clin. Invest. 51, 1287-1291 (1972).

III. Metabolism and Action of Fluoride

Chairmen: W. D. ARMSTRONG and I. VAN DER SLUYS VEER

Fluoride and its Relation to Bone and Tooth

J.A. WEATHERELL, D. DEUTSCH AND C. ROBINSON

Interest in the reaction of fluoride with the mineralizing tissues has increased because of proposals to fluoridate water supplies in the interests of dental prophylaxis and because of the hope that fluoride administration might help with the treatment of certain bone diseases. Such intentions necessitate a comprehensive knowledge of the uptake and distribution of fluoride in the mineralizing tissues and of possible hazards associated with its ingestion.

I. Occurrence, Absorption and Accumulation of Fluoride

All diets and most water supplies contain fluoride but the amount ingested varies considerably (WALDBOTT, 1963), depending upon local geography and dietary habits. In Britain, a major contribution is made by tea, an average infusion containing about 1 ppm (HARRISON, 1949). Tinned fish is a rich source of fluoride and it has been claimed that in some countries fish and sea-salt constitute a large proportion of the total fluoride intake (ELLIOTT and SMITH, 1960; HADJIMARKOS, 1962).

Unless the ingested fluoride is in an insoluble form (ERICSSON, 1958; WEDDLE and MUHLER, 1957) it is quickly absorbed (HODGE, 1956) and although excretion via the kidney is very efficient, up to half the absorbed fluoride is incorporated into the skeleton, where it accumulates with time (LARGENT and HEYROTH, 1949). In many experimental studies using relatively large dosage of fluoride, it was found that the initial rapid rate of incorporation into the skeleton gradually decreased, presumably due to saturation of the sites of skeletal uptake (ZIPKIN and McCLURE, 1952; WEATHERELL, 1969). At the restricted concentrations present in human diets, this does not occur and throughout life the concentration of fluoride in bone progressively increases (WEATHERELL, 1966). The bone of older individuals therefore contains a relatively high concentration of fluoride. Even in Leeds, where the concentration in the water supply is low (<0.1 ppm), the cortical bone in femora from elderly individuals can contain almost 3,000 ppm F on an ash basis. (Fig. 1.)

II. Concentration and Distribution of Fluoride within the Hard Tissues

The fluoride concentration is not the same in all parts of the skeleton and in general is highest at bone surfaces. This is perhaps the principal reason why cancellous bone, which has a very high surface/mass ratio contains 2-3 times as much fluoride as compact cortical bone (WEIDMANN and WEATHERELL, 1959; SINGER and ARMSTRONG, 1962). In the compact cortical bone itself, the surfaces usually contain the highest fluoride concentration and, in the human femur, the periosteal surface usually contains more fluoride than the endosteal surface (WEATHERELL,

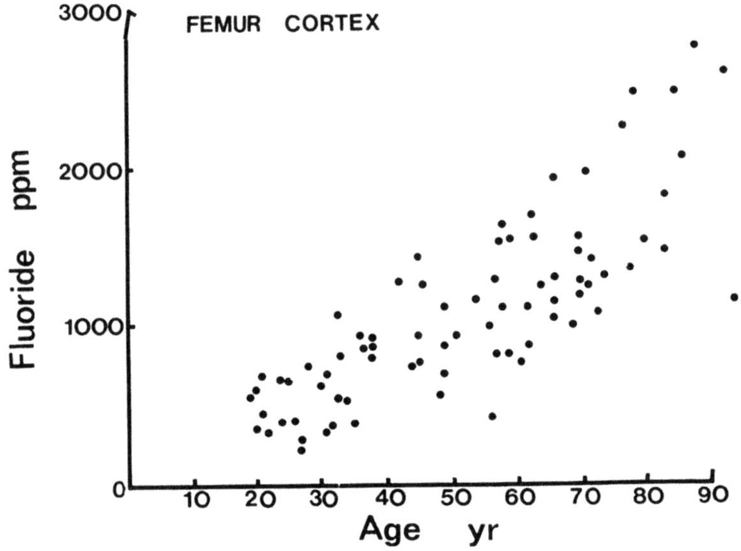

Fig. 1. Fluoride concentration in bone ash from the femoral diaphysis of males and females living in a district where the fluoride concentration in the water supply was <0.1 ppm

1969). This is probably because the amount of fluoride present in the surface region depends partly on the amount of bone formation at that surface and upon the amount of resorption taking place. In the femur, while bone formation occurs predominantly on the periosteal surface the endosteal surface is preferentially resorbed. With increasing age, this resorption gradually removes the relatively low-fluoride endosteal bone. Thus, although the femoral cortex undoubtedly accumulates fluoride with age, the rate at which the average concentration in the bone increases will be influenced by the pattern of resorption. This effect will be greater as the cortex becomes thinner with age. Resorption can also lower the average fluoride concentration of a tissue. In deciduous dentine, for instance, the average fluoride concentration increases as the tooth forms and decreases prior to shedding; because the resorption associated with the process of exfoliation removes the high-fluoride pulpal dentine (HARGREAVES and WEATHERELL, 1965) (Fig. 2a). It would be interesting to know whether all the fluoride removed during such resorptive processes finds its way into the general circulation or whether some of it is picked up locally by adjacent hard tissues. Local transfer of fluoride from one part of a bone to another has been reported from animal experiments (LIKINS et al., 1959). In the case of deciduous dentine, some of this fluoride lost by resorption from the pulpal surface might be taken up by the still mineralizing surface enamel of the underlying permanent teeth. Such fluoride transfer could make a significant contribution to the relatively high fluoride concentration of permanent enamel and the higher incidence of dental fluorosis found in permanent teeth, most of which are situated directly beneath a deciduous precursor and all of which, during eruption, come into close proximity with the overlying resorbing bone (Fig. 2b).

As in bone and dentine, the fluoride concentration in enamel is invariably highest in its surface region (JENKINS and SPEIRS, 1953; BRUDEVOLD et al., 1959). Most of this fluoride in the tooth surface

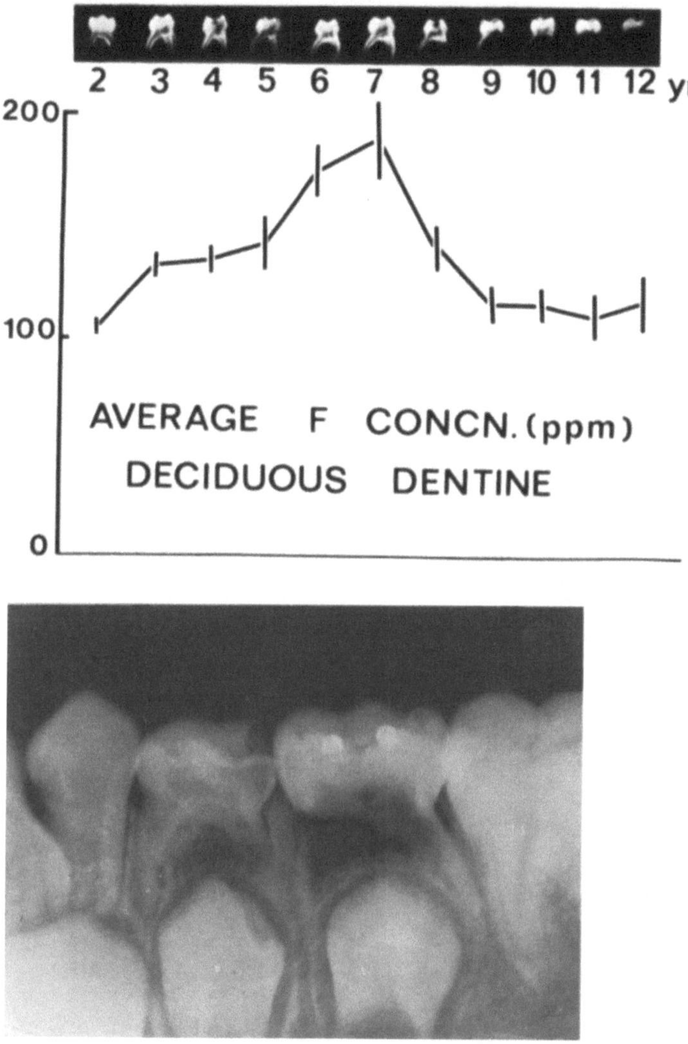

Fig. 2. (a) Variation with age in the average fluoride concentration of deciduous dentine. As the high-fluoride dentine of the pulpal surface is removed by resorption, the average fluoride concentration of the tissue falls until the tooth is exfoliated at about the age of 12. The vertical lines indicate standard deviations. (b) Radiograph of deciduous human molars, the resorbing dentine is situated directly over the mineralizing enamel of the future permanent dentition (From HARGREAVES and WEATHERELL, 1965)

appears to be acquired during enamel development and, in the fully mineralized tissue, fluoride uptake is very small. Once the tooth is erupted, some of the fluoride acquired during formation is removed by abrasion whereas at other sites, demineralisation brought about by a fall in pH at the tooth surface, increases fluoride uptake (WEATHERELL et al., 1973). (Fig.3.) The distribution pattern of fluoride in the surface of an erupted tooth is therefore extremely complex.

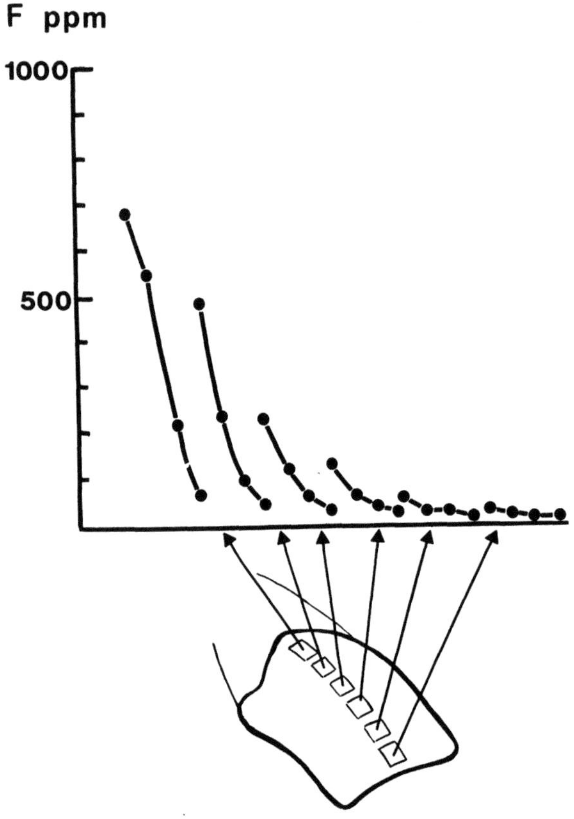

Fig. 3. Distribution of fluoride from surface to the interior of enamel in 6 regions on the labial surface of a 37-year-old human incisor. Towards the incisal region, much of the fluoride in the enamel surface has been removed by abrasion in the mouth. Near the cervical margin, where the enamel was covered by dental plaque, fluoride uptake probably continued throughout life

III. Pathological Effects of Fluoride Ingestion on Bones and Teeth

The overall concentration of fluoride in the hard tissues depends upon the amount of fluoride ingested and seems to correlate well with the concentration of fluoride in the drinking water supply (ZIPKIN et al., 1958). This can therefore usually be taken as an index for the level of ingestion.

Fluoride has an effect on bones and teeth which varies with the level of ingestion. At 1 ppm or less there is little or no visible effect, although the incidence of dental caries is reduced. If concentrations slightly higher than this are ingested during tooth development, however, changes begin to appear in the enamel surface (HODGE and SMITH, 1954). Sometimes these changes are not cosmetically objectionable, perhaps merely amounting to a patchy whiteness of the enamel due to slight increases in the porosity of the surface-region enamel. Gradually, however, with increasing concentrations of fluoride in the drinking water, the manifestations of dental fluorosis increase in severity. DEAN (1934) classified the symptoms into seven stages and FORREST (1965) more recently published excellent illustrations of the changes which occur in the permanent dentition. She describes how at about 2 ppm F in the drinking water, flecks can be detected in the enamel as a proportion of the teeth become affected by 'mild' fluorosis. These are cosmetically of little or no disadvantage and associated with increased resistance to dental caries. At 3-5 ppm some teeth become stained, a stage classified by DEAN (1934) as 'moderately severe' fluorosis. At 5-6 ppm the occasional unsightly pitted appearance of 'severe' dental fluorosis occurs.

At still higher levels of ingestion, about 8 ppm F or more in the drinking water, the skeleton can be affected, the earliest clinical changes being chiefly manifest in a slightly increased radio-density (LEONE et al., 1955) although, histologically, the occurrence of 'mottled' osteones seems to occur at an earlier stage of the condition (JOHNSON, 1964). The more severe chronic state, resulting sometimes from a life-long ingestion of over 20mg F/day can involve a crippling deposition of mineral into muscles and tendons, the development of bony out-growths and osteophytic nodules, increased resorption and, often, failure of the newly-formed tissue to mineralize (MØLLER and GUDJONSSON, 1932; ROHOLM, 1937; PANDIT et al., 1940; SINGH et al., 1962; SANKARAN and GADEKAR, 1964).

One of the most interesting features of skeletal fluorosis, especially from the point of view of bone therapy, is an apparent tendency for the balance of bone remodeling to shift towards formation, leading to a thickening of trabecular spicules in the cancellous structures and sometimes to an overall widening of the long bones. According to many authors, bones increased in density and hardness and, in 1961, RICH and ENSINCK tried to exploit this apparently beneficial property of fluoride. They administered fluoride to patients with osteoporosis in the hope that this might stimulate bone formation and increase the density of the porous bone. Subsequently, a large number of experimental and clinical studies have been carried out, from attempts to prevent disuse osteoporosis in animals (GEDALIA et al., 1966; MILICIC and JOWSEY, 1968), to clinical investigations in which doses of 30-70mg F/day were given to patients suffering from osteoporosis or various other skeletal pathologies (PURVES, 1962; RICH et al., 1964; COHEN and GARDNER, 1964; ROSE, 1965; AESCHLIMANN et al., 1966; BERNSTEIN and COHEN, 1967).

Unfortunately, it has not always been clear whether such fluoride therapy had actually stimulated new bone formation or had induced some of the less desirable features of skeletal fluorosis described above. There was also the possibility that any bone formed as a result of the fluoride administration would be too weak in structure or too poorly mineralized to be of any real advantage. Although strong, hard bone has undoubtedly been demonstrated in cases of chronic human skeletal fluorosis (ROHOLM, 1937; FRANKE et al., 1972) increases in <u>radio</u>density seem sometimes to have been interpreted as increases in physical density when, in fact, they merely reflected the production of more, but perhaps abnormal and weaker, poorly mineralized bone (JOHNSON, 1952; WEATHERELL, 1969; NORDIN, 1973). Fluorotic bone is not always hard. In experimental studies it has often proved to be porous and poorly mineralized with wide osteoid seams. Mineral can deposit in soft tissues and ligament insertions and numerous workers have described the greatly increased bone resorption, alleged by FACCINI (1969) to result from a fluoride-induced secondary hyperparathyroidism. Naturally, anxiety has been expressed that any stimulation of bone formation by fluoride administration might lead to undesirable bone pathology or to the mineralisation of the soft tissue surrounding bones.

Some recent investigations have tended to allay these fears to some extent. JOWSEY et al. (1972), for instance, seemed to have had considerable success with a twice weekly administration of fluoride to stimulate bone growth, together with vitamin D and calcium supplementation to facilitate mineralisation of the newly formed bone. Despite these fairly hopeful clinical findings, however, at the present time 'fluoride therapy' remains a controversial issue.

IV. Mechanism of Fluoride Action

Part of the uncertainty surrounding the clinical use of fluoride, is due to a lack of precise knowledge about its action on the hard tissues. Although the pathological changes in bones and teeth, and the established reduction in dental caries, are brought about by the absorption of a readily identifiable agent, the mechanism by which fluoride produces the effects is still not clear. The most favored current explanation is that fluoride replaces the hydroxyl group of the apatite lattice in bone and tooth mineral to produce a more stable crystallite, less easily dissolved by acid of the dental plaque and less easily resorbed by the osteoclast (NEUMAN and NEUMAN, 1958; JENKINS, 1962). While such a change in solubility might feasibly explain a reduction in dental caries, it does not so convincingly account for the more dramatic histological changes of skeletal and dental fluorosis often manifest in cellular changes and in alterations to the calcifying matrix. There is, moreover, no direct evidence to show that fluoridated apatite is less easily resorbed than non-fluoridated apatite or that the occurrence of fluorosis is directly related to the amount of fluoride in the bone mineral. Despite the one-time belief that there was indeed a 'critical threshold' concentration of fluoride in bone (McCLURE et al., 1958; JACKSON and WEIDMANN, 1958), very large amounts of fluoride can accumulate in bone without any visible pathological change (PEIRCE, 1939; ZIPKIN, 1960; WEIDMANN et al., 1963).

The factor which correlates most closely with the occurrence of fluorosis seems to be not the concentration in the hard tissues but the rate of fluoride ingestion (HODGE and SMITH, 1954; WEATHERELL, 1966). This has led some workers to suggest that the histological damage is due to an effect of the free-fluoride ion on enzymes, matrices or cells (JOHNSON, 1952; FLEMING, 1953). The chief difficulty with such a view has been the observation that the concentrations of free-fluoride ion available in the plasma are somewhat low even at relatively high levels of intake (ARMSTRONG et al., 1964), presumably because the absorbed fluoride is so efficiently excreted and quickly sequestrated by the skeleton.

However, perhaps the plasma concentration of fluoride does not correspond to that near the cells of the mineralizing tissues (WEATHERELL, 1969). When fluoride is absorbed, there will be a temporary increase in plasma fluoride concentration, so that a concentration gradient of fluoride from the plasma, across the cells to the apatite crystallites of the mineralizing tissues will be temporarily established. During this time, the small, highly hydrated crystallites in areas of active mineralisation and possibly also the inter-crystalline protein will trap relatively large amounts of fluoride. A short time after ingestion, however, the fluoride concentration in the plasma will fall, the concentration gradient will be reversed and the fluoride acquired at the mineralizing front will tend to diffuse back from the crystallites across the cell into the plasma (WEATHERELL, 1969; COSTEAS et al., 1970). In this way, the small crystallites and perhaps also the protein and water in the areas of active mineralisation will tend to trap and retain fluoride near the membrane of the mineralizing tissue cells at a level which is high relative to the fluoride concentration in the blood.

V. Fluoride Uptake During Mineralisation

Unfortunately, there is relatively little information about the concentration of fluoride in such forming areas of tissues. In bone, it is difficult to isolate a region which is exclusively forming and it would probably not be possible at present to measure any such effect directly. However, we have recently obtained some evidence from studies of forming enamel which suggests that the developing region contains a relatively high concentration of fluoride, and perhaps in a fairly labile form (DEUTSCH et al., 1973). Particles of enamel, approximately equal in length, were dissected from the forming region of the continuously-growing rat incisor and the fluoride/phosphorus ratio (F/P) of each sample determined. There was a relatively high F/P ratio in the late-forming or early-maturing region of enamel and, as the tissue mineralized, this decreased (Fig. 4.). Preliminary measurements in which an attempt has been made to relate fluoride to tissue volume, have suggested that a considerable amount of the fluoride present during this early phase of mineralisation was lost as the mineral content of the tissue increased. This, at least in part, probably explains the decrease in F/P ratio shown in Fig. 4. It has not been established whether this relatively high concentration of fluoride in the forming enamel is associated with mineral, protein or water. However, it appears from the above preliminary volume measurements that at a specific phase of development the enamel of the rat incisor accumulates a relatively high concentration of fluoride, which is fairly labile in the sense that much of it is lost again over a period of a few days. It seems possible that this loss is associated with the withdrawal of the protein matrix as the enamel mineralizes. The question is, whether or not this fluoride is free enough to have an effect on the enamel matrix or overlying ameloblasts, rather than merely on the mineral phase. If this were the case, it could explain how cell changes and fluorosis can occur at relatively low levels of fluoride absorption. In bone, a similar mechanism might operate; as in enamel some of the fluoride acquired by forming bone is certainly lost within hours of its incorporation (WALLACE-DURBIN, 1954). Fluoride seems to have less effect on dentine than on bone or enamel. This could be explained by the interposition of the non-mineralized predentine zone which, according to the above hypothesis, would provide a protective barrier between the crystallites at the mineralizing front of the tissue and the odontoblasts.

According to NEUMAN (1969), the concentrations of ions in the blood are not always the same as those in the milieu within the mineralizing tissues. The above observations made on forming enamel also point to the possible danger of regarding the concentration of fluoride in the plasma as necessarily relevant to the concentration of fluoride immediately adjacent to the cells of the enamel.

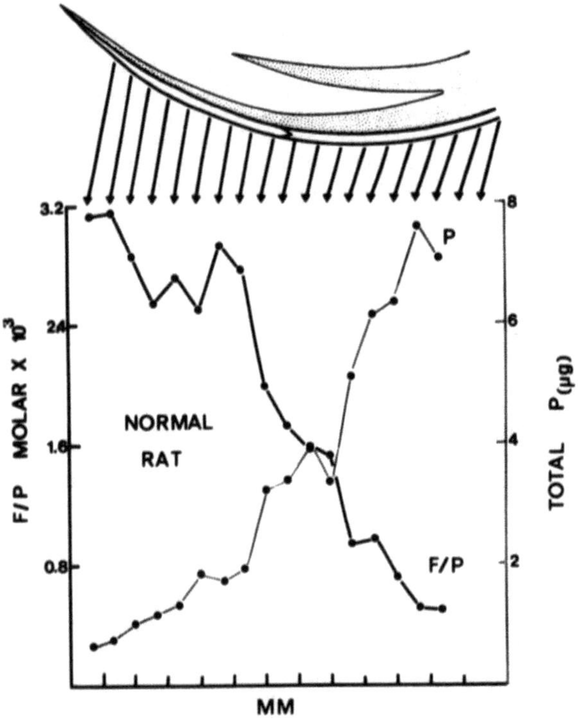

Fig. 4. Variation in F/P ratio in particles of forming enamel from the continuously-growing rat incisor. Although the width of the enamel increased over the first five particles, some idea of the change in mineralisation can be seen by the change in total P. As the amount of mineral in the enamel increased, the F/P ratio of the tissue fell

References

AESCHLIMANN, I., GRUNT, J.A., CRIGLER, J.F. Jr.: Effects of sodium fluoride on the clinical course and metabolic balance of an infant with osteogenesis imperfecta congenita. Metabolism, 15, 905-914 (1966).

ARMSTRONG, W.D., SINGER, L., ENSINCK, J., RICH, C.: Plasma fluoride concentrations of patients treated with sodium fluoride. J. clin. Invest. 43, 555 (1964).

BERNSTEIN, D.A., COHEN, P.: Use of sodium fluoride in the treatment of osteoporosis. J. Clin. Endocr. Metab. 27, 197-210 (1967).

BRUDEVOLD, F., GARDNER, D.E., SMITH, F.A.: The distribution of fluoride in human enamel. J. dent. Res. 35, 420-429 (1956).

COHEN, P., GARDNER, F.H.: Induction of subacute skeletal fluorosis in a case of multiple myeloma. New England J. Med. 271, 1129-1133 (1964).

COSTEAS, A., WOODARD, Q., LAUGHLIN, J.S.: Depletion of ^{18}F from blood flowing through bone. J. Nucl. Med. 11, 43-45 (1970).

DEAN, H.T.: Classification of mottled enamel diagnosis. J. Am. dent. Assoc. 21, 1421-1426 (1934).

DEUTSCH, D., WEATHERELL, J.A., ROBINSON, C.: Site of high fluoride concentration in developing enamel of rat incisor. Abstract, I.A.D.R. (Brit. Div. (1973).

ELLIOT, C.G., SMITH, M.D.: Dietary fluoride related to the fluoride content of teeth. J. dent. Res. 39, 93-98 (1960).
ERICSSON, Y.: The state of fluorine in milk and its absorption and retention when administered in milk. Acta odont. scand. 16, 51-77 (1958).
FACCINI, J.M.: Fluoride and bone. Calc. Tiss. Res. 3, 1-16 (1969).
FLEMING, H.S.: Effect of certain concentrations of fluoride on enamel and dentine as formed in transplants of tooth germs and related studies. J. dent. Res. 32, 469-485 (1953).
FORREST, Jean R.: Fluoridation of domestic water supplies: Mottled enamel. Br. dent. J. 119, 316-319 (1965).
FRANKE, J., DRESE, G., GRAU, P.: Klinische gerichtsmedizinische und physikalische Untersuchungen eines Falles von schwerer Fluorose. Krim. foren. Wissensch. 7, 107-122 (1972).
GEDALIA, I., SCHWARTZ, A., Sela, J., GAZENFIELD E.: Effects of fluoride intake on disuse atrophy of bone in rats. Proc. Soc. Exp. Biol. Med. 22, 657-660 (1966).
HADJIMARKOS, D.M.: The fluoride content of enamel in a group of Greek teeth. Archs. oral Biol. 7, 651-653 (1962).
HARGREAVES, J.A., WEATHERELL, J.A.: Variations of fluoride concentrations in human deciduous teeth. In: Advances in Fluorine Research and Dental Caries Prevention (ed. J.L. HARDWICK, Hans R. HELD, Klaus G. KÖNIG) vol. 3, 247-254. Oxford-London-Edinburgh-New York-Paris-Frankfurt: Pergamon Press Ltd. 1965.
HARRISON, M.F.: Fluorine content of teas consumed in New Zealand. Br. J. Nutr. 3, 162-166 (1949).
HODGE, H.C.: Fluoride metabolism: its significance in water fluoridation. J. Am. dent. Assoc. 52, 307-314 (1956).
HODGE, H.C., SMITH, F.A.: Some public health aspects of water fluoridation. In: Fluoridation as a Public Health Measure (ed. J. H. SHAW) pp. 79-109. Washington: Am Assoc. Advanc. Sci. 1954.
JACKSON, D., WEIDMANN, S.M.: Fluorine in human bone related to age and the water supply of different regions. J. Path. Bact. 451-459 (1958).
JENKINS, G.N.: The physiology of fluoride, Internat. dent. J. 12, 208-226 (1962).
JENKINS, G.N., SPEIRS, R.L.: Distribution of fluorine in human enamel. Proc. physiol. Soc. 121, 21 (1953).
JOHNSON, L.C.: Discussion in HODGE, H.C., The significance of the skeletal deposition of fluoride. In: Metabolic Interrelations (ed. E.C. REIFENSTEIN Jr.) Trans. 4th Conf., 250-260. New York: Josiah Macy jr. Found. 1952.
JOHNSON, L.C.: Morphologic analysis in pathology: The kinetics of disease and general biology of bone. In: Bone Biodynamics (ed. H.M. FROST) 543-654. Boston,USA: Little, Brown & Co. (Inc.) 1964.
JOWSEY, J., RIGGS, B.L., KELLY P.J., HOFFMAN, D.L.: Effect of combined therapy with sodium fluoride, vitamin D and calcium in osteoporosis. Am. J. Med. 53, 43-49 (1972).
LARGENT, E.J., HEYROTH, F.F.: Absorption and excretion of fluoride: III. Further observations on the metabolism of fluorides at high levels of intake. J. industr. Hyg. Toxicol. 31, 134-138 (1949).
LEONE, N.C., STEVENSON, C.A., HILBISH, T.F., SOSMAN, M.C.: A roentgenologic study of a human population exposed to high-fluoride domestic water. Am. J. Roentgenol. 74, 874-885 (1955).
LIKINS, R.C., SCOW, R.O., ZIPKIN, I., STEERE, A.C.: Deposition and retention of fluoride and radiocalcium in the growing rat. Am. J. Physil. 197, 75-80 (1959).
McCLURE, F.J., McCANN, H.G., LEONE, N.C.: Excessive fluoride in water and bone chemistry. Publ. Hlth Rep. (Wash.) 73, 741-746 (1958).
MILICIC, M., JOWSEY, J.: Effect of fluoride on disuse osteoporosis in cat. J. Bone Jt. Surg. 50-A, 701-708 (1968).

MØLLER, P.F., GUDJONSSON, S.V.: Massive fluorosis of bones and ligaments. Acta radiol. 13, 269-294 (1932).
NEUMAN, W.F.: The milieu intérieur of bone: Claude Berhard revisited. Fed. Proc. 28, 1846-1850 (1969).
NEUMAN, W.F., NEUMAN, M.W.: The chemical dynamics of bone mineral. Chicago: University of Chicago Press 1958.
NORDIN, B.E.C.: Metabolic bone and stone disease. London: Churchill Livingstone 1973.
PANDIT, C.G., RAGHAVACHARI, T.N.S., RAO, D.S., KRISHNAMURTI, V.: Endemic fluorosis in S. India. Indian J. med. Res. 28, 533-558 (1940).
PEIRCE, A.W.: Chronic fluorine intoxication in domestic animals. Nutr. Abstr. Rev. 9, 253-261 (1939).
PURVES, M.J.: Some effects of administering sodium fluoride to patients with Paget's Disease. The Lancet, pp. 1188-1189 (1962).
RICH, C., ENSINCK, J.: The effect of sodium fluoride on calcium metabolism of human beings. Nature (Lond.) 191, 184-185 (1961).
RICH, C., ENSINCK, J., IVANOVICH, P.: The effects of sodium on calcium metabolism of subjects with metabolic bone diseases. J. Clin. Invest. 43, 545-556 (1964).
ROHOLM, K.: Fluorine intoxication. London: H.K. Lewis and Co. Ltd. 1937.
Rose, G.A.: A study of the treatment of osteoporosis with fluoride therapy and high calcium intake. Proc. R. Soc. Med. 58, 436-440 (1965).
SANKARAN, B., GADEKAR, N.G.: Skeletal fluorosis. In: Bone and Tooth (ed. H.J.J. BLACKWOOD) pp. 357-362. London: Pergamon Press 1964.
SINGER, L., ARMSTRONG, W.D.: Comparison of fluoride contents of human dental and skeletal tissues. J. dent. Res. 41, 154-157 (1962).
SINGH, A., DASS, R., HAYREH, S.S., JOLLY, S.S.: Skeletal changes in endemic fluorosis. J. Bone Jt. Surg. 44-B, 806-815 (1962).
WALDBOTT, G.L.: Fluoride in food. Am. J. clin. Nutr. 12, 455-462 (1963).
WALLACE-DURBIN, P.: The metabolism of fluorine in the rat using ^{18}F as a tracer. J. dent. Res. 33, 789-800 (1954).
WEATHERELL, J.A.: Fluoride and the skeletal and dental tissues. In: Handbuch der experimentellen Pharmakologie (ed. FRANK A. SMITH) vol. XX, 141-172. Berlin-Heidelberg-New York: Springer 1966.
WEATHERELL, J.A.: Uptake and distribution of fluoride in bone and teeth and the development of fluorosis. In: Mineral Metabolism in Paediatrics (ed. D. BARLTROP, W.L. BURLAND) 53-70. Oxford-Edinburgh: Blackwell scientific Publications 1969.
WEATHERELL, J.A., HALLSWORTH, A.S., ROBINSON, C.: The effect of tooth wear on the distribution of fluoride in the enamel surface of human teeth. Archs oral Biol. 18, 1175-1189 (1973).
WEDDLE, D.A., MUHLER, J.C.: The metabolism of different fluorides in the rat: 1. Comparison between sodium fluoride, sodium silico-fluoride and stannous fluoride. J. dent. Res. 36, 386-390 (1957).
WEIDMANN, S.M., WEATHERELL, J.A.: The uptake and distribution of fluorine in bones. J. Path. Bact. 78, 243-255 (1959).
WEIDMANN, S.M., WEATHERELL, J.A., JACKSON, D.: The effect of fluoride on bone. Proc. Nutr. Soc. Engl. Scot. 22, 105-110 (1963).
ZIPKIN, I.: The effects of the absorption of fluoride: IV. The deposition of fluoride in human skeletal tissues as related to fluoride in the drinking water. Arch. industr. Hlth. 21, 329 (1960).
ZIPKIN, I., McCLURE, F.J.: Deposition of fluorine in the bone and teeth of the growing rat. J. Nutr. 47, 611-620 (1952).
ZIPKIN, I., McCLURE, F.J., LEONE, N.C., LEE, W.A.: Fluoride deposition in human bones after prolonged ingestion of fluoride in drinking water. Publ. Hlth Rep. (Wash.) 73, 732-740 (1958).

Absorption, Distribution in Body Fluids, and Bioavailability of Fluoride*

D. Henschler, W. Büttner and J. Patz

I. Introduction

The mechanisms by which fluoride acts in caries prevention and in the therapy of osteoporosis can be understood only in the light of the pharmacokinetic behavior of fluoride. More detailed studies of the dynamics of fluoride absorption, distribution, retention in hard tissues, and excretion in urine, saliva and sweat are now possible, since modern and more sensitive techniques of F^- analysis have become available, e.g. F^--specific electrometry and gas chromatography. They provide a continuous follow-up of F^- levels in body fluids. The concentrations of free F^- ions in plasma and saliva of man, and the rate and duration of F^- elevation in these body fluids following fluoride uptake and absorption are of special importance with regard to the availability of fluoride for the calcified tissues. Therefore, we evaluated fluoride levels in human plasma and saliva during 24-h periods under controlled dietary conditions.

II. Material and Method

1. Volunteers, Standardized Diet, and Sampling

For each experimental series 6 volunteers (dental students aged 22 to 28 years) were available. They received a standardized low-fluoride diet on the day preceding the experiment and the day of the experiment. Consumption of beverages was limited to 1,300ml/day. For the determination of the F^- concentration patterns in plasma, 8ml blood was taken by venepuncture at 7.30 a.m. in the fasting state, and was followed by the respective fluoride dose. Blood samples were then taken at intervals of 20 min to 6 h over the total 24-h period.

Simultaneously with the blood sampling, paraffin-stimulated whole mixed saliva was obtained (3-5 ml over 5 min). Before sampling, the volunteers brushed their teeth thoroughly with a toothbrush and tap water.

2. Fluoride Analysis

Fluoride in plasma was determined by the gas chromatographic method of FRESEN et al. (1968). Comparative, quantitative determinations of

+ The study was supported by the Deutsche Forschungsgemeinschaft, Sonderforschungsbereich 92, University of Würzburg. The authors wish to thank Mrs. H. BLEIFUSS and Mrs. I. STAHR for their skilful help during the study, as well as the dental students of the Würzburg University for their cooperation as volunteers.

F^- in pooled samples of plasma by the gas-chromatographic method and with the F^--sensitive ORION electrode gave equal values. Free F^- ions are determined with both methods.

Fluoride in saliva and sweat was analyzed with the ion-sensitive fluoride electrode, model 94-09, ORION Research Inc., according to the method of GRØN et al. (1968).

III. Results and Discussion

1. Absorption of Fluoride

Fluoride absorption is reflected in the speed of absorption and level of the F^- appearing in plasma. The absorption of soluble salts of F^- in man from stomach and gut varies by a factor of 3 to 4, depending on the kind and amount of ingesta. Likewise, peak levels of free plasma fluoride, determined by gas chromatography, are reached within 20 min to 2 h. After ingestion of the standardized diet and following controlled meal times, reproducible 24-h patterns of the plasma fluoride level are obtained. An example is shown in Fig. 1 (top), which gives the pattern after fluoride intake of 30mg NaF, 30mg F^-, and 10mg F^-, respectively. Under normal conditions of varying food habits, however, quite different patterns of the F^- level in plasma are observed. Fig. 1 also shows patterns in 3 adults after intake of 30mg F^-, and in 5 adults after intake of 10mg F^-. It can be seen that peak levels are sometimes reached after 20 min, in other individuals not before 2 h. Also, the height of the peaks differs. If milk was ingested instead of a standardized diet, the peak value was lower and the absorption time prolonged (Fig. 1, middle).

2. Distribution and Bioavailability of Fluoride

Most data available in the literature on the plasma fluoride concentration in man have been derived from spot samples (e.g. SINGER and ARMSTRONG, 1960, 1969; FRY and TAVES, 1970; VENKATESWARLU et al., 1971; ERICSSON et al., 1973). The plasma fluoride pattern following an intake of 10mg F^- as NaF over a period of 4 h was studied by COX and BACKER DIRKS (1968). Profile determinations over a period of 24 h, however, are desirable when studying the bioavailability of F^-. Therefore, 24-h patterns of the plasma fluoride concentrations were evaluated in 6 healthy young adult males, whose meal composition and meal times were strictly controlled. The concentrations of free F^- ions in plasma are shown in Figs. 2 and 3.

a) 24-h Patterns of Free Plasma Fluoride

Normal plasma F^- levels of individuals in nonfluoridated areas vary between 0.007 and 0.02 ppm. Following low F^- intakes of 1 or 2 mg, the peaks are reached after 30 min, followed by a rapid elimination over 4 to 6 h (Fig. 2). Fig. 3 shows on a semilogarithmic scale the elimination of F^- from plasma after single oral doses from 1 mg up to 100 mg. The plasma half-life was found to be between 4 and 8 h, depending on the height of the peak values; exponential elimination (with 1/2 t = 4 h) was found only from plasma concentrations of 0.1 ppm F^- upwards.

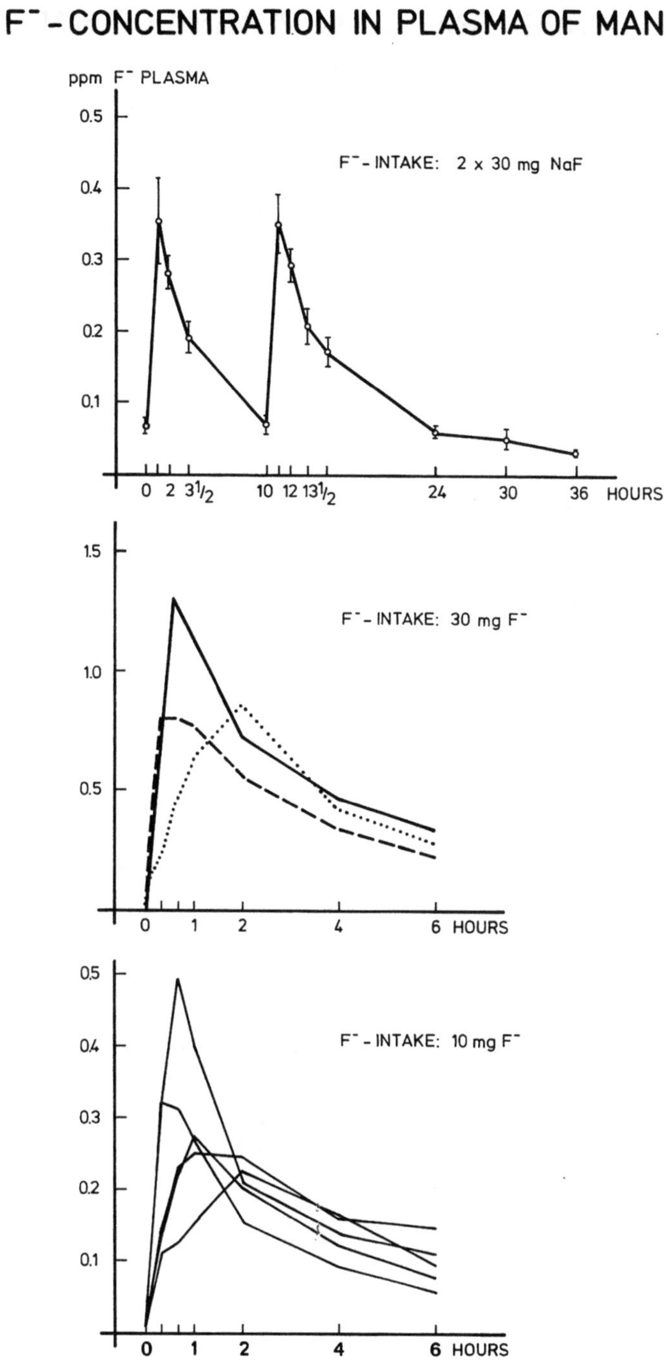

Fig. 1. Profiles of the plasma F^- concentration (free F^- ions) following intake of 2 x 30 mg NaF under standardized dietary conditions (above) and under uncontrolled conditions after intake of 30 and 10 mg F^- as NaF

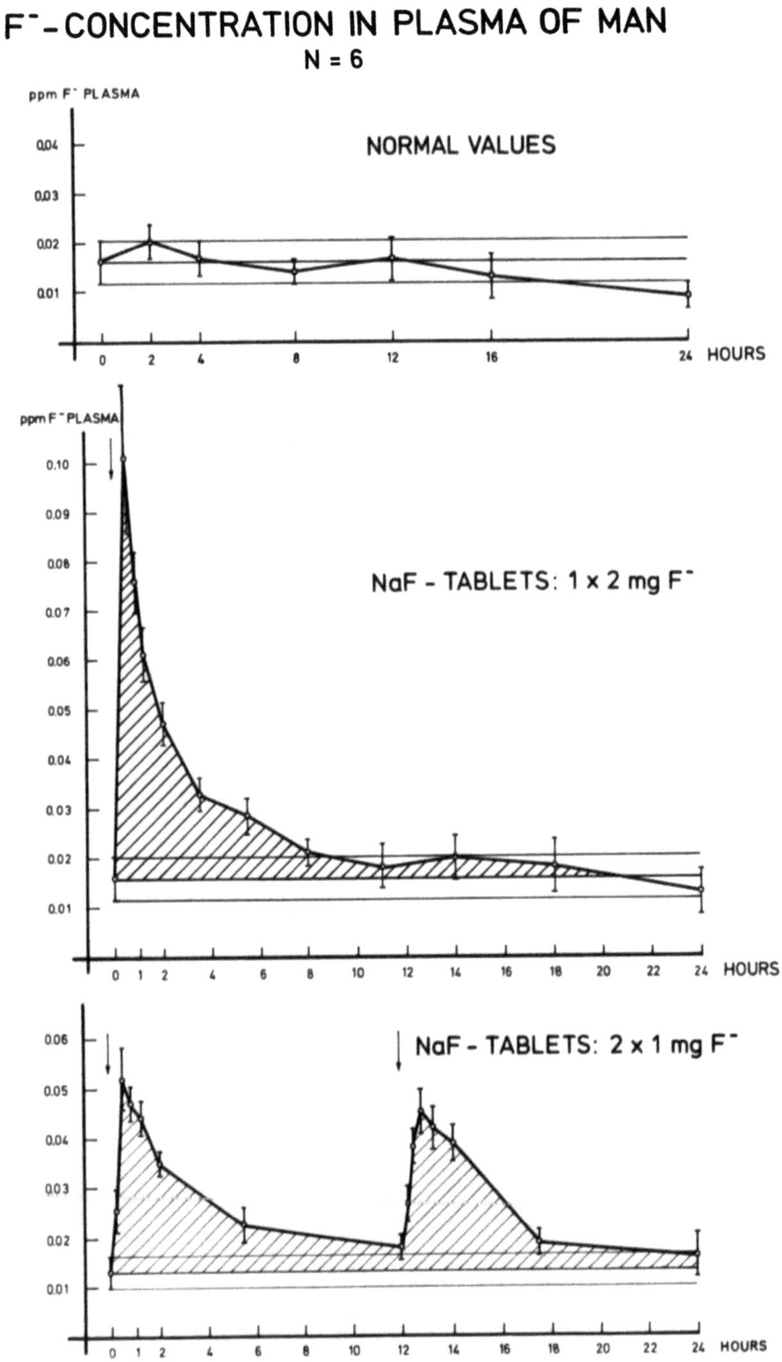

Fig. 2. Plasma F^- concentrations over 24 h under controlled dietary conditions. Mean values and standard deviations for 6 volunteers. Fluoride intake: none, 1 x 2 mg F^-, and 2 x 1 mg F^- as NaF in tablet form

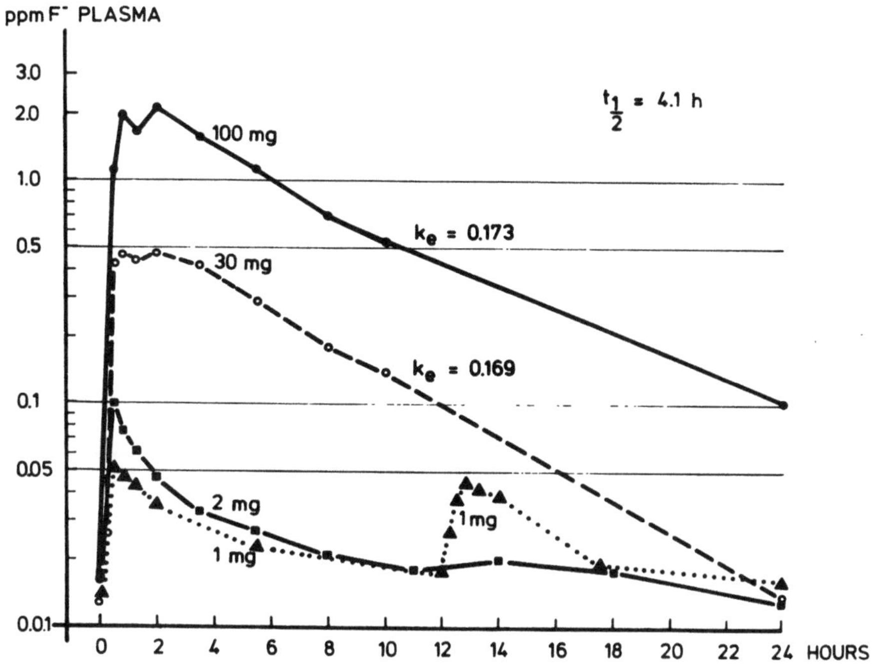

Fig. 3. Concentration peaks and elimination of F^- from human plasma during 24 h after a single oral dose of F^- as NaF

b) Correlation of Plasma and Salivary Fluoride

The 24-h profiles were also evaluated simultaneously for plasma and salivary fluoride concentrations in order to investigate the effect of fluoride intake on the speed of changes in the salivary fluoride levels. Again standardized conditions with regard to diet and meal times were used for the same volunteers. Fluoride in saliva was determined with the ORION fluoride electrode, thus ionized fluoride was recovered in saliva.

Fig. 4 shows the pattern of salivary fluoride following intake of 2 mg F^- in tablet form. The pattern in saliva is quite similar and the peak value appears nearly simultaneously with the peak in plasma 30 min after F^- intake. At 0.07 ppm, the salivary fluoride peak is slightly lower than that in plasma. A similar profile is found after intake of 1 mg F^-, the peak values in both plasma and saliva being about half those found with 2 mg F^-. The simultaneously obtained plasma and salivary F^- values of all series were nearly identical with an oral fluoride intake of 0.5, 1, 2, 10 or 30 mg, regardless of whether the analyses were carried out on increasing or decreasing F^- concentration. This observation leads us to assume that the free F^- ions in plasma can rapidly and easily diffuse accross the glandular membranes, since the free F^- ions in plasma and saliva seem to be in balance. It might thus be possible to rely on the much less complicated salivary F^- determination when predicting plasma F^- levels to be observed, e.g. during fluoride therapy in osteoporotic patients.

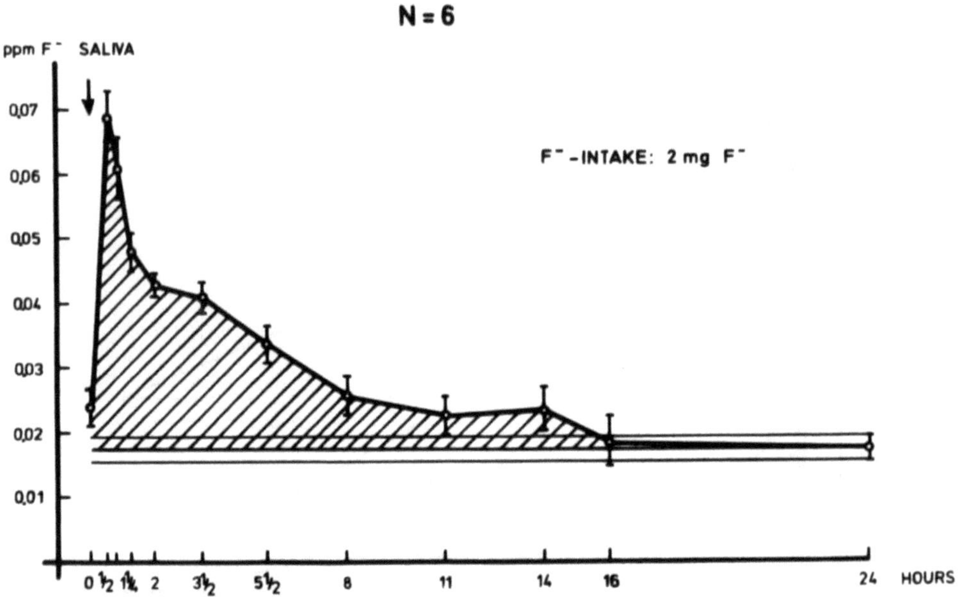

Fig. 4. F^- concentration pattern in saliva (whole stimulated mixed saliva) of man after intake of 2 mg F^- as NaF

3. Fluoride Levels in Plasma and Saliva Following Intake of Fluoride in Drinking Water

Frequent intake of fluoride in drinking water (e.g. 4 times 0.5 mg F^- in 250 ml water at intervals of 3.5 h) results in extremely different pharmacokinetic patterns as compared to intake of fluoride in tablet form. Minor elevations of F^- in plasma persist for 14 to 18 h (Fig. 5). This was also found in salivary-F^- concentrations for the same subjects. It seems, that as regards the cariostatic effect of fluoride, a continuously elevated salivary fluoride level is of advantage and is probably more important than concentration peaks of relatively short duration.

A sustained preparation of NaF in matrix pellets was therefore developed. As shown in Fig. 5, the pattern of plasma fluoride after intake in drinking water can be simulated by the use of sustained-release preparations.

4. Fluoride Pattern in Sweat

Equal concentrations of plasma and salivary fluoride were not observed when F^- levels in plasma and sweat were compared in the same volunteers. For these experiments doses of 10 and 30 mg F^- were used. As in the preceding series, peak plasma values were reached 30 to 40 min after intake, whereas no pronounced peaks were found in sweat. The F^- concentration in sweat increased steadily up to 0.12 and 0.18, respectively, after 2 h. Sweat was collected during the use of the sauna at 20-min intervals.

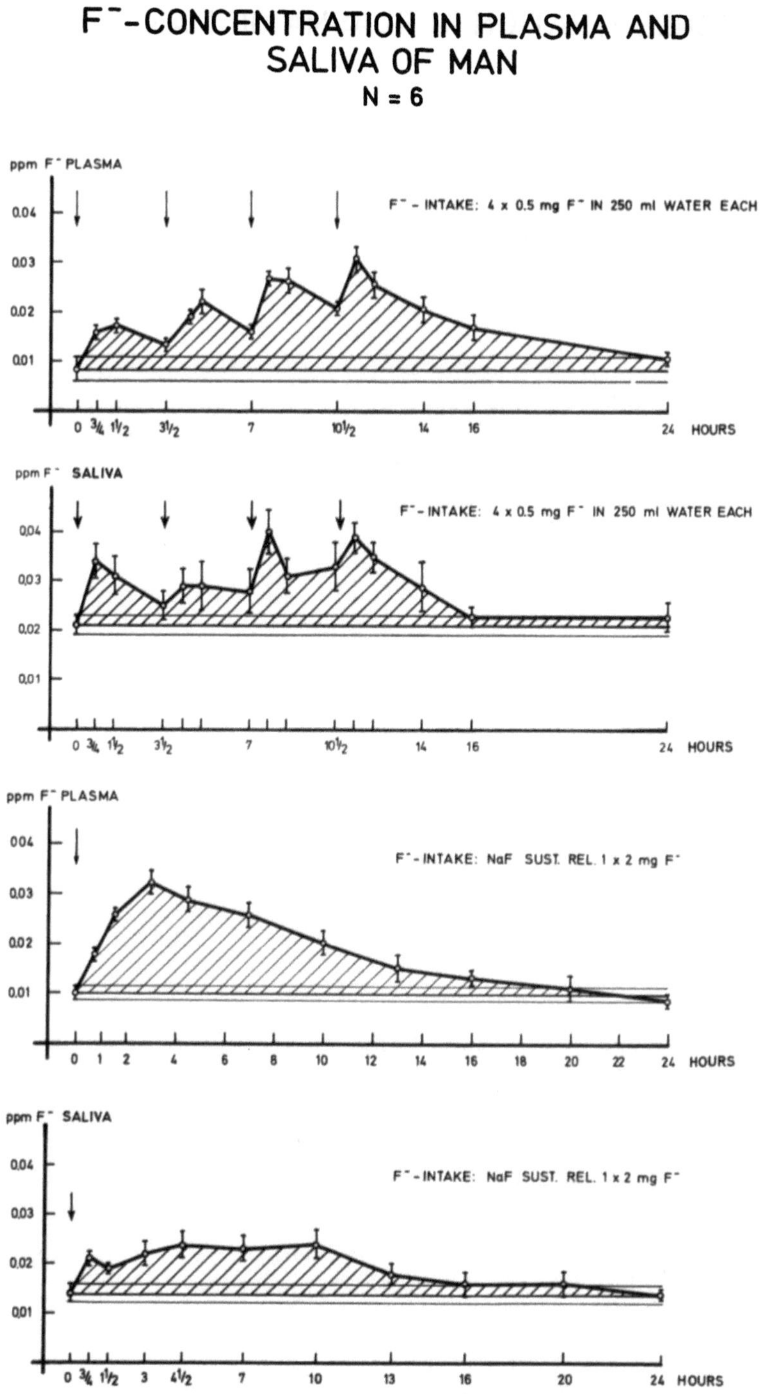

Fig. 5. Profiles of free F^- ions in plasma and saliva following F^- intake in drinking water or in sustained-release preparations of NaF

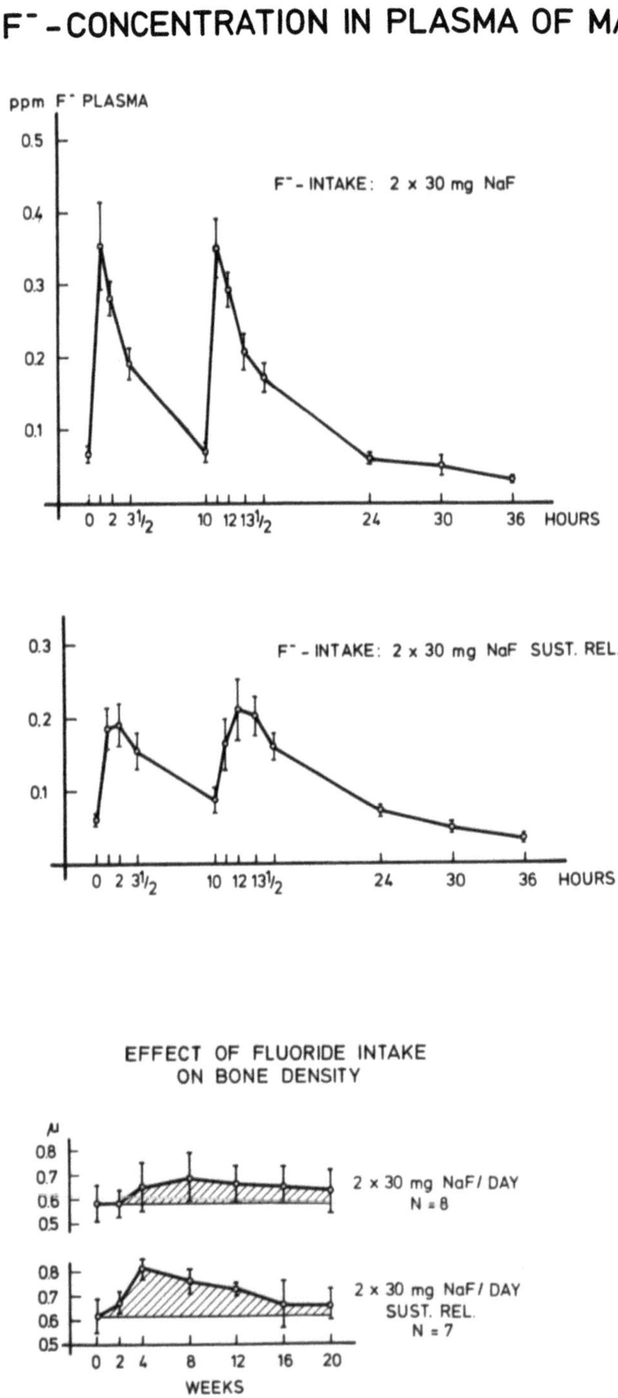

Fig. 6. Plasma F^- patterns in osteoporotic patients receiving twice daily 30 mg NaF as the salt or in sustained-release capsules. Measurements of bone density

Table 1. Side-effects in fluoride therapy of osteoporosis

Symptoms	Normal NaF (frequency)	Sustained-Release Prep. (frequency)
unpleasant feeling in the stomach after intake	0 / 11	1 / 11
loss of appetite, heartburn	4 / 11	0 / 11
prolonged nausea, necessitating cessation of medication	2 / 12	0 / 12

A similar pattern was observed when 10 mg F^- was ingested with 1000 ml of water in order to stimulate perspiration. Here too, sweat fluoride was still rising slightly after 2 h, although the F^- concentration in sweat was surprisingly low, being only one fifth of that in plasma.

5. Therapeutic Use of Fluoride in the Treatment of Osteoporosis

When high doses of fluoride are used in osteoporotic patients, extreme peaks of plasma fluoride sometimes produce side-effects. Therefore, a preliminary study with sustained-release capsules was conducted in aged people in osteoporotic states. Two series of 12 subjects were observed for 6 months, each group receiving 30 mg NaF twice daily, one group in the form of the normal salt in capsules, the other group in sustained-release capsules. The profiles of the plasma F^- levels recorded in the course of this type of medication are shown in Fig. 6.

Bone density of the compacta of the middle phalanx of the third finger was measured at 4-week intervals by the method of BÖRNER et al., (1970). As can be seen from Fig. 6, a larger increase is achieved by the sustained-release preparation as compared to the normal NaF-capsules. The slight reduction in the effect after 2 months of medication, seen in both groups, cannot be explained at present. One possibility is the influence of ambient heat during the months of June and July. More important is the advantage of the nearly total elimination of side-effects in the case of the sustained-release preparation. Table I is a synopsis of side-effects in the two groups.

IV. Summary

The 24-h patterns of the concentrations of free fluoride ions in plasma and saliva were evaluated in young healthy adults. After intake of 0.5, 1, 2, 10, 30, and 100 mg F^- as the NaF, the concentration peaks appeared approximately 30 min later, followed by elimination of F^- from plasma and saliva over 4 to 6 h. F^- concentrations in plasma and saliva were found to be nearly equal, suggesting that free F^- ions in plasma and saliva are easily diffusible and in balance. The administration of NaF in sustained-release preparations results in less pronounced F^- concentration peaks and prolonged F^- profiles in plasma and saliva. Especially for the therapeutic use of high doses of fluoride in osteoporotic patients, sustained-release preparations of NaF seem to give fewer side-effects and to improve bone density.

References

BÖRNER, W., MOLL, E. , RAUH, E., HEIEIS, G.: Ein empfindliches Messverfahren zur radiologischen Bestimmung der Mineralsalzdichte in Spongiosa und Kompakta des Fingerknochens. Z. Orthop. Grenzgeb. 108, 503-512 (1970).

COX, F.H., BACKER DIRKS, O.: The determination of fluoride in blood serum. Caries Res. 2, 69-78 (1968).

ERICSSON, Y., GYDELL, K., HAMMARSKJÖLD, T.: Blood plasma fluoride: an indicator of skeletal fluoride content. Intern. Res. Comm. 1 (1973).

FRESEN, J.A., COX, F.H., WITTER, M.J.: The determination of fluoride in biological materials by means of gas chromatography. Pharm. Weekblad 103, 909-914 (1968).

FRY, B.W., TAVES, D.R.: Serum fluoride analysis with the fluoride electrode. J. Lab. Clin. Med. 75, 1020-1025 (1970).

GRØN, P., McCANN, H.G., BRUDEVOLD, F.: The direct determination of fluoride in human saliva by a fluoride electrode. Fluoride levels in parotid saliva after ingestion of single doses of sodium fluoride. Archs. oral Biol. 13, 203-211 (1968).

SINGER, L., ARMSTRONG, W.D.: Regulation of human plasma concentration. Journal of Applied Physiology, 15, 508-510 (1960).

SINGER, L., ARMSTRONG, W.D.: Total fluoride content of human serum. Archs. oral Biol. 14, 1343-1347 (1969).

VENKATESWARLU, P., SINGER, L., ARMSTRONG, W.D.: Determination of ionic (plus ionizable) fluoride in biological fluids. Procedure based on absorption of fluoride ion on calcium Phosphate. Analyt. Biochem. 42, 350-359 (1971).

Bone-Body Fluid Fluoride Balance*

Y. ERICSSON

Fluoride (F) in urine has long been known to reflect intestinal F absorption as well as F liberation from a fluoride-rich skeleton, but blood plasma F has been believed only to show a limited absorption tide. Nobody seems to have observed plasma F values elevated due to the process of gradual F depletion of an F-rich skeleton. Some of the variations of plasma inorganic F may have been obscured by analytical methods, which have comprised the fraction of organically bound plasma F also. While the ionizable F (F^-) of a young person's fasting plasma generally is as low as 0.01 - 0.02 ppm, or 0.5 - 1 µM/l, the organically bound F may be 5 to 10 times higher. The nature of organic F bonding in plasma is not yet understood, but it is evident that the ionized or easily ionizable plasma F is the physiologically active fraction.

We observed that an osteoporosis patient, who had been treated with large doses of sodium fluoride for one and a half years, had a plasma inorganic F content as high as 0.2 ppm, or 10 µM/l, on a fasting stomach for many weeks after discontinuing the fluoride supply. This caused us to compare bone F and fasting plasma F^- contents in a number of persons with previous high F supply, most of them living in a community with a drinking water which had contained about 10 ppm F for 22 years. Bone F content had then already been determined for other purposes on de-fatted and dried biopsy samples from the iliac crest.

F in morning plasma was determined with the fluoride electrode after addition of 0.1 vol. of 2-M citrate buffer pH 5.6.

The results (Fig. 1) show the correlation of bone F to plasma F^-. In 21 cases all the fasting blood samples were taken on the same morning and under particularly rigorous control (open circles). These circles are well concentrated along a straight regression line (r = 0.93). The blood samples represented by the solid points, as well as all the bone biopsies, had been taken on different occasions during the previous months. The generally lower plasma F^- values of the samples represented by the solid points were probably due to less well-controlled fasting. We have compared fasting and non-fasting plasma F^- of some of our subjects, with the results appearing in Fig. 2. The explanation of the generally lower non-fasting values may be that an ordinary meal in this small high-F community largely consists of drinks and foods produced outside the community and hence with a low F content. This may "dilute" the intrinsically high plasma F^-.

The subjects with the highest bone F contents were older persons, of whom a few had roentgenographically observable osteosclerosis, some of them even with symptoms of reduced mobility. The subjects with low bone F and plasma F^- values were children down to 5 years of age.

+ These investigations were supported in part by Patentmedelsfonden för odontologisk profylaxforskning (Swedish Patent Revenue Research Fund).

The findings have been confirmed in several series of rat experiments. For example, while rats fed a diet containing about 0.5 ppm F and distilled water over the age period 35-75 days had a fasting plasma F^- level of 0.03 ± 0.003 ppm, a similar rat group fed the same diet and, for the last 10 days, water containing 40 ppm F had a plasma F^- level of 0.07 ± 0.014 ppm after a day on distilled water. When fasting rats of groups that had been on distilled water and 40 ppm F respectively for 40 days were given a peroral dose of 1 mg F 45 min before blood sampling they all had about the same high plasma F^- level, over 2 ppm. This huge plasma F^- elevation by a single F dose that has often been given in previous experiments is notable.

The determination of F^- in fasting plasma can be used diagnostically in cases of suspect osteofluorosis - most certainly also in domestic animals after establishing the regression lines for different species - and this parameter will also be useful for the control of F dosage in the treatment and prevention of osteoporosis.

Since plasma F probably mirrors the F content of cancellous bone rather than that of compact bone, the correlation between plasma F and cancellous bone ash F will probably be still better than that shown by the open rings of the diagram in Fig. 1, which were based on the total bone biopsies. The plasma F^- test may thus be refined.

Salivary F has been found to follow plasma F^- alimentary tides but to be lower than the latter (ADLER et al., 1970; GRØN et al., 1968). In our cases with high fasting plasma F^-, simultaneous samples of parotid saliva F showed a weak positive correlation with the plasma values (r = 0.37) but were only about 60% of the latter.

A number of analyses (ADLER et al., 1970; BACKER DIRKS et al., 1974) indicate that human milk F is as low as fasting plasma F^- and is little influenced by plasma F tides, which agrees with results obtained in

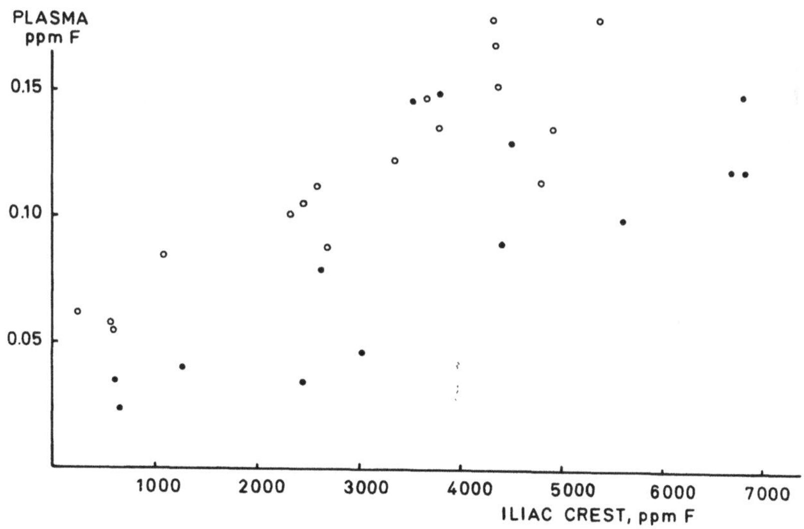

Fig.1. Bone F - plasma F^- relationship
o plasma samples taken under well-controlled fasting conditions
● plasma samples taken under less well-controlled fasting conditions

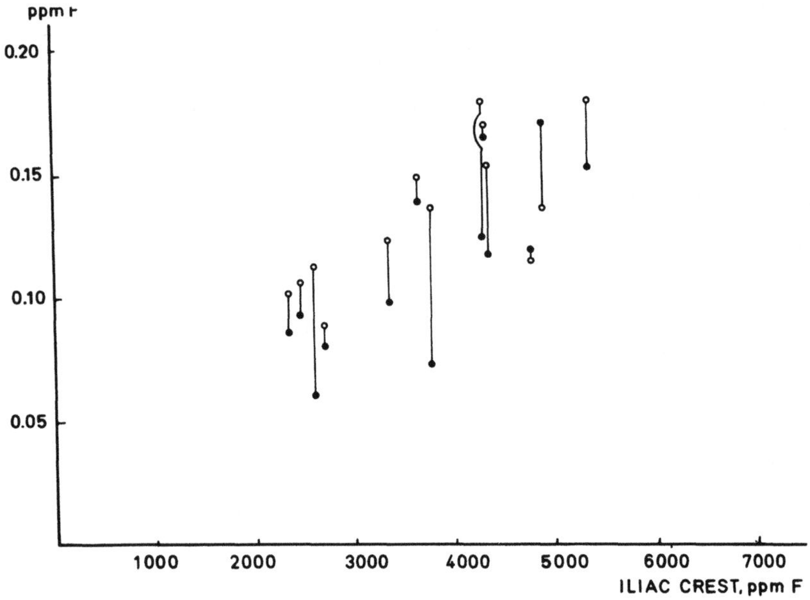

Fig. 2. Relationship fasting - non-fasting plasma F^-

○ fasting, ● non-fasting } values in the same individual

cows. Infant feeding with water-diluted dry-milk formulas will supply many times greater F doses than breastfeeding; the implications of this fact have been investigated and discussed in some recent papers (ERICSSON and RIBELIUS, 1971; ERICSSON et al., 1972; ERICSSON, 1973).

In conclusion some requisite knowledge on fasting plasma F^- levels may be summarized, together with some tentative data:

Purpose	Maximal F^-	Optimal F^-
Formation of normal enamel in children, resistant to caries by F incorporation	< 10 µM (0.2 ppm) in cattle. (SUTTIE et al., 1972)	....
Formation of normal bone in adolescents and adults, resistant to osteoporosis by F incorporation	< 15 µM (0.3 ppm) in rats. (TAVES, 1970)	Increase with age up to ∾ 7 µM (0.13 ppm)? (ERICSSON et al., 1973)
Formation of increased bone mass in osteoporotics, with maximal F incorporation but without impairment of bone quality	....	5-10 µM (0.1-0.2 ppm) in humans? (TAVES, 1970)

With a clearer concept of the problems and with the analytical tools now available it should not be long before we have the data necessary for a scientifically based control of the plasma F levels.

References

ADLER, P. et al.: Fluorides and human health. W.H.O. Monograph Ser. No. 59 (1970)

BACKER DIRKS, O., JONGELING-EIJNDHOVEN, J.M.P.A., FLISSEBAALJE, T.D., GEDALIA, I.: Total and free ionic fluoride in human and cow's milk as determined by gas-liquid chromatography and the fluoride electrode. Caries Res. 8, 181 (1974).

ERICSSON, Y.: Effect of infant diets with widely different fluoride contents on the fluoride concentrations of deciduous teeth. Caries Res. 7, 56 (1973).

ERICSSON, Y., GYDELL, K., HAMMARSKIÖLD, T.: Blood plasma fluoride: an indicator of skeletal fluoride content. J. internat. Res. Communic. 1, 33 (1973).

ERICSSON, Y., HELLSTRÖM, I., HOFVANDER, Y.: Pilot studies on the fluoride metabolism in infants on different feedings. Acta Paediat. Scand. 61, 459 (1972).

ERICSSON, Y., RIBELIUS, U.: Wide variations of fluoride supply to infants and their effect. Caries Res. 5, 78 (1971).

GRØN, P., McCANN, H.G., BRUDEVOLD, F.: The direct determination of fluoride in human saliva by a fluoride electrode. Archs. oral Biol. 13, 203 (1968).

SUTTIE, J.W., CARLSON, J.R., FALTIN, E.C.: Effects of alternating periods of high- and low-fluoride ingestion on dairy cattle. J. Diary Sci. 55, 790 (1972).

TAVES, D.R.: New approach to the treatment of bone disease with fluoride. Fed. Proceed. 29, 1185 (1970).

Possible Systemic Effects Following the Ingestion of Low Doses of Fluoride

R. L. SPEIRS

I. Introduction

FERGUSON (1971) observed that the ingestion by human subjects of drinking water containing about 1 ppm fluoride, or the daily ingestion of 5 mg sodium fluoride, was accompained by a small, transient reduction in the activity of serum alkaline phosphatase. As this was not due to a direct inhibition of the enzyme by the plasma fluoride levels, it was suggested that fluoride might restrict the release of phosphatase from or synthesis within cells. The enzyme originates from liver, intestinal mucosal cells and osteoblasts. Since fluoride ingestion at higher dosage can affect skeletal structure, it seemed possible that the reduction in serum alkaline phosphatase following fluoride ingestion in human subjects and in animals (RIEKSTNIECE et al., 1965; FERGUSON, 1972) might be indicative of altered bone metabolism.

The present study was an attempt to test this hypothesis by determining the effect of fluoride intake on the output of certain urinary constituents which reflect bone metabolism. Also, the susceptibility of the various alkaline phosphatase isoenzymes to fluoride was studied.

II. Methods and Materials

In the first of these investigations, 21 healthy, young adult males on relatively uncontrolled diets participated. Three to eight control 24 h urinary collections were made over a 3-4 week period and preceded those made over a similar experimental period during which 1.5 or 3.0 mg NaF were ingested daily as tablets. The daily dosage of fluoride was similar to that ingested by adults in a water fluoridated area.

The urinary constituents which were studied, the methods of analysis and the rationale for their inclusion are listed in Table 1. A non-parametric test, the Wilcoxon-signed rank test, was used in determining the statistical significance of the overall differences between experimental and control means for all the subjects. The degree of dependence between the output of any two constituents was measured both as the Spearman rank-correlation coefficient and as the correlation coefficient. In the second study, 16 students, 20 rats and 6 rabbits provided weekly serum samples for alkaline phosphatase estimations during control and experimental periods. The fluoride concentration in drinking water of the animals was 25 ppmF. Total serum alkaline phosphatase levels and the contribution made by various isoenzymes to this total activity were determined by the methods of KING and WOOTON (1956) and HORNE et al. (1968).

Table 1.

Urinary constituent	Significance	Analytical method
Fluoride	Related to intake and degree of retention in skeleton	NEEFUS et al. (1970)
Total hydroxyproline (mostly peptide bound)	Reflects degradation and synthesis of bone collagen	KIVIRIKKO et al. (1967)
Free hydroxyproline	Usually considered to be more specific for breakdown of mature bone collagen	ibid
Non-dialyzable hydroxyproline	Degradation of new bone collagen (only included in 1 subject)	KRANE et al. (1970)
Citrate	Related to plasma levels and to bone mineral resorption	JONES (1967)
Creatinine	Crude check on completeness of 24 h urinary collections	BOEHRINGER - based on Jaffé reaction

III. Results

The overall mean urinary fluoride excreted during the control period was 0.64 mg F/day. The mean retention of the ingested dose in 20 subjects was 50%, but extended from 76% to only 30%. Subjects were subsequently ranked according to their fluoride retention results. Over the short duration of the experimental period, there was no obvious change in the amount of fluoride retained.

Although there was a tendency towards lower citrate levels during the experimental period in 10 of 15 subjects (Fig. 1), no statistical significance could be attached to this. The differences between experimental and control means for citrate were positively related to the calculated fluoride retention values ($r_s = 0.65$, $p = 0.05$). The results

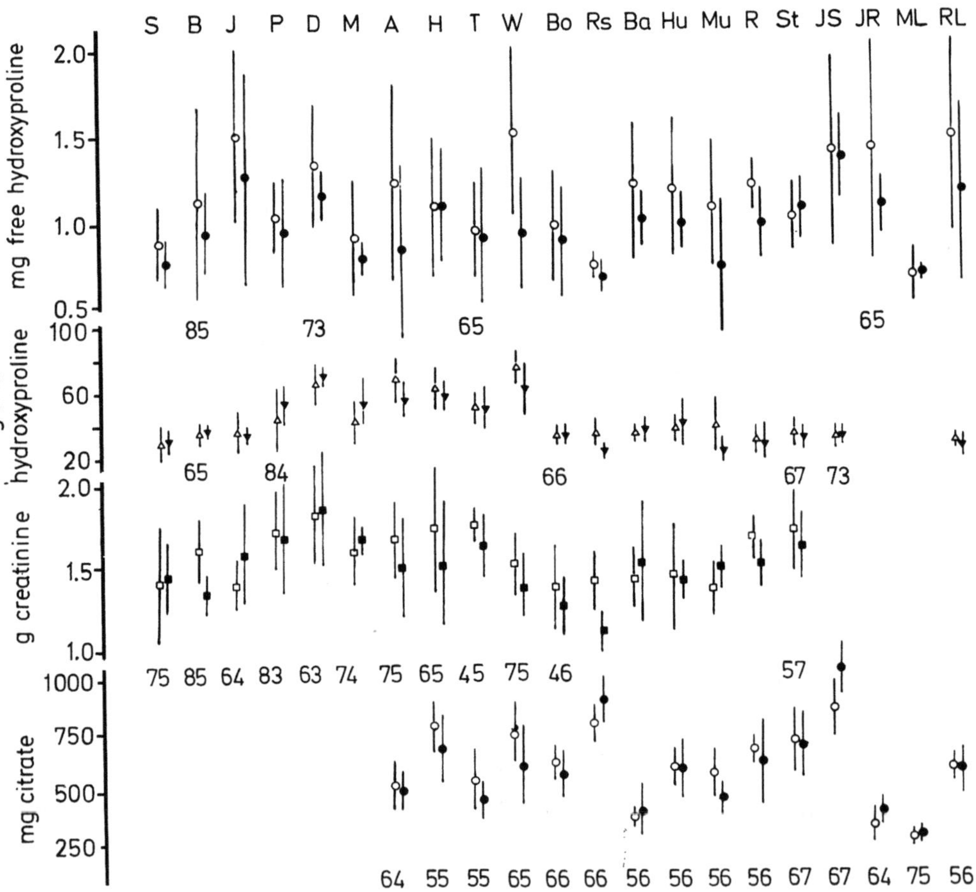

Fig. 1. Daily output of free OHP, citrate, total OHP and creatinine during control (o) period and period of fluoride ingestion (•). The points represent the mean values and vertical lines ± 1 standard deviation. The figure at the base of each column gives the number of 24-h collections upon which determinations were made. The daily intake of NaF was 1.5 mg for subjects S and B, and 3.0 mg for all others

for the excretion of total hydroxyproline (OHP) revealed no statistically significant influence of fluoride. A similar conclusion was reached when the results were expressed in terms of mg total OHP per g creatinine excreted.

However, there was a strong tendency for the output of free OHP to be depressed by fluoride. The signed-rank test showed a highly statistically significant overall reduction during the experimental period ($P < 0.01$), although in only 4 subjects did the difference between the means of control and experimental periods reach statistically significant values. When the results were expressed as mg free OHP per g creatinine, there was no statistically significant trend associated with fluoride ingestion. This was attributed to the concomitant reduction in creatinine excretion in many subjects during the experimental period. Even so, when the difference between the experimental and control mean values for both free OHP and creatinine in each subject were correlated, they were found to be independent of each other. Wilcoxon analysis showed that overall the apparent tendency towards lowered excretion of creatinine during the experimental period was not statistically significant.

In both rabbit and man, fluoride ingestion caused no significant change in total serum alkaline phosphatase levels or in the percentage contribution of the isoenzymes. In the rat too, there was no overall effect, though in 9 of the 20 animals a reduction in total activity was observed which was associated in 6 with a reduction in an isoenzyme probably of liver origin.

IV. Discussion

Fluoride ingestion caused a significant overall reduction in the output of only the free imino acid. This effect was observed within several days following the ingestion of fluoride tablets in some subjects. These results must, however, be examined in relation to the reduction in excretion of other constituents, particularly creatinine, and to the apparently high fluoride retention figures seen in many subjects during the experimental period.

The possibility that these effects are caused by incomplete collections of urine during the period of fluoride ingestion has been examined and considered to be very improbable for the following reasons:

1. The student volunteers understood the design of the experiment.
2. The study was undertaken on small groups of subjects over a period of some 15 months.
3. If it is assumed that most, but not all, of the samples provided during the experimental period were inaccurate as regards volume, the variance for creatinine results, and other results too, within the experimental period, should be greater than during the control period. This has not been observed. In fact, the variance is no greater than that reported for a small group of patients under supervised hospital conditions (EDWARDS et al., 1969).
4. The reductions in creatinine output during the experimental period do not correlate well with those for free OHP, total OHP, and citrate or with the percentage retention values for fluoride. Indeed, several subjects showed a decrease in free OHP excretion accompanied by an increase in creatinine output.

The finding that the fluoride retention figures correlated well with the reductions in citrate excretion implies an effect of fluoride on bone metabolism and the decreased output of free OHP would support this view. An alternative explanation for the latter is that the catabolism of OHP in the liver might be increased by fluoride ingestion. In this regard, it is of interest to note that FERGUSON (1973) has recently obtained results, which though falling short of statistical significance, suggest that fluoride ingestion in the rat and rabbit might inhibit the release into the serum of liver alkaline phosphatase and other enzymes and thus allow their accumulation in hepatic cells. However, in the present work, no statistically significant difference has been found in the rates of utilization of OHP by liver homogenates from 11 control and 11 of the fluoride-fed rats.

References

EDWARDS, O.M., BAYLISS, R.I.S., MILLEN, S.: Urinary creatinine excretion as an index of the completeness of 24-h urine collections. Lancet 2, 1165-1169 (1969).

FERGUSON, D.B.: Effects of low doses of fluoride on serum proteins and a serum enzyme in man. Nature, New Biology 231, June 159-160 (1971).

FERGUSON, D.B.: Effect of low doses of fluoride on blood enzymes in the rabbit. J. dent. Res. 51, 1260-1261 (1972).

FERGUSON, D.B.: Effect of low doses of fluoride on tissue enzyme levels in the rat. Proc. 21st Int. Ass. Dent. Res. (British Div.) Abstract 193. (1973).

HORNE, M., CORNISH, C.J., POSEN, S.: Use of urea denaturation in the identification of human alkaline phosphatases. J. Lab. Clin. Med. 72, 905-915 (1968).

JONES, G.C.: Estimation of microgram quantities of citrate in biological fluids. Anal. Biochem. 21, 286-292 (1967).

KING, E.J., WOOTON, I.D.P.: Micro-analysis in Medical Biochemistry, 3rd edition, 101-103. London: Churchill 1956.

KIVIRIKKO, K.I., LAITINEN, O., PROCKOP, D.J.: Modifications of a specific assay for hydroxyproline in urine. Anal. Biochem. 9, 249-255 (1967).

KRANE, S.M., MUNOZ, A.J., HARRIS, E.D.: Urinary polypeptides related to collagen synthesis. J. Clin. Invest. 49, 716-729 (1970).

NEEFUS, J.D., CHOLAK, J., SALTZMAN, B.E.: The determination of fluoride in urine using a fluoride-specific ion electrode. J. Amer. Indust. Hyg. Ass. 31, 96-99 (1970).

RIEKSTNIECE, E., MYERS, H.M., GLASS, L.E.: In vivo effects of sodium fluoride on serum proteins and enzymes as studied with starch gel electrophoresis. Archs oral Biol. 10, 107-117 (1965).

Chairmen: J. JOWSEY and J. MENCZEL

Effects of Bone Fluoride on Bone Resorption and Metabolism

W. D. ARMSTRONG, H. MESSER AND L. SINGER

Calvaria of 5-day-old mice with ash percentage of in vivo incorporated fluoride of 0.007 ± 0.004 (low-F), 0.041 ± 0.004 (moderate-F), and 0.107 ± 0.008 (high-F) were used in tissue-culture experiments to investigate bone resorption as affected by metabolically incorporated fluoride and some roles of parathyroid hormone (PTH) and of calcitonin (CT) on bone metabolism. Calcium increments of the media rather than ^{45}Ca release were used to assay bone resorption to avoid complications of calcium exchange.

In media not containing PTH, low-F bones were more labile than those of higher F content. Bone resorption was increased by PTH (0.1 or 0.5 units per ml) but in presence of PTH bones of low-F content exhibited greater resorption than moderate-F or high-F bones. In media containing 0.5U/ml PTH, absorption of bones of high-F content was less than that of low-F bones in absence of PTH. Moderate-F and high-F bones showed a reduced rate of absorption in the first 24 h (compared with the second 24 h) but this did not occur with low-F bones. 0.4 mU/ml/CT reduced calcium release by all bones and promoted calcium uptake from the media by bones of moderate-F and high-F groups.

Lactate and citrate production by all bones was increased by 0.5 U/ml PTH; this was not affected by the F content of the bones. Acid phosphatase in fresh bones of the two higher F groups was increased over that of low-F bones; alkaline phosphatase of fresh bones was not altered by F content. Incubation with PTH increased acid phosphatase of bones of low-F and moderate-F contents, and 0.4 mU/ml/CT produced a greater increase of bone acid phosphatase than was found with PTH. Alkaline phosphatase of all bones was increased by CT.

Hydroxyproline (Hyp) release to culture media was decreased by F in the bones. 0.5 U/ml PTH increased Hyp release, but this effect was reduced with bones of higher F contents and 0.4 mU/ml/CT reduced Hyp release. With ^{14}C-labelled proline it was found that F had no effect on the conversion of proline to Hyp but increased incorporation of newly formed collagen into bone matrix in presence or absence of PTH. PTH reduced collagen synthesis and incorporation of ^{14}C-Hyp into bone matrix; CT had no influence on either process.

References

MESSER, H.H., ARMSTRONG, W.D., SINGER, L.: Fluoride, Parathyroid Hormone and Calcitonin: Inter-Relationship in Bone Calcium Metabolism, Calc. Tiss. Res. 13, 217 (1973).

MESSER, H.H., ARMSTRONG, W.D., SINGER, L.: Influence of Calcitonin on Bone Phosphatases and Phosphate Release in Organ Culture, Proc. Soc. Exp. Bio. and Med. 143, 690 (1973).

MESSER, H.H., ARMSTRONG, W.D., SINGER, L.: Influence of in Vitro-Incorporated Fluoride on Collagen Metabolism by Mouse Calvaria in Organ Culture, Arch. Oral Biol. 18, 1393 (1973).

MESSER, H.H., ARMSTRONG, W.D., SINGER, L.: Characteristics of Calcium Uptake by Calcitonin-Treated Mouse Calvaria in Vitro, Calc. Tiss. Res. (In Press).

MESSER, H.H., ARMSTRONG, W.D., SINGER, L.: Calcitonin: Stimulation of Bone Phosphatases in Organ Culture, Calc. Tiss. Res. (In Press).

Urine Fluoride Levels Following Ingestion of Fluoridated Domestic Salt over Three Years

K. TÓTH AND E. SUGÁR

Many authors state that, provided the air and soil are free of F pollution, the most important source of fluorine for humans up to a level of 5 mg/l F is drinking water; the urine F level depends on the drinking water; the fluoride content of the water drunk and of the urine is almost identical (BREDEMANN, 1956; DOMAZALKA and LASSOCINSKA, 1969; W.H.O., 1970; GEDALIA, 1958; JIRASKOVA and RUZICKA, 1959; LIKINS et al., 1956; McCLURE and KINSER, 1944; SAVAGE, 1967; SMITH, 1966; TRUSWELL, 1966; ZIPKIN et al., 1956).

The amount excreted in the urine gives information not only about the fluoride consumed in the drinking water but about any other source (e.g. NaF tablets) consumed in some soluble form (WEATHERELL, 1966; ZIPKIN and LEONE, 1957). Thus, if fluoridated salt is consumed in an area with F-deficient drinking water, the F content of the urine reflects the amount of fluoride ingested with the salt (ERICSSON, 1971; MÜHLEMANN, 1965; WESPI and BÜRGI, 1969). We have been marketing fluoridated domestic salt in two settlements where no other salt is on sale in the shops. Thus, provided consumption of this salt has continued for long enough, it is possible to check ingestion through the determination of urine F levels.

I. Method

The fluoride analyses were made by means of the Coleman Ionselective Fluoride Electrode 3-803 at pH 5.2. McCLURE and Kinser (1944), ZIPKIN, LIKINS et al. (1956), GEDALIA (1958, 1970), WHO and SMITH (1966) showed that in determining the mean value of urine F level of population groups it is not necessary to take 24-h samples. Within the experimental error of F determination, a good result can be obtained by analyzing single urine fractions taken from the members of the group and calculating the mean value. Another possibility is to pool equal volumes (e.g. 5-20 ml) taken from the members of the group at one time and to analyze this mixture. Our preliminary investigations show that the above mentioned statements are also valid in the circumstances prevailing in Hungary.

Accepting this as a starting point, and adding only that we always collected the urine samples about noon (between 11 and 12 o'clock), we present our analyses as follows.

II. Number of Persons Examined and Origin of Samples

In both the experimental and control settlements we took urine samples from adults and children of school and kindergarten age. A total of 359 persons were examined.

1. Experimental group

We took urine samples from 89 persons in one village (Deszk) whose population had been consuming exclusively 25o mg/kg of F in fluoridated domestic salt for 4 years, and from 54 persons at another village (Röszke) whose population had been consuming only 200 mg/kg of F in fluoridated domestic salt for 2 years.

2. Control group

Urine samples were taken in one city (Szeged) from 118, in two villages (Szóreg, Tápé) from 41 and 57 persons. In these places fluoridated salt is not sold; the only fluoride source is the food the inhabitants eat; the drinking water is F-deficient.

III. Results (Fig. 1)

1. In all 5 places the drinking water is F-deficient and the fluoride level varies between 0.06 and 0.20 pp. Where ingestion was not increased by fluoridated domestic salt (Tápé, Szóreg, Szeged), the fluoride level in the drinking water and the urine samples was almost the same.
2. We did not find any difference between the urine F level in men and women.
3. At Röszke and Deszk, where fluoridated domestic salt had been consumed for years, the urine F level was higher in all age groups than in the control settlements.

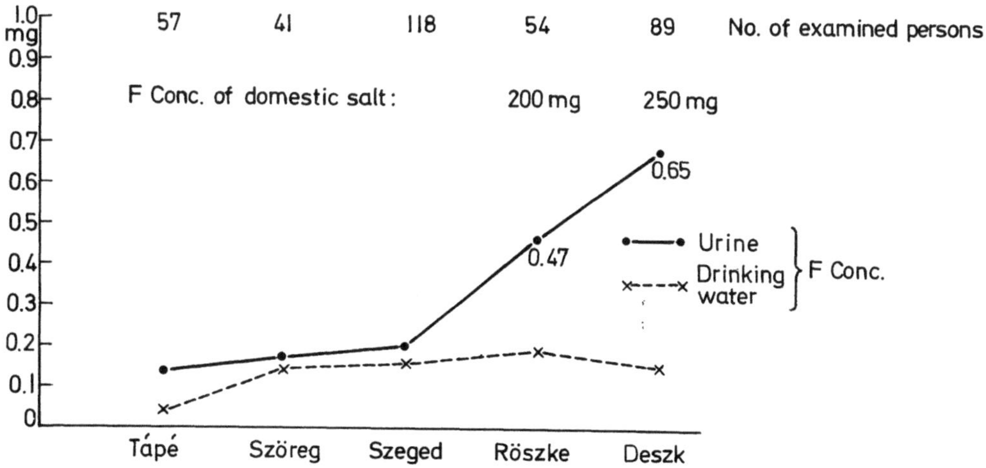

Fig. 1. F concentration in drinking water and urine

IV. Evaluation

1. In agreement with the data published abroad, at places in Hungary with F-deficient drinking water the human urine F level is almost identical with that of the water drunk.
2. At Röszke, fluoridated domestic salt consumption had increased the urine F level to 2.5 times above the original level. The 200mg/kg F consumption corresponds to drinking water with about 0.45 ppm F.
3. At Deszk, the 250 mg/kg F domestic salt regularly consumed increased the level of urine concentration more than 3 times, an addition corresponding to drinking water with a 0.65 ppm F.
4. According to our data concerning domestic salt ingested per kg body weight (TÓTH and SUGÁR, in Press), in Hungary 350-400 mg F has to be mixed with 1 kg of salt to give the optimal level. The urine determinations recently presented above support this opinion. We have recently begun research with domestic salt containing 350 mg/kg F.

V. Summary

The authors analyzed the F level of urine fractions collected from 359 persons living in an F-deficient area. Of these 359, 216 persons did not consume fluoridated domestic salt, while 54 persons had been consuming 200 mg/kg F fluoridated domestic salt for 2 years and 89 persons 250 mg/kg F fluoridated domestic salt for 4 years. The urine F level of persons who had not consumed fluoridated domestic salt was almost identical with that of the water they drank, whereas the urine F level of those who had consumed 200 mg fluoride was 0.47 ppm, and of those who had consumed 250 mg fluoride 0.65 ppm. On the basis of these data, the authors suggest a further increase in the F content of domestic salt.

References

BREDEMANN, G.: Biochemie und Physiologie des Fluors und die industriellen Fluor-Rauchschäden. II. Ausgabe. Berlin: Akademie-Verlag 1956.

DOMAZALKA, E., LASSOCINSKA, A.: Age differences affecting the fluoride content in human urine, saliva and blood. Csas. Stomat. 22, 317-323 (1969).

ERICSSON, Y.: Urinary estimation of optimal fluoride dosage with domestic salt. Acta odont. Scand. 29, 43-51 (1971).

GEDALIA, I.: Urinary fluorine levels of children and adults. J. dent. Res. 37, 601-604 (1958).

JIRÁSKOVÁ, M., RUZICKA, J.: Vztah mezi koncentraci fluoru v pitné vode a v moci. /The relation between the fluorine concentration in drinking water and urine. Ceskoslov. Stomatol. 59, 166-170 (1959).

LIKINS, R.C., McCLURE, F.J., STEERE, A.C.: Urinary excretion of fluoride following defluoridation of a water supply. Pub. Health Rep. 71, 217-220 (1956).

McCLURE, F.J., KINSER, C.A.: Fluoride domestic waters and systemic effects. II. Fluorine content of urine in relation to fluorine in drinking water. Pub. Health Rep. 59, 1575-1591 (1944).

MÜHLEMANN, H.R.: Dietary fluoride and dental caries. Int. dent. J. 15, 209-217 (1965).

SAVAGE, R.H.: The medical importance of fluorine. Guy's Hosp. Gaz. 81, 159-172 (1967).

SMITH, F.A.: Metabolism of inorganic fluoride. Handbuch der experimentellen Pharmakologie. Vol. XX. Berlin, Heidelberg, New York: Springer 1966.

TÓTH, K., SUGÁR, E.: Über die tägliche Zufuhr der Speisesalz-Menge auf Körpergewicht-Kilogramm errechnet. Dtsch. zahnärztl. Z. (in Press).

TRUSWELL, A.S.: Is fluoridation harmful? Roy. Soc. Health J. 86, 121-125 (1966).

WEATHERELL, J.A.: Fluoride and the skeletal and dental tissues. Handbuch der experimentellen Pharmakologie. Vol. XX.Berlin, Heidelberg, New York: Springer-Verlag 1966.

WESPI, H.J., BÜRGI, W.: Fluoridausscheidung im Urin unter Fluorprophylaxe. Ztschr. für Präventivmed. 14, 291-292 (1969).

WORLD HEALTH ORGANISATION: Fluorides and human health. Geneva: 1970.

ZIPKIN, I., LEONE, N.C.: Rate of urinary fluoride output in normal adults. Amer. J. Pub. Health 47, 848-851 (1957).

ZIPKIN, I., LIKINS, R.C., McCLURE, F.J., STEERE, A.C.: Urinary fluoride levels associated with use of fluoridated water. Pub. Health Rep. 71, 767-772 (1956).

Fluoridation of Drinking Water and Bone Mineral Content, Analyzed by Monochromatic (^{125}I) Radiation Absorptiometry*

J. VAN DER SLUYS VEER, A. P. M. MELMAN, T. POT, B. HOUWINK AND O. BACKER DIRKS

The effect on bone density of drinking water with fluoride content has been studied before, either directly or as part of studies analyzing potential hazards of fluoride.

BERNSTEIN et al. (1966) reported that roentgenographic evidence of osteoporosis was substantially higher in 715 persons living in low-fluoride areas in North Dakota, than in 300 persons living in an area where the natural fluoride content of drinking water was high (4 - 5.8 ppm). These results confirmed earlier findings of LEONE et al., (1955, 1960), who reported that osteoporosis was rare in the high-fluoride (8 ppm) area of Bartlett, Texas, while a much higher incidence of osteoporosis was found in Framingham, where the water has a very low fluoride content (0.04 ppm). The studies of BERNSTEIN et al. and LEONE et al. suggest a favorable effect on bone density of relatively high natural concentrations of fluoride (over 4 ppm) in domestic water. Much less is known, however, of the influence in this respect of drinking water which has been artificially fluoridated to only 1 ppm.

ANSELL and LAWRENCE (1965) for instance doubt whether the extra fluoride was responsible for the lower incidence of osteoporosis as judged by hand and cervical radiographs in women in the fluoridated town of Watford in England, compared with female inhabitants of Leigh where the water supply contains less than 0.2 ppm F.

In Elmira, N.Y., according to GOGGIN et al. (1965) no difference was noted in senile fractures of the femur in postmenopausal females before and after 5 years fluoridation.

KORNS et al. (1969) also found no significant difference in frequency of osteoporosis or its sequellae as crush fractures of vertebrae and of femoral- or other fractures, between persons over 40 years of age, who lived in Kingston (0.05 ppm F) or in Newburgh, where the water has been fluoridated to 1.0 ppm since 1945.

Fractures may be late complications of mineral loss, and differences of bone density as judged from radiographs are only noticed when large amounts of bone mineral have disappeared.

Methods which are based on absorptiometry of monochromatic gamma radiation provided by ^{125}I of ^{241}Am sources, as have been developed by CAMERON and SORENSEN (1963), allow detection of differences of bone mineral content of 3 per cent or less.

+We thank Miss L. BAUER and Drs.G.W. KWANT and A. GROENEVELD for active participation in this study, and Mrs.M.L. BOUWHUIS-HOOGERWERF for the statistical analysis.

Direct photon beam absorptiometry was therefore used in the present study in the town of Tiel (21.900 inhabitants) in the Netherlands, where the water supply has been fluoridated to 1 ppm since March 1953, and in the town of Culemborg (15.600 inhabitants) where drinking water contains only 0.01 ppm fluoride, to analyze the possible existence of small differences in bone mineral content.

I. Materials and Methods

Bone mineral content was analyzed according to the method described by SORENSEN and CAMERON (1967). ^{125}I was used as photon source (27.4 KeV). of 5 years), born and living in Tiel. The results were compared with data obtained from 245 men and 255 women, born and living in the non-fluoridated town of Culemborg. All subjects were working in their place of residence and they had never left it for a consecutive period of more than three months. Careful studies revealed that both towns were sociologically matched. Their inhabitants had been subjected to a continuous study of the influence of fluoride on caries frequency since the beginning of fluoridation in 1953.

Bone mineral content was analyzed according to the method described by SORENSEN and CAMERON (1967). ^{125}I was used as photon source (27.4 KeV). The shaft of the right radius at 1/3 from the distal end was transversely scanned at a constant speed of 0.3 mm per second. With a spectrometer and automatic scaler-timer, the number of detected photons was counted and every two seconds recorded by a paper tape punch. The data were consequently processed by a computer. A "bone standard" was measured twice daily for reference and control purposes.

The mass of mineral Mb was calculated according to the equation:

$$Mb = \Delta x \; \Sigma \ln \frac{I_o^*}{I} \; \frac{\rho B}{\mu B \rho B - \mu S \rho S} \qquad (g/cm)$$

where μB (2.05) and μS (0.373) are mass absorption coefficients in cm^2/g and ρB (1.8) and ρS (1.0) are densities (g/cm^3), Δx is the interval (0.6 mm) between steps across the bone. I_o^* is the initial beam intensity after passing only through soft tissue and I is the intensity after passing through soft tissue and bone.

II. Results

In Table 1 the results of the measurements of mineral content of males and females, living respectively in Culemborg and Tiel are presented with the number of subjects investigated. Since comparison is greatly facilitated when the bone mineral content (BMC) is expressed in g/cm per cm width of the radius, the values are thus presented in Table 1 and in Fig. 1. All BMC values of men are higher than those of women. No distinction between the two towns was found.

In the age groups of 25 - 50 years constant BMC values are found and the difference between men and women is about equal. In women of 50 - 74 the average BMC is 14 per cent lower than in the age group of 25 - 50. In men this difference in only 5 per cent. Also in this respect there was no difference between the two towns.

Table 1. Calculated bone mineral content (BMC) values of the men and women studied in Culemborg (control) and Tiel (+F)

		CULEMBORG (control)			TIEL (+ F)		
	age	n	BMC ($\frac{g/cm}{width}$)[a]	sd	n	BMC ($\frac{g/cm}{width}$)[a]	sd
women	25-29	24	0.80	0.07	25	0.85	0.05
	30-34	28	0.83	0.07	26	0.84	0.08
	35-39	25	0.84	0.07	24	0.83	0.06
	40-44	27	0.80	0.08	29	0.84	0.07
	45-49	28	0.81	0.07	25	0.83	0.07
	50-54	27	0.81	0.07	27	0.82	0.06
	55-59	27	0.73	0.09	25	0.75	0.08
	60-64	27	0.68	0.09	27	0.72	0.09
	65-69	20	0.69	0.09	31	0.68	0.08
	70-74	22	0.62	0.15	23	0.69	0.10
		255			262		
men	25-29	23	0.93	0.08	29	0.94	0.07
	30-34	22	0.94	0.07	25	0.95	0.07
	35-39	29	0.91	0.07	21	0.95	0.08
	40-44	24	0.91	0.07	31	0.93	0.06
	45-49	24	0.91	0.07	24	0.93	0.05
	50-54	25	0.91	0.08	27	0.90	0.08
	55-59	22	0.90	0.07	25	0.90	0.08
	60-64	24	0.86	0.09	21	0.90	0.06
	65-69	25	0.84	0.08	29	0.86	0.07
	70-74	27	0.85	0.11	24	0.86	0.10
		245			256		

[a] To compare these data with Cameron's normal data (1969) our values have to be multiplied by 0.9, due to the use of other mass absorption coefficients in the U.S. studies.

In 8 of 10 age groups, in women as well as in men, mean BMC-values are higher in the fluoridated town. The difference varies from 1 - 5 per cent. It is only significant in one group, but for all groups of men together ($P_D < 0.02$) and for all groups of women together ($P_D < 0.01$) differences are significant.

III. Discussion

As has been shown by absorptiometric (CAMERON, 1969; JOHNSTON and GOLDSMITH, 1970) and other methods (MEEMA and MEEMA, 1968) men have a higher bone mineral content than women. The present study confirms this observation. There is no difference between the towns in this respect. The rate of decrease of bone mineral content with age, which after the age of 50 is more impressive in females than in males, is also similar in both areas. Apparently fluoride has no influence on this phenomenon. It is known that the caries frequency was similar in both towns in 1953 (and since then has decreased by 60 per cent in the 15 year-olds). No data on bone mineral content of the period before the fluoridation are available. The possibility therefore cannot be excluded that the difference in BMC between the two populations has not existed since the beginning of fluoridation. There is no explanation for that, however, since such conditions as social status, dietary habits and work were and are well-matched in both towns. At present no significant difference in weight and height existed between the subjects of both towns. No explanation is therefore available at this moment to account for distinction between the two towns and it seems reasonable to suggest that the difference in BMC-values is related to the fluoridation of the water supply in Tiel. Further studies are necessary, however, to prove this supposition.

IV. Summary

Bone mineral content (BMC) was analyzed by direct photon beam absorptiometry of the shaft of the right radius in 256 men and 262 women of 25 - 75 years of age (about 25 of both sexes per age group of 5 years) living in a Dutch town where the drinking water has been fluoridated since 1953. The results were compared with data obtained from 245 men and 255 women living in a non-fluoridated town. All BMC values of men are higher than those of women. The difference between men and women in the fluoridated area is equal to that in the non-fluoridated area.

In women of 50 - 74 the mean BMC is 14 per cent lower than in the age groups of 25 - 50. In men this difference is only 5 per cent. In this respect also no distinction is found between the two towns. In 8 out of 10 age groups in men as well as in women, mean BMC values are higher in the fluoridated area. The difference is only significant in one age group, but for all groups of men together (P_D 0.02) and for all groups of women together (P_D 0.01) differences are significant. Although no data on bone mineral content from the period before fluoridation are available, it is suggested that the differences in BMC values between the two towns are related to the fluoridation of drinking water.

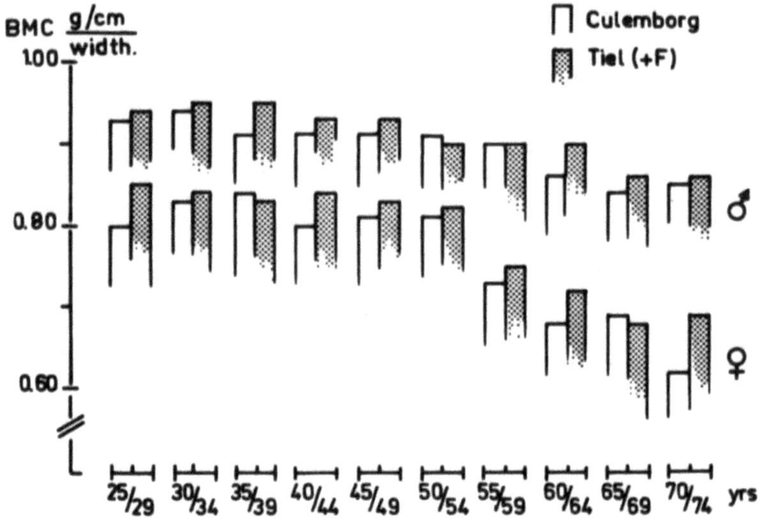

Fig. 1. Representation of BMC values per age group of 5 years of men and women from the non-fluoridated (Culemborg) and fluoridated area (Tiel)

References

ANSELL, B.M., LAWRENCE, J.S.: Fluoridation and the rheumatic diseases. A comparison of rheumatism in Watford and Leigh. Anw. Rheum. Dis. 25, 67 (1965).

BERNSTEIN, D.S.,SADOWSKY, N., HEGSTED, D.M., GURI, C.D., STARE, F.J.: Prevalence of osteoporosis in high- and low-fluoride areas of North Dakota. J.A.M.A. 198, 499 (1966).

CAMERON, J.R.: Progress Rep. COO-1422-41. Summary of data on the bone mineral of the radius in normals. Dept. of Radiology, Univ. of Wisconsin, Madison, Wisc. (1969).

CAMERON, J.R., SORENSEN, J.A.: Measurements of bone mineral in vivo. An improved method. Science 142, 230 (1963).

GOGGIN, J.E. et al.: Incidence of femoral fractures in postmenopausal women. Publ.Hlth.Rep.Wash. 80, 1005 (1965).

JOHNSTON, J.O., GOLDSMITH, N.F.: A clinical trial of the I-125 isotope scanning device as a scanning method for the early detection of osteoporosis. Bone Measurement Conf. Chicago, I-11. May 22-23, 1970. Proc. Conf. 700515, p. 340 (1970).

KORNS, R.F. et al.: Relationship of water fluoridation to bone density in two N.Y. towns. Publ.Hlth.Rep. 84, 9, 815 (1969).

LEONE, N.C., STEVENSON, C.A., HILBISH, T.F., SOSMAN, M.C.: A roentgenologic study of human population exposed to high-fluoride domestic water: a ten year study. Amer. J. Roentgen. 74, 874 (1955).

LEONE, N.C. et al.: The effects of the absorption of fluoride: II. A radiological investigation of 546 human residents of an area in which the drinking water contained only a minute trace of fluoride. Arch.Industr.Hlth. 21, 326 (1960).

MEEMA, H.E., MEEMA, S.: Mineral mass and mineral density in human radius: a roentgenologic-densitometric study. Conf. on progr. in methods of bone mineral measurements, Bethesda, Febr. 15-17, 1968, CONF - 680211, p. 135 (1968).

SORENSEN, J.S., CAMERON, J.R.: A reliable in vivo measurement of bone mineral content. J. Bone Jt. Surg. 49A, 481 (1967).

Effects of Supply and Withdrawal of Fluoride on the Concentration of Fluoride and Glycosaminoglycans in Cortical Bone from Rabbits

R. LEMPERG AND J. ROSENQUIST

Fluoride effects on bone have been extensively studied but little attention has been paid to possible alterations of the organic constituents. The reversibility of manifest skeletal fluorosis has also been scarcely investigated. As acid glycosaminoglycans (GAG:s) have been associated with bone mineralization, and skeletal fluorosis is characterized by excessive bone formation and resorption, this investigation was carried out with the aims of studying the relationship between morphological alterations and concentrations of fluoride and organic constituents (GAG:s and hydroxyproline) during fluoride ingestion and after its withdrawal. Furthermore age-related changes were studied and an attempt was made to isolate keratan sulphate.

Rabbits, aged 48 - 52 days, were fed nutritionally equivalent diets containing 12 ppm (group I:B) or 2 ppm of fluoride and given distilled water (ROSENQUIST, 1973).

Series I		Effect of fluoride ingestion	
Group	I:O	No fluoride supplement	(total of 34 animals)
	I:A	0.5 mg of F^-/kg/day	(total of 35 animals)
	I:B	10 mg of F^-/kg/day	(total of 32 animals)

Observations were made at 2, 4, 7 and 14 weeks.

Series II		Effect of withdrawal of fluoride after completed growth	
Group	II:O	No fluoride supplement	(total of 21 animals)
	II:A	0.5 mg of F^-/kg/day for 14 weeks	(total of 24 animals)
	II:B	10 mg of F^-/kg/day for 14 weeks	(total of 22 animals)

After 14 weeks the fluoride supplement was withdrawn and observations were made at 4, 12 and 24 weeks.

Cortical diaphyseal bone from the left femur and tibia was analyzed for its content of hydroxyproline, total hexosamines, CPC-precipitable GAG:s and fluoride. Microfractionation of the acid GAG:s was carried out using the CPC precipitation technique. Fractions from the microcolumns were further characterized by electrophoresis on acetate membranes in different buffer systems. Identical parts of the right femur and tibia were used for morphological studies and for morphometry on microradiographs. GAG:s from the CPC supernatant were precipitated and chemically analyzed in order to demonstrate keratan sulphate (LEMPERG and ROSENQUIST, 1974; ROSENQUIST, 1973).

The animals from groups I:O and II:O showed no morphological alterations and will be called normal. The changes of the fluoride concentration in bone ash during treatment are shown in the figure. During the same period, enormous, mainly periosteal, bone formation occured in the lateral part of the tibia. This newly formed bone had a higher degree of porosity than normal cortical bone (Table 1). The degree of mineralization, however, appeared to be normal (ROSENQUIST, 1973). The excessive bone formation was accompanied by an increase of bone resorption. In contrast, the morphological changes in the femur were generally slight.

Table 1. Relative "porosity" of cortical bone expressed as non-mineralized areas per surface unit

Obs.time		0 (21)		4 (25)		12 (33)		24 (45)
Group	n		n		n		n	
Control	5	3.6$\pm$1.1[b(*)]	5	3.4$\pm$1.2[b(*)]	5	3.5$\pm$1.0[a(*)]	5	3.5$\pm$1.1[a(*)]
Exptl.	6	30.0$\pm$3.5	6	11.6$\pm$2.2[b(+)]	6	7.0$\pm$1.3[a(+)]	6	6.4$\pm$1.3

The precision of the determinations s_e = 1.0 per cent. Observation times calculated in weeks after withdrawal of fluoride supplement. Age in brackets. M $\pm$ S.D. The statistical significance of differences between the groups is indicated by (*) and between different observation times in the experimental group by (+).

[a] $0.01 > p > 0.001$

[b] $0.001 > p$

In the normal animals the femur showed significantly lower hydroxyproline concentrations than the tibia; while total hexosamines and CPC-precipitable GAG:s were equal when calculated as percent of dry weight. The 1 percent CPC fraction (glycopeptides and keratan sulphate) was significantly smaller in the normal tibia than in the femur, whereas the sum of the $MgCl_2$ fractions (0.25 M to 0.80 M - chondroitin sulphate) was significantly higher. In the tibia there were no consistent changes in the mutual distribution of the different GAG:s either with age or fluoride ingestion. In contrast, the normal femur showed an age-dependent increase of the CPC fraction from 20.3$\pm$0.6 percent at 14 weeks of age to 25.9+3.5 percent at 45 weeks of age. There was a corresponding decrease of the sum of the $MgCl_2$ fractions (0.25 M to 0.80 M) from 67.2$\pm$0.9 percent to 63.0$\pm$3.2. In the group given 10 mg of F^-/kg/day these alterations were not seen. The solubility profiles of the CP-chondroitin sulphate complexes from the tibia showed the peak in the 0.45 M to 0.60 M $MgCl_2$ fractions at the age of 14 and 21 weeks of age almost all animals showed the peak in the 0.45 M $MgCl_2$ fraction, thus a marked shift towards smaller/lower sulphated chondroitin sulphate molecules at higher age was demonstrable. In the femur, most of the profiles showed the peak in the 0.60 M $MgCl_2$ fraction without age-related alterations. No effect of fluoride ingestion was observed.

Upon electrophoresis chondroitin-4-sulphate was found in all $MgCl_2$ fractions and traces of hyaluronic acid in the 0.3 M NaCl fraction. Neither age-related nor fluoride-induced alterations were demonstrable.

Withdrawal of the fluoride supplement after an ingestion period of 14 weeks did not cause any reduction of the fluoride concentration in the bone ash during the following 24 weeks (Fig. 1), and it showed no further decrease even after one year (ROSENQUIST and LEMPERG, unpublished). Nevertheless, normalization of the bone morphology was apparent. By morphometry a highly significant reduction of the degree of bone porosity was ascertained (Table 1). The only chemical alteration in the fluoride-supplement animals was the lack of age-related changes in the mutual distribution of the different GAG:s in the femur.

Material isolated from the CPC supernatant showed on chemical analyses and on electrophoreses characteristics of keratan sulphate. Also low sulphated chondroitin was present.

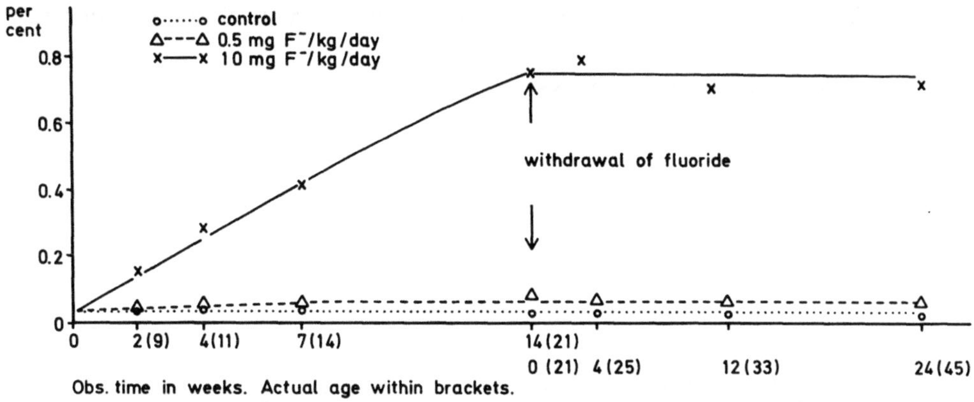

Fig. 1. Diagram showing the fluoride concentration in cortical diaphyseal bone of rabbits during fluoride ingestion and after withdrawal of the fluoride supplement. Expressed as per cent of ash

In this investigation on cortical bone it was thus not possible to demonstrate quantitative or qualitative changes in acid GAG:s during development of skeletal fluorosis. Only minor alterations limited to the femur were demonstrable in the mutual distribution of the GAG:s after withdrawal of fluoride and during regression of skeletal fluorosis. The observations in this study indicate that the ability of the osteoblastic cells to synthesize normal amounts of GAG:s was not affected.

References

LEMPERG, R., ROSENQUIST, J.: Concentration of acid glycosaminoglycans and hydroxyproline in cortical bone. Acta path. microbiol. scand. 82 A, 435-444, 1974

LEMPERG, R., ROSENQUIST, J.: Long Term Observations on the reversibility of skeletal fluorosis. Acta path. microbiol. scand. 82 A, 495-498, 1974.

ROSENQUIST, J.: Effects of supply and withdrawal of fluoride. Experimental studies on growing and adult rabbits. Umeå University Medical Dissertations. No. 12, (1973).

Histological Bone Changes after Long-Term Treatment with Sodium Fluoride

A. J. OLAH, F. W. REUTTER AND R. K. SCHENK

I. Introduction

In a previous study, patients with senile osteoporosis were treated with 50 - 75 mg of sodium fluoride daily, combined with vitamin D, over a period of 1 - 2 years. The quantitative histological results of repeated iliac crest biopsies from these patients are reported in an earlier paper (SCHENK et al., 1970). We observed a stimulation of osteoblastic bone formation, an increase in osteoid seam thickness due to accelerated matrix formation and delayed mineralization of the newly-formed osteoid, and a slight increase in osteoclastic bone resorption. These changes in bone turnover led to increased volume of cancellous bone as an expression of the positive skeletal balance, which was also confirmed by clinical, biochemical and radiological findings (REUTTER et al., 1970).

II. Material and Methods

In view of the biopsy findings and of the clinical improvement, the treatment of these patients was continued over several additional years. The present study is based on data from 27 osteoporotics, who have been under sodium fluoride for more than 2 years. From the 2nd year on, they received 50 mg of sodium fluoride daily and in addition vitamin D_3 at 800 IU/day. On iliac crest biopsies from these patients, 14 of which were taken after 3 to 5 or more years of treatment, histomorphometric analysis was done according to our method of point counting and line intersectioning (MERZ, 1967). To evaluate bone turnover and bone structure, the following parameters were used:

1. Surface density of osteoid-osteoblast interface (Sv_{Ob}), which represents the surface area of trabeculae lined by osteoblasts. It is expressed as mm^2 surface per cm^3 total volume of cancellous bone.
2. Mean thickness of osteoid seams ($\bar{s}$) is calculated from the volume and the surface extent of osteoid tissue and is given in mμ.
3. Osteoclast index (OI) results from direct counting of osteoclasts present in the test area. Their number is related to the unit surface area in mm^2 per cm^3 total bone.
4. Volume density cancellous bone (V_V) represents the percentage of bone tissue compared to the total volume of cancellous bone.
5. Mean diameter of trabeculae ($\bar{D}_{trab}$, mμ) is not measured directly, but calculated from the total volume and total surface of all trabeculae present in the test field (paper in preparation).

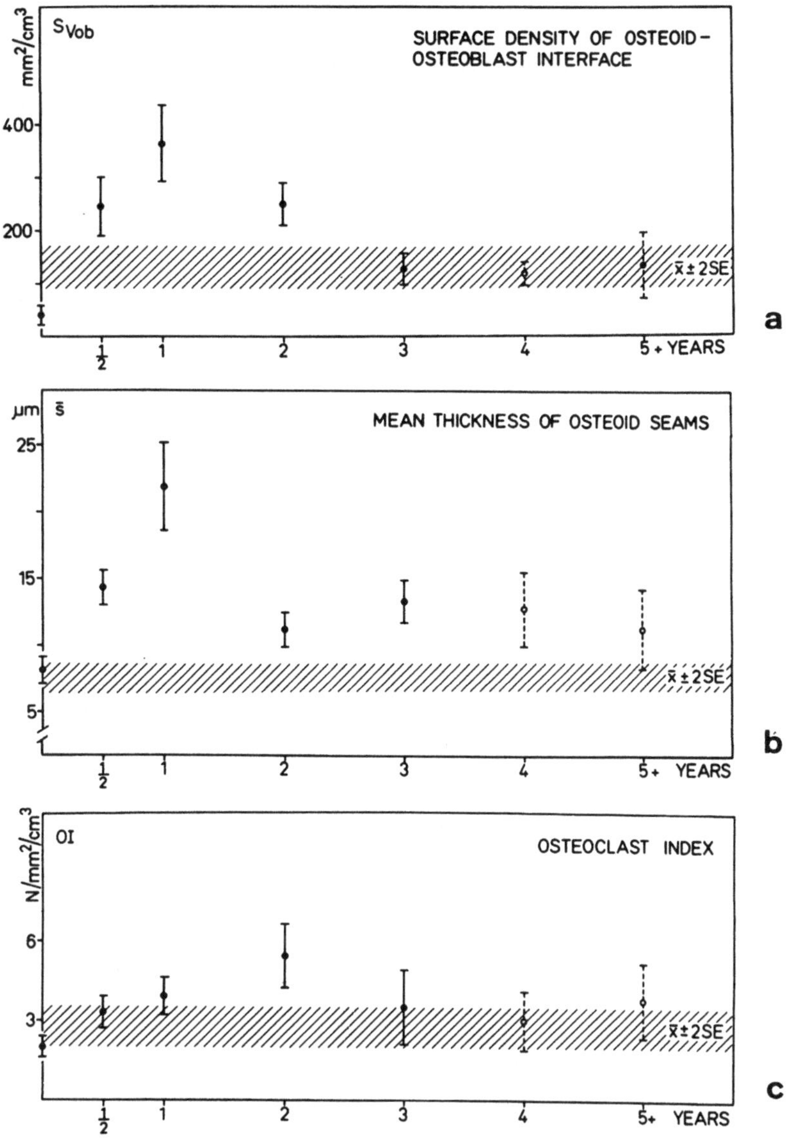

Fig. 1 a-c

III. Results

The histomorphometrical results in bone turnover are given in Fig. 1. The hatched band represents the normal range for the respective age group. The mean values of the different groups are plotted with ± SE. The groups after 4 and 5 years of treatment consist only of 4 cases each, and are therefore indicated by open circles and dashed lines. Prior to treatment, the surface density of osteoid-osteoblast interface (i.e. the surface area of the trabeculae lined by osteoblasts) was clearly below the norm, a typical feature of senile osteoporosis

(Fig. 1). Under treatment, the number of osteoblasts increases enormously during the first year. In spite of continuous treatment, the osteoblasts begin to decrease in the second year, and reach the normal range in the third year. The cases examined in the fourth year of treatment or thereafter exhibit, on the average, normal osteoblast numbers, but they never return to the extremely low values present before therapy.

In contrast to these changes, the number of osteoclasts (Fig. 1) remains close to the upper limit of the norm except in the second year of treatment, when their number is significantly elevated.

The increase in bone formation and the normal or only slightly elevated resorption lead to a positive skeletal balance. It is, however, primarily confirmed by changes in bone structure as measured in the cancellous bone of the iliac crest. Volume density (Fig.2) exhibits a significant increase as early as half a year after the initiation of therapy which continues until the beginning of the third year. Thereafter, the volume density sinks a little, but remains throughout the entire period within the normal range. In view of this increase, the question arises as to whether this is the product of an increase in the thickness of already-existent trabeculae or whether the formation of new trabeculae is responsible. This question is answered by computing the mean diameter of trabeculae (Fig. 2). We found that beginning from subnormal values before therapy, this parameter increases conspicuously in the first year of treatment. In addition, the thickening of the trabeculae continues during the second and third years and remains at this high level. On the basis of these results, the conclusion is justified that administration of sodium fluoride in osteoporosis leads to a significant net gain in bone substance, which is especially marked in the first year and which can be maintained for many years thereafter. This increase in bone density is, moreover, not only histologically demonstrable, but can be observed on x-rays, and is accompanied by a considerable clinical improvement. It is an important fact that none of the patients who was treated for longer than two years had any pathological fractures.

On the other hand, hesitation has been expressed respecting the success of fluoride therapy. Among other things, it has been suggested that, although under sodium fluoride therapy new bone is produced in excess, this fluoride-loaded bone calcifies incompletely and is inferior in its substructure. Indeed, in the initial months of therapy, one does observe a great increase in the volume of osteoid. In addition, both incompletely mineralized cement lines as well as enlarged osteocyte lacunae are demonstrable (JOWSEY et al., 1968). For these reasons, we have focused special attention on a more precise analysis of fluoride-induced mineralisation disturbances in our material. From the standpoint of histomorphometry, the most revealing parameter is the thickness of the osteoid seams (Fig. 1). This parameter already increases prodigiously at the beginning of treatment, with a maximum after one year. During the second year, it decreases but without ever completely sinking into the normal range. In continued treatment, the osteoid seam thickness remains for years at this elevated, relatively constant level. Thus we must conclude that under sodium fluoride, the osteoid does mineralize slowly but in the course of months and years, the lamellae formed under treatment attain the same mineral density as the preexisting bone. On the other hand, the enlarged osteocyte lacunae and the incompletely mineralized cement lines persist up to 8 years after the onset of treatment, regardless of the period in which this bone moiety has been formed.

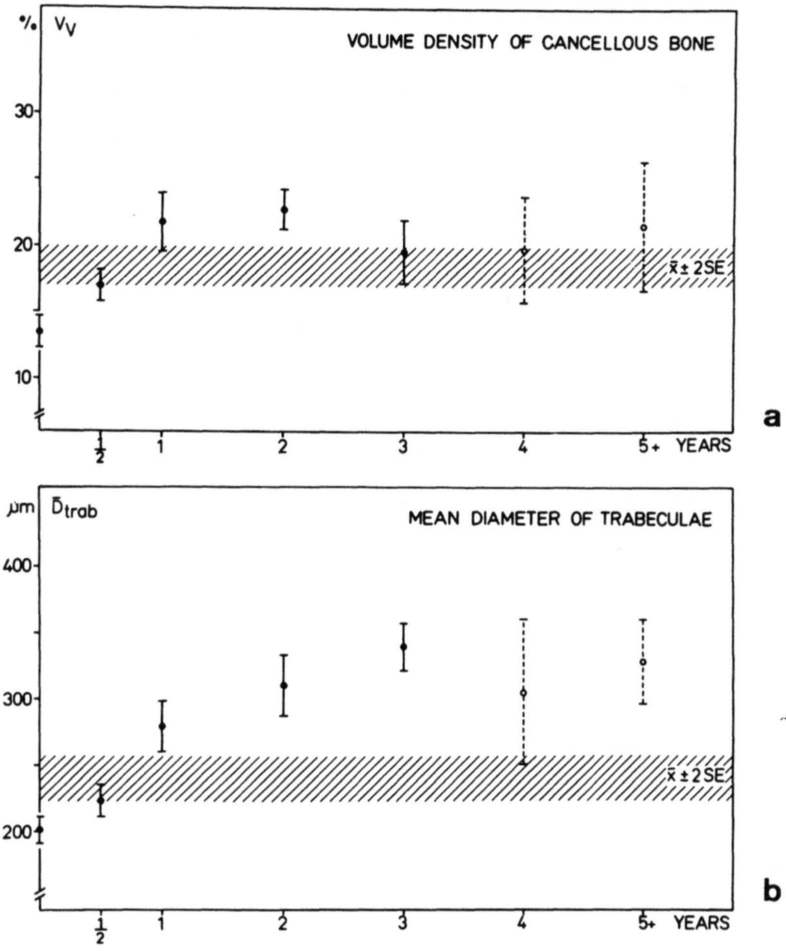

Fig. 2a and b

According to our histomorphometrical and qualitative-histological findings, the effect of prolonged treatment of osteoporosis with sodium fluoride can be divided into two phases:

1. During the first two years, a vigorous stimulation of osteoblastic activity dominates and leads to a positive skeletal balance and results in a hypertrophy of the trabeculae in the formerly porotic cancellous bone.
2. If treatment is continued for more than two years, bone turnover normalizes and reaches a new steady state at a higher level which is maintained for years.

In view of these facts, the benefit of a fluoride treatment which lasts for more than two years becomes questionable. Up to the moment, we dispose of only 2 cases where the fluoride application was interrupted after two years, and a histological control was carried out 2 and 4 years later. In spite of these long, fluoride-free intervals, the results are identical with the group under continuous treatment, and the gain in bone substance was maintained for the whole period. We realize that with only these two cases, it is too early to come to any definite conclusion and that further investigation is necessary. But from the present histological findings, it seems likely that application of fluoride in the treatment of osteoporosis could be restricted to repeated periods each lasting not longer than 1 to 2 years.

References

MERZ, W.A.: Die Streckenmessung an gerichteten Strukturen im Mikroskop und ihre Anwendung zur Bestimmung von Oberflächen-Volumen-Relationen im Knochengewebe. Mikroskopie 22, 132-142 (1967).

JOWSEY, J., SCHENK, R.K., REUTTER, F.W.: Some results of the effect of fluoride on bone tissue in osteoporosis. J. clin. Endocrin. 28, 869-875 (1968).

REUTTER, F.W., SIEBENMANN, R., PAJAROLA, M.: Fluoride in osteoporosis. In: Fluoride in Medicine, 143-152. Ed. Th. Vischer. Bern: Hans Huber 1970.

SCHENK, R.K., MERZ, W.A., REUTTER, F.W.: Fluoride in osteoporosis. Quantitative histological studies on bone structure and bone remodeling in serial biopsies of the iliac crest. In: Fluoride in Medicine, 153-168. Ed. Th. Vischer. Bern: Hans Huber 1970.

Long-term Experience with Fluoride and Fluoride Combination Treatment of Osteoporosis

J. JOWSEY, B.L. RIGGS AND P.J. KELLY

I. Osteoporosis

An osteoporotic individual is defined symptomatically as one with pain and disability resulting from fractures, generally a femoral neck fracture or one or more fractures of the spine. The disease can be defined by the degree of bone loss using quantitative radiological techniques such as bone cortical thickness or the Singh Index (BARNETT and NORDIN, 1960; SINGH et al., 1972).However, an arbitrary decision must be made as to precisely what level of bone loss constitutes osteoporosis and ultimately the definition remains dependent on the presence or absence of symptoms resulting from fracture. A symptomatic patient with a recent fracture is expressing evidence of a long history of the process of bone loss while a non-symptomatic, and therefore "normal" individual, may have the same degree of bone loss as a diagnosed osteoporotic patient and has only failed to sustain a fracture because no stress has occurred. It is obvious therefore that in any population there is going to be considerable overlap between individuals with symptomatic osteoporosis and controls as far as bone mass, bone metabolism and hormonal abnormalities are concerned. The disease or pain reflects rate of bone loss, duration of bone loss and recent stress. A fracture results from both a decreased amount of bone and stress; if the stress does not occur, there is no fracture.

Decrease in bone mass, which is the significant feature of osteoporosis, appears to be caused by an increase in resorption of tissue compared with its deposition. (BORDIER and TUN CHOT, 1972; FROST, 1961; JOWSEY, 1966). Bone biopsy indicates a decrease in the number of trabeculae in the spongy bone, while cortical bone becomes more porotic and tends to disappear on the endosteal surface. Bone loss appears to be related to need, in that bone which has important weight-bearing functions tends to be protected. The outer third of the cortex of long bones and the vertically oriented trabeculae of the vertebrae and ends of the long bones remain and appear to be resistant to resorption.

II. The Treatment of Osteoporosis

For successful treatment of osteoporosis there must be an increase in bone mass and as a result, a decrease in fracture incidence. Many forms of treatment have been suggested, the majority having no beneficial effect on the progress of the disease although both calcium and vitamin D, and estrogens and androgens have been shown to slow the progress of bone loss (Table 1).

It is possible that some combination of two agents might be successful, such as calcitonin and calcium. However, since the proposed usefulness of calcitonin lies in an ability to decrease bone resorption,

Table 1. Therapeutic agents for the treatment of osteoporosis

Agent	Mechanism	Effect
Hormones		
Estrogens and Androgens	Mediate the PTH effect on osteoclasts and decrease bone resorption.	Slows bone loss
Calcitonin	Lowers serum calcium and causes secondary hyperparathyroidism.	Accelerates bone loss
Growth hormone	Increases bone periosteally.	No clear effect
Chemicals		
Phosphate supplements	Stimulates PTH secretion and increases resorption.	Accelerates bone loss
Diphosphonates	Elevates serum phosphorus and increases PTH secretion. Increase in osteoid.	No clear effect
Calcium and Vitamin D	Suppresses PTH and bone resorption.	Slows bone loss
Fluoride	Stimulates PTH and increases bone resorption. Increase in osteoid.	Accelerates bone loss

even combined with calcium to prevent the secondary effect due to the hypocalcemia, no increase in mass would result.

1. Fluoride

Many investigators have established that fluorine will stimulate osteoblastic activity and cause the change in cellular behavior in bone which could achieve the increased mass essential for correction of osteoporosis. (CASS et al., 1966; COHEN et al., 1969). However, studies with fluoride as the therapeutic agent in the treatment of osteoporosis were disappointing. Although a more positive calcium balance tended to occur (RICH et al., 1964), this was not accompanied by radiological increase in bone density and cessation of symptoms in all cases (COHEN et al., 1969; SHAMBAUGH et al., 1968). The reason became evident in animal studies in which high doses of fluoride resulted in a depression of serum calcium. This led to stimulation of parathyroid gland activity and increased release of the hormone and increased resorption of bone. Osteomalacia was also apparent on biopsy (Table 2).

2. Fluoride and Calcium

However, it has been demonstrated experimentally that the secondary hyperparathyroidism and hyperosteoidosis can be prevented by the administration of added amounts of calcium as an oral supplement (BURKHART and JOWSEY, 1968). In man, increased bone mass was seen in individuals with multiple myeloma treated with a combination of fluoride

Table 2. Effect of fluoride on bone

Fluoride → increases in osteoblasts → increase in matrix formation
↓
[+] increase in demand for Ca
↓
Hyper-osteoidosis ← failure of mineralization ← lowering of serum Ca
↓
bone loss ← stimulation of osteoclasts ← increase in PTH

and calcium while cortisone-induced osteoporosis also responded to fluoride and calcium treatment with increased radiological density (COHEN et al., 1969). It appears that the additional need for calcium resulting from stimulation of new bone tissue can be fulfilled by oral supplements[+]. In recent studies vitamin D has been added to the combined fluoride and calcium regime to produce an increase in the absorption of calcium (JOWSEY et al., 1972).

There were two major questions; the first concerned the optimal level of fluoride which would produce a significant increase in bone formation, while the second was related to toxicity. In order to answer the first question patients were given different levels of fluoride varying from 15 mg to 45 mg of fluoride per day; given in the form of sodium fluoride. The level of supplemental calcium was also varied to find the amount necessary for calcification of new bone and for suppression of any secondary hyperparathyroidism. Preliminary studies suggested that 25 mg of fluoride per day and a gram of calcium per day as an oral supplement (the carbonate form appears satisfactory) is an effective dose. 50,000 units of vitamin D was also given twice a week. Further investigation is needed at this time to decide if this is the optimum dose or if vitamin D is required at all. (JOWSEY et al., 1972). Of the three components of the treatment regime the vitamin D is potentially the most toxic since it can produce hypercalcemia and hypercalcuria and endanger renal function.

a) Radiological findings. The x-ray appearance of fluoride-treated individuals may be described as 'fluorosis' in that there is coarsening of trabeculae and increased density of bone as a result of increased amounts of bone. In the present study more than 50% of the patients showed improved radiological appearance in the spine at the end of three years of treatment.

b) Bone biopsy results. The iliac crest biopsies taken before treatment and after five years of therapy showed a clear increase in the cortical thickness and in the width of trabeculae in all cases. In the majority the organization of bone lamellae and, to a lesser degree, the mineralization of the bone formed under the influence of fluoride, calcium and vitamin D was appreciably different from normal.

c) Side effects. A periosteal exostosis has occurred in one patient at the site of the bone biopsy. It is probable that this phenomenon resulted from a combination of the trauma due to biopsy and the fluoride, which over-stimulated healing at the biopsy site. The two most frequent adverse effects were gastric discomfort and joint pain. Gastric pain was generally overcome by giving the fluoride with meals; while swelling and pain of the joints, particularly at the knees, was reversible on withdrawal of the fluoride and did not recur when treatment was continued except in one patient.

III. Conclusions

Further long-term studies are needed, particularly to evaluate relief of symptoms and frequency of fracture. At this time the combination of fluoride and calcium, perhaps with the addition of vitamin D, is the only treatment regime that appears to increase bone mass and result in a reversal of the pathology of osteoporosis and of the bone disease.

References

BARNETT, E., NORDIN, B.E.C.: The radiological diagnosis of osteoporosis. A new approach. Clin. Rad. XI, 166-174 (1960).

BORDIER, Ph. J., TUN CHOT, S.: Quantitative histology of metabolic bone disease. Clinics in Endocrinology vol. 1, 197-215 (1972).

BURKHART, J.M., JOWSEY, J.: Effect of variations in calcium intake on the skeleton of fluoride-fed Kittens. J. Lab. Clin. Med. 72, 943-950 (1968).

CASS, R.M., CROFT, J.D., Jr., PERKINS, P., NYE, W., WATERHOUSE, C., TERRY, R.: New bone formation in osteoporosis following treatment with sodium fluoride. Arch. Intern. Med. 118, 111-116 (1966).

COHEN, P., NICHOLS, G., BANKS, H.H.: Fluoride treatment of bone rarefaction in multiple myeloma and osteoporosis. Clin. Orthop. 64, 221-249 (1969).

FROST, H.M.: Postmenopausal osteoporosis: a disturbance in osteoclasia. J. Am. Geriat. 9, 1078-1085 (1961).

JOWSEY, J.: Quantitative microradiography: a new approach in the evaluation of metabolic bone disease. Am. J. Med. 40, 485-491 (1966).

JOWSEY, J., RIGGS, B.L., KELLY, P.J., HOFFMAN, D.L.: Effect of combined therapy with sodium fluoride, vitamin D and calcium in osteoporosis. Am. J. Med. 53, 43-49 (1972).

RICH, C., ENSINCK, J., IVANOVICH, P.: The effects of sodium fluoride on calcium metabolism of subjects with metabolic bone diseases. J. Clin. Invest. 43, 545-556 (1964).

SHAMBAUGH, G.E., Jr., PETROVIC, A.: Effects of sodium fluoride on bone. J. American Medical Association 204, 969-973 (1968).

SINGH, M., RIGGS, B.L., BEABOUT, J.W., JOWSEY, J.: Femoral trabecular-pattern index for evaluation of spinal osteoporosis. Annals of Internal Medicine. 77, 63-67 (1972).

IV. Collagen Structure and Calcification

Chairmen: A. Ascenzi and A. S. Posner

Collagen Structure and Calcification*

G. L. MECHANIC

All normal mammalian collagens whether from fetal or mature tissues, exhibit the same physical chemical properties such as shrinkage temperature, molecular weight, etc. The protein, no matter what its tissue origin, exhibits the same x-ray diffraction pattern as well as the same electron microscopic appearance. The amino acid compositions have characteristic contents such as 1/3 glycine approximately 1/3 amino acids and in all the combined hydroxylysine, lysine totals are the same. Some do however vary in the ratio of hydroxylysine to lysine such as fetal tissues, bone and cartilage. There exist also variations in glycosylation. In all cases the collagen fibrils consist of collagen molecules arranged in a specific three dimensional array, although the collagens in the different tissues serve varied physiological functions. It therefore might be possible that despite specific array another order of specificity might exist that directs the organization and packing in a particular tissue that serves that tissue's particular function.

The mineralized organic matrices of tooth and bone are the end results of the calcification process. The ability of collagen to nucleate and mineralize is highly dependent upon its macromolecular structure, organization and packing. All must eventually be stabilized to provide a cohesive tissue. This stabilization is now known to occur via intermolecular covalent cross-links. The other collagenous tissues of the body are stabilized in the same manner. The loci, on the collagen molecules, of the cross-link precursors must attain specific juxtapositions in order for condensation to occur for the formation of a cross-link. The cross-links from only when the collagen molecules exist as 640 Å fibrils as opposed to unstructured fibrils. (TANZER and MECHANIC, 1970).

The type and number of the cross-links vary from tissue to tissue (MECHANIC, 1971; MECHANIC and TANZER, 1970; MECHANIC et al., 1971) as well as with the degree of maturation of a single tissue. For example, the difference between fetal bone and mature bone is remarkable (MECHANIC et al., 1971). More cross-links were evident in the mature bone as well as additional aldehydes.

The ratios of δ, δ'-dihydroxylysinonorleucine (DHLNL) δ-hydroxylysinonorleucine (HLNL) in the fetal and mature bone was 5.2 and 1.0 respectively. The ratio of the above cross-links in fetal and mature dentin was 5.5 and 5.0 respectively (MECHANIC et al., 1971). A rationale for the unaltered change in cross-link pattern for the dentin may be that it is a collagenous tissue that does not turn over and therefore its maturation is due to chronological age alone. The change in the cross-link patterns and ratio of the cross-links in bone could then be due to maturational phenomena caused by the turnover of collagen.

The quantitative and qualitative varity of cross-link chemistry that collagen displays when obtained from different tissues indicates:

+ Supported by PHS Grant No. DE-02668 and RR-05333 from the National Institutes of Health.

1. Some fine biochemical control mechanisms are operating that may be characteristic of the tissue function.
2. Different genomes might control the types of collagen that are formed in each tissue.

Following this we can therefore propose that alterations in micromolecular sequences may exist that direct the formation of the intermolecular cross-link precursors which form the cross-links that stabilize the macromolecular fibrous matrix of a particular tissue. Therefore there may be regions of the collagen molecule that contain these particular micromolecular alterations which carry the tissue-specific information and modulate the processes of macromolecular structure, organization and packing. These would vary with the type of support structure formed depending on the physiological function served by a particular connective tissue.

Since the intermolecular cross-links might reflect organization and packing, it is therefore important to obtain detailed information concerning the type of cross-links and their sites on the collagen molecules in different tissues. Data such as this might lead to more knowledge concerning differences in organization and packing of soft and hard tissue collagens leading to insights dealing with mineralization of collagen matrices. Following reduction of insoluble collagen from bovine tendon and dentin with NaB^3H_4, the most abundant radioactive compound in both cases is the intermolecular covalent cross-link dihydroxylysinonorleucine (MECHANIC and TANZER, 1970; MECHANIC et al., 1971). This derives from the Schiff base $\Delta^{6,7}$-dehydro-dihydroxylysinonorleucine, which is formed by the condensation of the carbonyl group of α-amino-δ-hydroxy-adipic-δ-semialdehyde on one collagen molecule, with the ε-amino group of hydroxylysine on another.

Since these tissues have quite different structural properties and functions, it might be expected, despite the fact that dihydroxylysinonorleucine is the most abundant reduced cross-link in both tissues, that differences in the distribution of the cross-link would occur which might indicate that the organization and packing of the collagen macromolecular fibrous matrix in tendon and dentin might not be the same. Studies such as this correlated with Katz's findings, that the intermolecular distances between molecules in soft and hard tissues vary, might lead to insights as to why certain fibrous matrices are predisposed to mineralization phenomena.

Unreduced and NaB^3H_4-reduced bovine insoluble tendon and dentin collagens were cleaved with CNBr. CM-cellulose chromatography indicated the CNBr peptides of both unreduced proteins were homologous with $\alpha1(I)_2$, α-2 type collagens. The data agreed extremely well with previously reported results (VOLPIN and VEIS, 1973). The chromatographic profiles obtained from the CNBr peptides derived from the NaB^3H_4-reduced tendon and dentin did not differ significantly from each other or with the unreduced collagens. The radioactivity was distributed differently when compared to each other. This result was expected because of the increased number of radioactive substances present in reduced tendon (MECHANIC and Tanzer, 1970) when compared with dentin (MECHANIC et al., 1971). The N-terminal portions of the reduced collagens were not altered upon reduction indicating a lack of <u>in vitro</u> reducible Schiff bases in the N-terminal regions of insoluble tendon and dentin. A reducible cross-link involving the N-terminal portion of collagen has been shown to occur in soluble rat tail tendon collagen upon reconstitution (KANG, 1972). The cross-link peptide α1-CB (1x6) has been isolated from unreduced insoluble skin and dentin collagen (VOLPIN and VEIS, 1973) while the α1-CB (1x6) from NaB^3H_4-reduced soluble rat tail tendon fibrils has been shown to contain the radioactive cross-link hydroxylysinonorleucine (KANG, 1972). It is possible that all of the

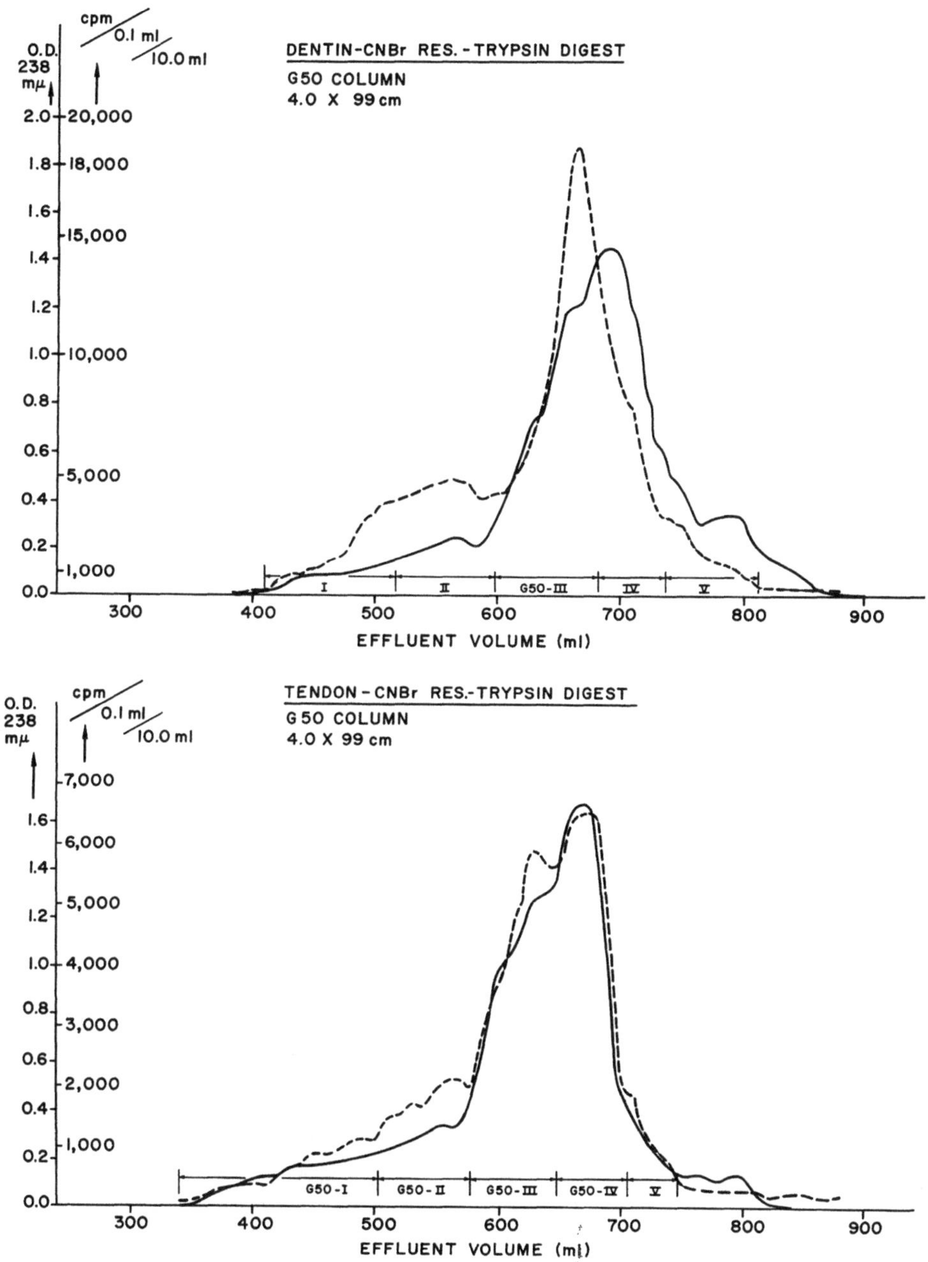

Fig. 1

Schiff base precursor of hydroxylysinonorleucine in the N-terminal portion from bovine insoluble tendon and dentin may have been reduced in vivo. In vivo reduction of Schiff base cross-links has been shown to occur in bovine insoluble dentin and bone collagen (MECHANIC et al. 1971).

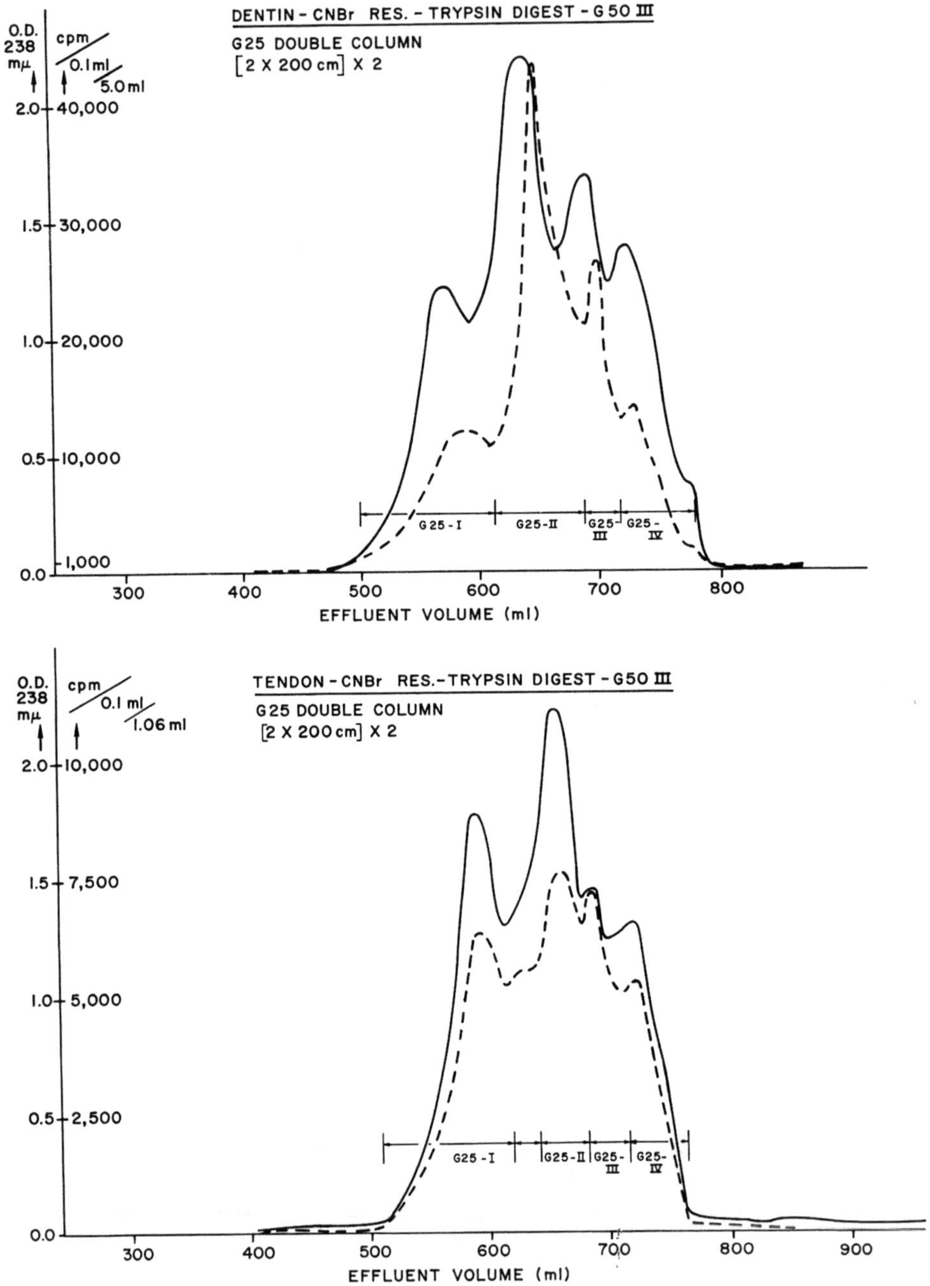

Fig. 2

The CNBr products from the NaB^3H_4-reduced insoluble tendon and dentin collagens were digested with trypsin. These peptides were chromatographed on Sephadex-G-50 (fine). The elution profiles are shown in Fig. 1. and indicate that both products contain the same molecular

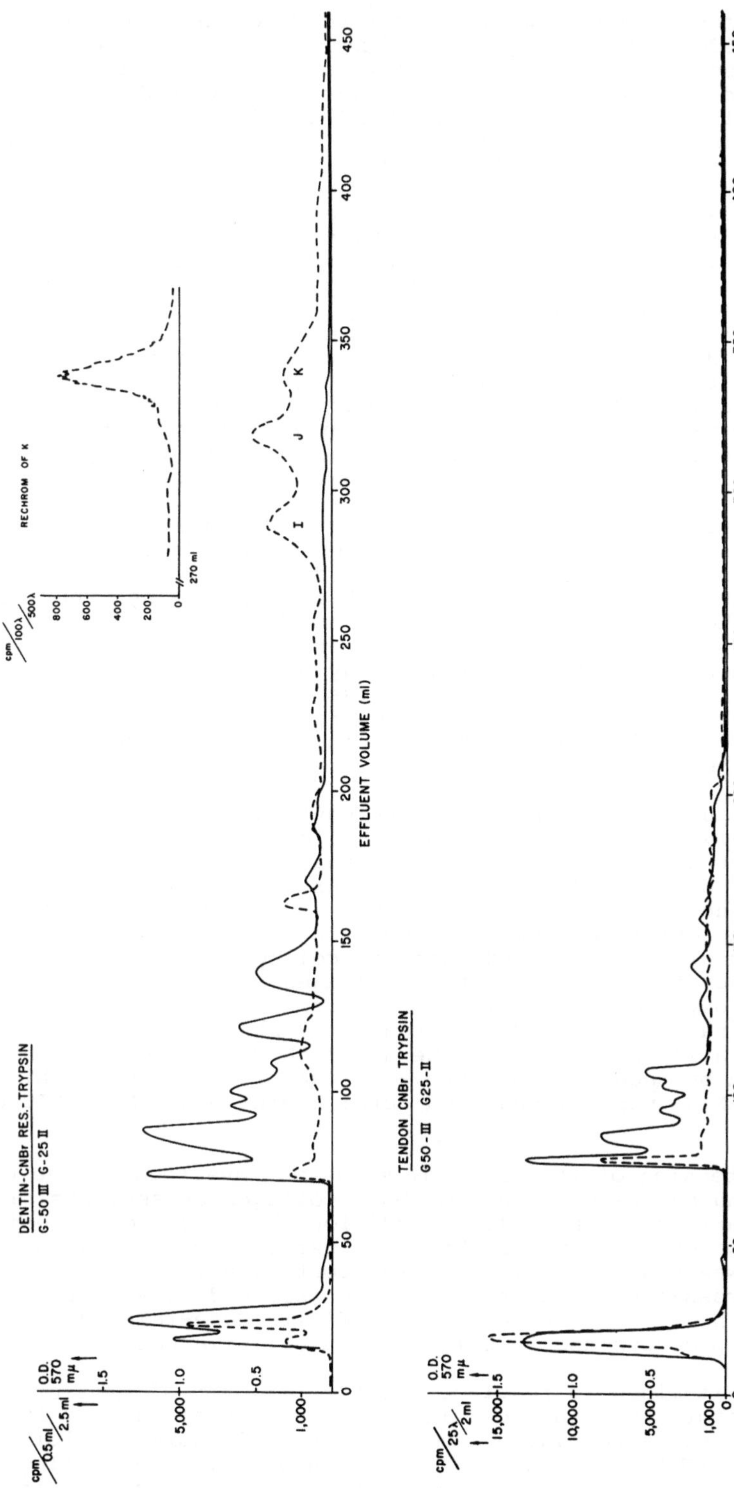
DENTIN-CNBr RES.-TRYPSIN
G-50 III G-25 II
cpm/0.5ml/2.5ml
O.D. 570 mμ
RECHROM OF K
cpm/100λ/500λ
270 ml
I
J
K
EFFLUENT VOLUME (ml)
TENDON CNBr TRYPSIN
G50-III G25-II
cpm/25λ/2ml
O.D. 570 mμ
EFFLUENT VOLUME (ml)

Fig. 3

weight distribution as would be expected because their respective CNBr peptides are homologous. As expected also the radioactivity was distributed differently. The most abundant radioactive peak (G-50III) from the dentin peptides and the corresponding peak from the tendon were refiltered further on Sephadex G-25 (Superfine) (Fig. 2). The G-25II peak contained the largest amount of radioactive substances in the dentin. This fraction and the corresponding fraction from the tendon were subjected to chromatographic analyses for reduced aldehydic cross-link precursors and cross-links. They were shown to contain dihydroxylysinonorleucine and hydroxylysinonorleucine. This fraction (G-50III-G-25II) was then chromatographed on a cation-exchange column using the pyridine-formate, pyridine-acetate volatile buffer system. The results are shown in Fig. 3. The dashed lines represent the radioactive profiles. These differ in that peaks I, J and K which appear in dentin do not appear in tendon. Chromatographic analysis indicated that all three radioactive peaks (I, J and K) derived from dentin contain dihydroxylysinonorleucine. Two of the peptides purified from peaks J and K have molecular weights of 3984 and 2850 daltons and consist of 38 and 26 amino acids respectively (KUBOKI et al., 1973). Neither involves the N-terminal portion of collagen.

Radioactive dihydroxylysinonorleucine is the most abundant intermolecular cross-link in NaB^3H_4-reduced tendon and dentin collagen but it is clear from these results that the distribution of this reduced cross-link is different in the two tissue collagens. The loci containing the cross-link precursors must therefore also be different. Thus some of the differences in properties and function between tendon and dentin-insoluble collagen may be due to different distribution of the same cross-links in these soft and hard tissues. It is clear that the necessary conditions required for the formation of cross-links and therefore orientational packing must be under very fine biochemical control.

Earlier studies in this laboratory have shown that the NaB^3H_4-reduced dihydroxylysinonorleucine from rachitic chick bone has an increased ratio over hydroxylysinonorleucine than is present in normal bone (MECHANIC et al., 1972). In the latter we had suggested that vitamin D deficiency did not allow synthesis of the normal complement of cross-links that exist in a normally maturing bone collagen. Subsequently BARNES et al. (1973) found that the collagen in chicks and rat rachitic bone contain $\alpha 1(I)_2$-$\alpha 2$ type collagen and both $\alpha 1$ and $\alpha 2$ chains are hyperhydroxylated when compared to normal bone. This is reflected in the original study (MECHANIC et al., 1972) where dihydroxylysinonorleucine ratios in rachitic bone are higher than normal. The high ratios exhibited by rachitic bone are also exhibited by fetal chick bone (G. MECHANIC, unpublished) and fetal bovine bone (MECHANIC et al., 1971).

In an extension of the original study we investigated the effect of growth and excess vitamin D on bone collagen cross-links. The data are summarized in Table 1. In addition there was no appreciable change in body weight or bone weight in either the -D, or high D groups at 1 and 2 weeks when compared to the control animals. Therefore the changes in the cross-link patterns cannot be ascribed to either inhibition or growth. The data indicate that a definite dose-related response relationship occurs and is effective at all time periods.

Perhaps vitamin D is required for the turnover of immature collagen to a more mature type that calcifies in the normal manner. The collagen structure seems to play an important role in normal calcification and it may be supposed that fetal collagen or a younger collagen does not contain the structure necessary for full normal calcification.

Table 1. Ratio of dihydroxylysinonorleucine to hydroxylysinonorleucine in chick bone collagen

	1 wk.	2 wk.	3 wk.	4 wk.
-D	4.2	3.9	6.6	7.6
Control	3.6	3.0	2.2	1.8
+D	3.2	2.2	1.8	1.8

Averages of 3 determinations.

References

BARNES, M.J., CONSTABLE, B.J., MORTON, L.F., KODICEK, E.: Biochem. J. 132, 113 (1973).

KANG, A.H.: Biochemistry, 11, 1828 (1972).

KUBOKI, Y., TANZER, M.L., MECHANIC, G.L.: Arch. Biochem. Biophys. 158, 106 (1973).

MECHANIC, G.: Israel J. Med. Sci. 7, 458 (1971).

MECHANIC, G., GALLOP, P.M., TANZER, M.L.: Biochem. Biophys. Res. Commun. 45, 644 (1971).

MECHANIC, G., TANZER, M.L.: Biochem. Biophys. Res. Commun. 41, 1597 (1970).

MECHANIC, G.L., TOVERUD, S.U., RAMP, W.K.: Biochem. Biophys. Res. Commun. 47, 760 (1972).

TANZER, M.L., MECHANIC, G.: Biochem. Biophys. Res. Commun. 32, 885 (1968).

VOLPIN, D., VEIS, A.: Biochemistry 12, 1452 (1973).

Calcification of Collagenous Material in vitro as a Model System for Biological Calcification

B. N. BACHRA

Many investigators have shown that collagen preparations can be used as a model system to study nucleation catalysis for apatite precipitation in metastable solutions. This process has been studied with reconstituted acid soluble collagen (SOBEL and BURGER, 1954; STRATES et al., 1957; GLIMCHER et al., 1957; BACHRA et al., 1959, BACHRA and SOBEL, 1959), with collagen from dentine (SOLOMONS et al., 1960) skin and Achilles tendon (BACHRA et al., 1959; STRATES and NEUMAN, 1958) It has been inferred from these studies that the ability to catalyze the nucleation of apatite is a property intrinsic to collagen, irrespective of its source (GLIMCHER, 1959; GLIMCHER, 1960; GLIMCHER and KRANE, 1968).

In recent years, other investigators have suggested that cells in calcifying tissues play a role in biological calcification, either directly by participating in the formation of the earliest mineral deposits or indirectly by conferring calcifiability on the extracellular matrix or by destroying this property. Elsewhere (BACHRA, 1970a), we have proposed the term "matrix modulation" for the latter two postulated cellular activities.

The main support for the concept that collagen can act as the nucleation catalyst in vivo has been drawn from the close morphological relationship between the collagen fibrils and the crystallites of the mineral phase as observed in bone and dentine (e.g. ROBINSON and WATSON, 1955; CARLSTRÖM and ENGSTRÖM, 1956), but also from studies with model systems, such as described above.

In order to settle the possible role of collagen as a nucleation catalyst we decided that more information was required on the characteristics of metastable calcium phosphate solutions at 37° and their interactions with solid apatitic material and collagenous nucleation catalysts. We have studied this in the past ten years and also the effects of various cations and anions on the stability of such solutions. The results of these studies have been published in a series of publications dealing with the effects of carbonate (BACHRA et al., 1963; BACHRA et al., 1965a; BACHRA et al., 1965b), Mg^{++} (BACHRA et al., 1965b; BACHRA and FISCHER , 1969), pyrophosphate (BACHRA et al., 1963; BACHRA et al., 1965a; BACHRA et al., 1965b), Sr^{++} and F^{-} (BACHRA et al., 1965b; BACHRA and FISCHER, 1969), a variety of polyvalent metal ions in trace amounts (BACHRA and VAN HARSKAMP, 1970) and of tetracycline and oxytetracycline (BACHRA and VAN HARSKAMP, 1973). These results can be summarized as follows.

Carbonate ions have a pronounced effect on both the stability of the solutions and the properties of the solids which can be deposited in these solutions. Raising the total carbonate concentration from 22 to 110 mM causes a contraction of the metastable concentration region of the solution, while the solid deposited in this region under the influence of apatitic seeding material or spontaneously in the region of instability becomes very poorly crystallized or even amorphous. When 3 mM Mg^{++} is added, in addition to the 110 mM total carbonate, the formation of apatitic material is completely suppressed, as judged from the x-ray diffraction patterns. Such "amorphous" calcium carbonate phosphate, formed in the presence or absence of Mg^{++} is stable at 37° for at least 3 days, when in contact with the supernatant solution (BACHRA et al., 1963; BACHRA et al., 1965a; BACHRA et al., 1965b; BACHRA, 1967). It should be emphasized that the term amorphous is used operationally in this context. It pertains to very poorly crystallized Mg^{++}-free or -containing calcium carbonate phosphate which does not show discrete x-ray diffraction lines of apatite, but only a broad and diffuse diffraction ring. It does not imply that the material should be devoid of considerable short range order. This point is of importance, since in recent years it has been claimed that considerable amounts of non-crystalline calcium phosphate are present both in young and mature bone mineral, that such amorphous material may be secreted as such by the cells of calcifying tissues and that further mineral deposition may occur not under the influence of a biological nucleation catalyst, but rather by accretion of this material and its crystallization into apatite (HARPER and POSNER, 1966; TERMINE and POSNER, 1966; TERMINE and POSNER, 1967; TERMINE et al., 1967).

Although solutions metastable with respect to calcium phosphates were easily obtained, metastability with respect to calcium carbonates was never observed under our conditions. Catalyzed nucleation of biological calcium carbonates therefore is not to be expected at 37°. Small concentrations of phosphate suppressed the formation of crystallized calcium carbonates and trace amounts of pyrophosphate suppressed also the formation of solid calcium phosphates. It seems therefore that biological mechanisms to remove phosphate from and to add carbonate and perhaps calcium ions to sites of physiological calcium carbonate deposition are mandatory (BACHRA et al., 1963; BACHRA et al., 1965a; BACHRA et al., 1965b), while pyrophosphate removal from sites of bone mineral deposition may be required (BACHRA, 1967; FLEISCH and BISAZ, 1964).

The effects of Mg^{++}, Sr^{++} and F^- and of combinations of these were studied, because it has been claimed that the inhibitory effect of these and other ions on the calcification <u>in vitro</u> of rachitic epiphyseal rat bone cartilage may be due to a direct effect on the catalytic system for apatite formation (SOBEL, 1950; SOBEL, 1952; SOBEL, 1955; SOBEL, 1965; SOBEL et al., 1958; SAMACHSON et al., 1959). Our studies (BACHRA and FISCHER, 1969) showed that these ions affected both crystal nucleation and growth and the combinations of Mg^{++} with Sr^{++} of F^- ions showed additive effects. Crystal growth was less susceptible to inhibition than nucleation. No specific effects of these inhibitors occurred on the nucleation catalysis for apatite formation by demineralized bone collagen different from their effects on the nucleation of apatite as such. This implies that the effects found by SOBEL et al. are not specifically operative in the catalyzed nucleation as such, but an effect on cellular process or on extracellular mechanisms involving membranous structures or enzymatic activities cannot be excluded. In recent work (BACHRA and VAN DER MEULEN-VAN HARSKAMP, 1973) it was shown that the effects of oxytetracycline and tetracycline in our system were similar to that of Mg^{++}, Sr^{++} and F^-. Although these antibiotics have been shown to be preferentially bound <u>in vivo</u> to the mineral phase of the earliest deposits in calcified tissues and can even inhibit calcification (JACOBS et al., 1964), these effects thus do not seem to involve a specific effect on the biological nucleation catalyst, although these antibiotics also become bound to the collagen (JACOBS et al., 1964). That far lower concentrations of these drugs inhibit mineralization in organ culture, as compared to those required in our system (BACHRA and VAN DER MEULEN-VAN HARSKAMP, 1973), may be connected to their known inhibitory action on protein synthesis. This could impair "matrix modulation" by the cells which may be required preparatory to calcification.

The effect of polyvalent metal ions on our metastable calcification buffer was found to be pronounced (BACHRA and VAN HARSKAMP, 1970). At concentrations of 1 µM or less, Pb^{2+}, Fe^{2+}, Fe^{3+}, Cr^{3+}, Bi^{3+} and Al^{3+} ions were found to destabilize the buffer. These results suggested that insoluble phosphates formed by these ions are able to initiate the precipitation of calcium apatite. The effect of these ions implies that meticulous care must be taken to exclude such ions from the solutions in order to prepare reliable metastable solutions. They further imply that the pathological liberation of such ions may give rise to pathological calcification and that topical calciphylaxis due to such ions (SELYE, 1962) may be caused by a similar mechanism. Furthermore, this phenomenon may also be involved in many cases of indirect systemic calciphylaxis, observed experimentally by SELYE (1962).

In the studies described in the previous paragraphs a metastable calcification buffer was employed with the following compositon:

salt	NaCI	$NaHCO_3$	$Na_2HPO_4.12H_2O$	$CaCI_2$
mM	120	22	1.67	3.75 or 3.38

The ionic strength was 0.16, the pH: 7.30 and the temperature on incubation 37°.

While this solution did not form a spontaneous precipitate of apatite within 3 days, although a thin film of such material usually formed at the liquid-air interface, demineralized bone collagen or apatite seeds were able to initiate the formation of considerable amounts of apatite (BACHRA and FISCHER, 1968a). It is clear therefore that such solutions are to be defined as metastabile, i.e. supersaturated, but not giving rise to a spontaneous precipitate throughout the solution within a reasonable period of time (BACHRA, 1967). Recently, Dr. Walter E. BROWN has calculated the ion activity products for five different calcium phosphates in our buffer: $CaHPO_4.2H_2O$ (brushite), $CaHPO_4$ (monetite), $Ca_4H(PO_3)_3.2.5H_2O$ (octacalcium phosphate), $Ca_3(PO_4)_2$ (whitlockite) and $Ca_{10}.(PO_4)_6(OH)_2$ (calcium hydroxyapatite). When comparing these calculated activity products to the known solubility products of these salts, he concluded that the ionactivity product considerably exceeded the solubility product for all five salts, even if due allowance was made for a certain degree of complexing of Ca^{++} ions by the carbonate ions. It is clear that this situation may contribute to the poor crystallinity of the apatitic material which can be formed in this solution.

In our studies, we observed that in our metastable calcification buffer demineralized compact sheep bone collagen (CSBC) or added apatite seeds were able to initiate the formation of considerable amounts of apatite. Reconstituted acid-soluble rat tail tendon collagen (RTTC) was a poor catalyst for apatite formation in this system (BACHRA and FISCHER, 1969; BACHRA and FISCHER, 1968a; BACHRA, 1970a and b; BACHRA, 1972; BACHRA and FISCHER, 1968b). The mineralization of the bone collagen was not due to the presence of residual mineral after decalcification. This was effectively excluded by our drastic decalcification procedure which involved stirring of 10 g of bone powder in the cold room with 1 l of 0.5 M EDTA, pH 8.1, for 2 weeks (the decalcifying solution was renewed after 1 week). Only traces of calcium remained in the material and no detectable phosphate, while no remaining crystals were observed with the electron microscope. The difference in behavior of these two types of collagen may have a physiologically significant meaning, because CSBC and RTTC originate from normally calcifying and normally non-calcifying tissues, respectively.

The mineralization of the bone collagen was studied chemically and by electronmicroscopy. Mineral deposition showed a rapid initial phase, followed by an appreciable slowdown of further deposition. At the end of the rapid initial phase, the amount of mineral deposited was proportional to the amount of collagen used, while the percent mineralization was independent of the amount of collagen. These findings suggested that the slowdown in mineral deposition was caused by locally impaired diffusion. The time course of mineralization was independent of the size of the bone collagen particles. This excluded the possibility that mineral deposition occurred mainly on the surface of the particles.

EM studies showed that the mineralization begins in discrete island-like regions within the particles. The apatite crystallites accentuated the 640 Å banding pattern of the collagen, which suggests that the collagen played an active role in apatite nucleation. At an overall mineral content of 12.5 percent, ellipsoid islands were present having dimensions usually smaller than 1 μ. At a mineral content of 32 percent they had reached dimensions of up to 4 x 6 to 7 μ, while many of the islands had become confluent. Large regions of the collagenous matrix were still devoid of crystallites, even at the higher mineral content which is at a stage near to the end of the rapid phase of mineral deposition. These findings indicated that the catalytically active regions in the bone collagen were localized at discrete sites. These give rise to numerous small islands of mineral. Upon further growth of the deposits by crystal growth and by epitaxy of new crystals on existing ones the collagenous matrix is compressed and more and more free water is removed from the mineral islands. This causes a slowdown in ion diffusion and of further mineral deposition within the islands.

The close morphological relationship observed between the 640 Å banding pattern of the bone collagen and the position of the mineral crystallites suggests that the collagen is involved in the nucleation catalysis. This involvement should be either direct as the catalyst or by being part of it, or indirect, by binding the actual non-collagenous catalyst in some relationship to the 640 Å banding pattern. If non-collagenous substances are involved, they or some of them should be bound strongly to the collagen (probably covalently). This was concluded from experiments in which the demineralized bone matrix was denatured by heating for 6 hours in 10 percent NaCl in boiling water. When placed in the calcification buffer, the residual non-soluble material renatured and the 640 Å banding pattern of the collagen was restored, as was its catalytic activity.

Since soluble substances are removed upon demineralization of bone, this may also remove substances involved in the nucleation catalysis _in vivo_. This could have caused the rather patchy mineralization pattern observed upon subsequent mineralization _in vitro_, as compared to the more even distribution pattern of bone mineralized _in vivo_.

It should be realized that we shall not attempt to explain tissue calcification by one generalized mechanism. In recent years, electron microscopy of various calcifying cartilages, dentine and embryonic bone (BONUCCI, 1967; SCHERFT, 1968; ANDERSON, 1969; ALI et al., 1970; BERNARD, 1972) has shown the presence of mucopolysaccharide-containing vesicles in the extra-cellular matrix which accumulate apatite crystallites and which have no morphological relationship with the collagen fibrils. Their possible function and their mechanism for apatite formation are unclear, at the present time. Their possible relationship to the calcium-containing granules on the endoplasmatic reticulum and in the mitochondria of epiphyseal chondrocytes has not been elucidated (MATUKAS and KRILOS, 1968; MATTHEWS et al., 1968; MARTIN and MATTHEWS, 1969). On the role of the calcium- and phosphate-containing extracellular PPL fractions in the epiphyseal plate (HOWELL et al., 1968 and 1969), either as initiators or as inhibitors of biological calcification, only speculations can be given at the moment. We feel, however, that at some stage of tissue calcification the nucleation of a solid phase will have to occur and we think that it is likely that this is a process actively catalyzed by biopolymers, be they collagen, enamel-proteins or the organic matrix of calculi etc. In our opinion, this process must be involved, irrespective of all the other tissue-specific phenomena which may precede the nucleation.

The following model is proposed for the nucleation catalysis in apatite formation by biopolymers (BACHRA, 1970; BACHRA, 1974). Active biopolymers are able to bind critical and subcritical clusters and in this way, to stabilize these sufficiently long to enable them to grow out fully-fledged apatite crystals which remain situated close to the original nucleation site. It seems likely that charged groups of the amino acid side chains of catalytically active proteins are involved in the cluster binding and stabilization. In the catalytically active bone matrix, collagen seems to be part of the active nucleation sites, at least beyond the stage of vesicle calcification.

This could, for example, be the result of material originating from the vesicles being bound to the collagen after its release from these vesicles.

References

ALI, S.Y., SAJDERA, S.W., ANDERSON, H.C.: Proc. Natl. Acad. Sci. 67, 1513 (1970)

ANDERSON, H.C.: J. Cell Biol. 41, 59 (1969).

BACHRA, B.N.: Clin. Orthop. and Rel. Res. 51, 199 (1967).

BACHRA, B.N.: Int. Rev. Conn. Tiss. Res. 5, 165 (1970a).

BACHRA, B.N.: Calc. Tiss. Res. 4 (suppl.), 31 (1970b).

BACHRA, B.N.: Calc. Tiss. Res. 8, 287 (1972).

BACHRA, B.N.: In: Biological Mineralization, (I. Zipkin,Ed.). New York: John Wiley and Sons, Inc. 1974.

BACHRA, B.N., FISCHER, H.R.A.: Calc. Tiss. Res. 2, 343 (1968a).

BACHRA, B.N., FISCHER, H.R.A.: Proc. of the Fifth Europ. Symp. on Calc. Tiss. p. 53. Bordeaux: 1968b.

BACHRA, B.N., FISCHER, H.R.A.: Calc. Tiss. Res. 3, 348 (1969).

BACHRA, B.N., VAN HARSKAMP, G.A.: Calc. Tiss. Res. 4, 359 (1970).

BACHRA, B.N., VAN DER MEULEN-VAN HARSKAMP, G.A.: Calc. Tiss. Res. 11, 95 (1973).

BACHRA, B.N., SOBEL, A.E.: Arch. Biochem. Biophys. 85, 9 (1959).

BACHRA, B.N., SOBEL, A.E., STANFORD, J.W.: Arch. Biochem. Biophys. 84, 79 (1959).

BACHRA, B.N., TRAUTZ, O.R., SIMON, S.L.: Arch. Biochem. Biophys. 103, 124 (1963).

BACHRA, B.N., TRAUTZ, O.R., SIMON, S.L.: Adv. in Fluorine and Dental Caries Prevention 3, 101 (1965a).

BACHRA, B.N., TRAUTZ,O.R., SIMON, S.L.: Arch. Oral. Biol. 10, 731 (1965b).

BERNARD, G.W.: Ultrastr. Res. 41, 1 (1972).

BONUCCI, E.: J. Ultrastr. Res. 20, 33 (1967).

CARLSTRÖM, D., ENGSTRÖM, A.: In: The Biochemistry and Physiology of Bone (G.H. Bourne, Ed.), Chapter 6, p. 149. London and New York: Academic Press 1956.
FLEISCH, H., BIZAS, S.: Proc. First Europ. Bone and Tooth Symp., p. 249. Oxford: Pergamon Press 1964.
GLIMCHER, M.J.: Revs. Mod. Phys. 31, 359 (1959).
GLIMCHER, M.J.: In: Calcification in Biological Systems (R.F. Sognnaes, Ed.), p. 421. Publication No. 64 of the A.A.A.S., Washington: D.C. 1960.
GLIMCHER, M.J., HODGE, A.J., SCHMITT, F.O.: Proc. Natl. Acad. Sci. 43, 860 (1957).
GLIMCHER, M.J., KRANE, S.M.: In: G.N. Ramachandran and B.S. Gould (Eds.): Treatise on Collagen,Vol. 2 B, Chapter 2. London and New York: Academic Press:1968.
HARPER, R.A., POSNER,A.S.: Proc. Soc. Exp. Biol. and Med. 122, 137 (1966).
HOWELL, D.S., PITA, J.C., MARQUEZ, J.F., GATTER, R.A.: J. Clin. Invest. 48, 630 (1969).
HOWELL, D.S., PITA, J.C., MARQUEZ, J.F., MADRUGA, J.E.: J. Clin. Invest. 47, 1121 (1968).
JACOBS, R., HARRIS, W.H., KATZ, E.P., GLIMCHER, M.J.: Biochim. Biophys. Acta 86, 579 (1964).
MARTIN, J.H., MATTHEWS, J.L.: Calc. Tiss. Res. 3, 184 (1969).
MATTHEWS, J.L., MARTIN, J.H., LYNN, J.A., COLLINS, E.J.: Calc. Tiss. Res. 1, 330 (1968).
MATUKAS, V.J., KRIKOS, G.A.: Cell Biol. 39, 43 (1968).
ROBINSON, R.A., WATSON, M.L.: Ann. N.Y. Acad. Sci. 60, 596 (1955).
SAMACHSON, J., NOBEL, S., SOBEL, A.E.: J. Dental Res. 38, 253 (1959).
SELYE, H.: Calciphylaxis. The University of Chicago Press: 1962.
SCHERFT, J.P.: J. Ultrastr. Res. 23, 333 (1968).
SOBEL, A.E.: In: E.C. Reifenstein, Jr., (Ed.): Metabolic Interrelations,Josiah Macy Jr. Found. 2, 113 (1950).
SOBEL, A.E.: In: E.C. Reifenstein, Jr., (Ed.): Metabolic Interrelations, Josiah Macy Jr. Found. 4, 113 (1952).
SOBEL, A.E.: Ann. N.Y. Acad. Sci. 60, 713 (1955).
SOBEL, A.E.: Proc. Second Europ. Symp. on Calcified Tissues (L.J. Richelle and M.J. Dallemagne Eds.), p. 291. Université de Liège: 1965.
SOBEL, A.E., BURGER, M.: Proc. Soc. Exper. Biol. and Med. 87, 7 (1954).
SOBEL, A.E., BURGER, M., NOBEL, S.: Proc. Soc. Exper. Biol. and Med. 99, 341 (1958).
SOLOMONS, C.C., IRVING, J.T., NEUMANN, W.F.: In: Calcification in Biological Systems (R.F. Sognnaes Ed.), p. 203. Publication No. 64 of the A.A.A.S., Washington: D.C. 1960.
STRATES, B., NEUMANN, W.F.: Proc. Soc. Exper. Biol. and Med. 97, 688 (1958).
STRATES, B.S., NEUMANN, W.F., LEVINSKAS, G.J.: J. Phys. Chem. 61, 279 (1957).
TERMINE, J.D., POSNER, A.S.: Nat. 211, 268 (1966).
TERMINE, J.D., POSNER, A.S.: Calc. Tiss. Res. 1, 8 (1967).
TERMINE, J.D., PULLMAN, I., POSNER, A.S.: Arch. Biochem. Biophys. 122, 318 (1967).

Chairmen: H.J. Höhling and G.L. Mechanic

Further Investigations on the Function of the Osteonic Lamellae According to Collagen and Crystallite Orientation*

A. Ascenzi and E. Bonucci

Our previous articles on the mechanical properties of single osteons (ASCENZI and BONUCCI, 1968, 1971, 1972) and that describing a method for isolating osteonic lamellae (ASCENZI et al., 1973), all serve to support the view that the compactness of haversian systems is strengthened by the presence of lamellae whose fiber bundles have an Almost transversal spiral course. Accepting this as a premise, we have now developed a research method able to yield direct evidence on the tightening function of these same lamellae through measurement of the internal pressure which eventually leads to the fracture of cylindrical osteon samples.

I. Material and Methods

The American Society for testing Materials has described a very simple hydrostatic method for measuring the resistance of the walls of cylindrical tubes to an internal pressure (ASTM Designations: D 1598 - 58 T and D 1599 - 58 T). Unfortunately, this method is apparently inapplicable to a microscopic sample as small as a single osteon. A different procedure therefore had to be developed.

Cross sections 100 μ thick were prepared by grinding human femoral shafts from 8 subjects aged between 20 and 25. Cylindrically shaped samples were obtained from cross-sectioned osteons, using a specially designed device which has been described in detail elsewhere (ASCENZI and BONUCCI, 1968). A careful selection of the osteons was made in order to test osteonic units having a known degree of calcification and a known orientation of collagen bundles. The degree of calcification was determined microradiographically, the aim being to select fully calcified osteons or osteons at the initial stage of calcification. Among the various arrangements produced by differences in fiber bundle and crystallite direction in successive lamellae, those characteristic of two types of osteon were chosen. In the first (called here Type I) the fibers and crystallites in one lamella have a marked longitudinal spiral course, while in the next the fibers and crystallites have an almost transversal spiral course, so that the fibers and crystallites in two successive lamellae form an angle of nearly 90°. Under the polarizing microscope, osteons of this type reveal an alternation of dark and bright lamellae in cross-section. In the second type (called Type II) fibers and crystallites have a marked longitudinal spiral course, with the pitch of the spiral changing so slightly that the angle of the fibers and crystallites in one lamella is practically the same as that of the fibers and crystallites in the next lamella. Under the polarizing microscope osteons of Type II appear uniformly dark in cross-section, although they are often bordered by a bright lamella

+ This work was supported by the grant No 72.00631.04 of the National Research Council of Italy.

both at the periphery and around the central canal. In every case the peripheral lamella was removed in cutting the osteon samples.

Each osteon sample was loaded along its axis by progressive penetration of a steel cone within the central canal until fracture occured (Fig. 1). This process was observed and controlled under the microscope. The angular width of the cone was constant (35°), which allowed the most accurate comparison of results. As the cone was gradually loaded with additional weights, it pressed with increasing force on the upper circumference of the haversian canal, first producing deformation and then progressive fracture of the lamellae, starting with those nearest the haversian canal.

The repeated forward movement of the cone until the ultimate strength was reached, was measured using a microwave micrometer based on cavity and pulse techniques (see ASCENZI et al., 1966; ASCENZI and BONUCCI, 1968,1972). The cone was fixed to the lower end of a push cylinder whose upper end was a disc. The disc was the lower plane of a cylindrical cavity, which functioned as a resonator for electromagnetic waves. When weights were added to a nylon thread attached to the push cylinder, the cone penetrated into the sample and the lower plane of the cylindrical cavity fell, thus increasing the height of the cavity. The forward movement of the cone was exactly equal to the increase in height of the cavity, so that changes in the position of the cone were easily deduced from changes in the resonant frequency of the cavity.

Forty samples were tested in all. All were kept wet.

II. Results and Conclusions

The main results and conclusions are as follows.

Fig. 1. Osteon sample ready to be tested. The steel cone is seen penetrating into the haversian canal. The top of the cone was cut off to allow it to move forward freely until the ultimate strength was reached. x 100

1. Osteons whose fiber bundles and crystallites have a marked longitudinal spiral course in successive lamellae (Type II) are those least able to resist the dilating loads produced by a conic punch moving forward along the haversian canal. This finding supports the view that in the osteon samples whose fiber bundles and crystallites change direction in successive lamellae through an angle of about 90° (Type I), the compactness of bone is increased by the lamellae whose fiber bundles and crystallites have a marked transversal spiral course.

2. The breaking strength of the osteons rises as calcification proceeds.

3. The stress-strain curves can be divided into two distinct portions. The first, very short portion bends slightly. The second, much longer portion approximates to a straight line (Fig. 2).

4. Microscope examination of the osteon samples before and after loading reveals that only the innermost lamellae - those surrounding the haversian canal - become greatly deformed.

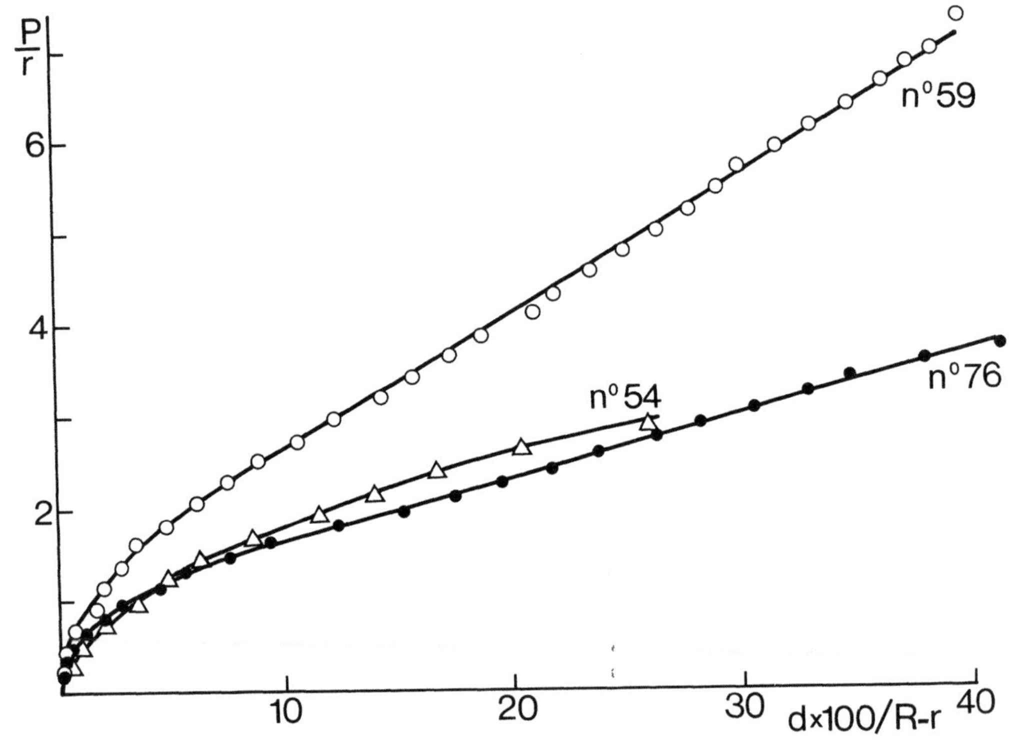

Fig. 2. Stress-strain curves of three osteon samples: No 59, a fully calcified osteon of Type I; No 76, an osteon of Type I at the initial stage of calcification; No 54, a fully calcified osteon of Type II. P = load; R = external radius of osteon samples; r = radius of the haversian canal; d = forward movement of the cone

References

ASCENZI, A., BONUCCI, E.: The compressive properties of single osteons. Anat. Rec. 161, 377-392 (1968).

ASCENZI, A., BONUCCI, E.: A micromechanic investigation on single osteons using a shearing strength test. Israel J. Med. Sci. 7, 471-472 (1971).

ASCENZI, A., BONUCCI, E.: The shearing properties of single osteons. Anat. Rec. 172, 499-510 (1972).

ASCENZI, A., BONUCCI, E., CHECCUCCI, A.: The tensile properties of single osteons studied using a microwave extensimeter. In: Studies on the Anatomy and Function of Bone and Joints (Edited by F.G. Evans), 121-141. Heidelberg: Springer 1966.

ASCENZI, A., BONUCCI, E., SIMKIN, A.: An approach to the mechanical properties of single osteonic lamellae. J. Biomechanics 6, 227-235

The Lability of Aldimine Crosslinks on Dissolution of Chicken Bone Collagen by Protein Denaturants

D. R. EYRE AND M. J. GLIMCHER

Bone collagen is highly insoluble in dilute solutions of salt or acid in comparison to the collagens of most unmineralized connective tissues. However, it has been shown that a large proportion of collagen in demineralized chicken bone can be solubilized as gelatin by protein denaturants such as LiCl, KCNS, $CaCl_2$, guanidine HCl or guanidine thiocyanate (GLIMCHER and KATZ, 1965; EYRE and GLIMCHER, 1972). It was thought that the extraction of the collagen was mediated by the rupture of certain crosslinking bonds that were peculiar to the bone collagen and were responsible for its extreme insolubility in non-denaturing solvents. It has been suggested however that the bone gelatin that was extracted from chicken bone by guanidine HCl may be extensively degraded, presumably by cleavage of peptide bonds (MILLER et al., 1967; PIEZ, 1968).

In the present study the effect of denaturing solvents on the stability of the known crosslinking compounds of chicken bone collagen was investigated. Recently much information has been published on the structure and role of aldimines (Schiff bases) as crosslinks in collagen (TRAUB and PIEZ, 1971; GALLOP et al., 1972). One aldimine compound, dehydro-hydroxylysinohydroxynorleucine (dehydro-Hyl OHNle), is especially abundant in the mineralized collagens of bone and dentine (MECHANIC et al., 1971; DAVIS and BAILEY, 1971), and may contribute to the unique physical properties of collagen in the mineralized tissues. Recent studies indicate that in bone collagen this aldimine rearranges to a keto-amine, a more stable crosslink, (EYRE and GLIMCHER, 1973; ROBINS and BAILEY, 1973).

Metatarsal bones from 12 - 14 week old chickens were cleaned, dried and then powdered in a mill that was cooled by solid CO_2 (GLIMCHER and KATZ, 1965). The bone powder was demineralized in several changes of 0.2M EDTA, pH 7.9 at 4° C, washed thoroughly with distilled water and freeze-dried. Samples of this material were extracted in 5M KCNS, 5M LiCl or 4M $CaCl_2$ at 4°C or 25°C. Some samples were reacted with sodium borohydride in 0.1M Na-phosphate, pH 7.4, prior to extraction (EYRE and GLIMCHER, 1973).

Treatment of the matrix of chicken bone with sodium borohydride markedly inhibited the extraction of the collagen by protein denaturants (Table 1). Now borohydride chemically reduces the aldimine crosslinks of collagen to secondary amines (e.g. dehydro-HylNle→HylNle). This stabilization of reducible bonds may explain why the dissolution of the bone collagen was inhibited by the borohydride. Consequently, the results suggest that the reducible crosslinks were breaking during extraction of the chicken bone matrix in certain protein denaturants, thus allowing the extraction of the collagen.

The denatured collagen that was extracted from the untreated bone matrix was examined by molecular sieve chromatography on a column (2.5 cm x 150 cm) of Bio-Gel Agarose A15 m (100 - 200 mesh) (Bio-Rad Laboratories), and by disc electrophoresis in SDS-polyacrylamide gels

Table 1. Effect of $NaBH_4$-Reduction on the Extractability of Collagen from Demineralized Chicken Bone by Protein Denaturants

	% of Total Collagen Extracted [a]					
	5M KCNS		5M LiC1		4M $CaC1_2$	
Serial Extracts	Control	$NaBH_4$ Treated	Control	$NaBH_4$ Treated	Control	$NaBH_4$ Treated
1 (1 week, 4°)	22	6	9	5	39	6
2 (1 week, 4°)	5	3	3	3	18	4
3 (1 week, 25°)	6	2	47[b]	4[b]	23	3
TOTAL	33	11	59	12	80	13

[a] Calculated from the hydroxyproline content of the extracts. Hydroxyproline was estimated by the method of STEGEMANN (1972) adapted to the Technicon Autoanalyser.

[b] Extraction carried out for 2 weeks at 25° C.

(FURTHMAYR and TIMPL, 1971). A large proportion of the solubilized collagen (50-75%) was eluted from the Agarose column as α component. Disc electrophoretic analysis of those fractions that contained polymeric chains (β-s, γ-s, and larger crosslinked polymers) also revealed a considerable amount of α chains. These α chains were virtually absent from the polymeric fractions, however, if the bone gelatin was treated in solution with sodium borohydride before chromatography on the Agarose column. Also, when samples of acid-soluble collagen from rat tail tendon were analyzed by a similar procedure α chains were not detected in the column fractions that contained β chains and larger polymeric chains.

It appeared that the solubilized collagen from bone included some polymeric chains which were held together during molecular sieve chromatography by relatively labile bonds. These labile polymers apparently dissociated under the conditions of disc electrophoresis to release free α chains. This finding can be explained if the labile bonds were aldimine crosslinks. Borohydride would eliminate the dissociable polymers by reducing the aldimines to stable secondary amines. Any free aldehyde groups in the soluble bone collagen would also be reduced, thereby preventing more aldimine bonds, and hence mor of the labile polymers, from forming.

The speculation that the reducible crosslinks were broken during solubilization of the collagen could be tested by using tritiated sodium borohydride to label the reducible crosslinking compounds. Demineralized bone, and a solution of the denatured collagen derived from it, were treated with NaB^3H_4 in 0.1M Na-phosphate, pH 7.4. Denatured collagen that was extracted in 4M $CaC1_2$, which in this experiment represented 40% of the total collagen in the bone matrix, was dialyzed exhaustively against distilled water, then against 0.1M Na-phosphate, pH 7.4. The collagen remained in solution under these conditions and was adjusted to a concentration of 1 mg/ml by addition of phosphate buffer. The labeled collagen was hydrolysed in 3N HC1 (TANZER et al., 1970), and the radioactive derivatives of the crosslinking compounds were detected after column chromatography on the amino acid analyzer. Fig. 1 compares the elution profile of tritiated compounds derived from the bone matrix with that derived from the solution of denatured collagen. The elution profile for the intact matrix is typical of similar analyses of bone collagen reported by BAILEY et al., (1969) and MECHANIC et al., (1971). The two main peaks of activity are the reduced crosslinks hydroxylysinohydroxynorleucine (Hy1OHN1e) and hydroxylysinonorleucine (Hy1N1e). The two smaller peaks, eluting earlier in the chromatogram are the alcohols dihydroxynorleucine (diOHN1e) and hydroxynorleucine (OHN1e), which before reduction were the aldehydes α-amino δ-hydroxy adipic δ-semialdehyde (hydroxyallysine) and α-amino adipic δ-semialdehyde (allysine). In the elution profile of the solubilized bone collagen the major peaks are clearly the reduced aldehydes, hydroxynorleucine and dihydroxynorleucine. The reduced crosslinking compounds are almost completely absent. When the residue of bone matrix that remained after extraction by 4M $CaCl_2$ was analyzed for reducible crosslinking components, reduced aldimines and reduced aldehydes were both detected but the concentration of aldimines was somewhat lower than that found in the intact tissue.

The results show that most of the reducible crosslinks are absent from the bone collagen that was solubilized by the $CaCl_2$. The aldimine, dehydro Hy1N1e, appeared to cleave quantitatively to the aldehyde, α amino adipic δ-semialdehyde, judging by the recovery of hydroxynorleucine from the solubilized collagen. But the major reducible component of the bone collagen dehydro-Hy1OHN1e, was apparently not converted quantitatively to its aldehyde precursor, α-amino δ-hydroxy adipic δ-semialdehyde. If it had been, then dihydroxynorleucine would be the major reduction product in the solubilized bone collagen. Nevertheless, the absence of Hy1OHN1e suggests that this crosslink was destroyed when the collagen was denatured and solubilized. Perhaps the crosslink is converted into products other than aldehyde and amine, which are not reducible and hence not detected. Indeed, as mentioned above, the aldimine, dehydro-Hy1OHN1e, apparently adopts the more stable, keto-amine configuration in bone collagen (EYRE and GLIMCHER, 1973; ROBINS and BAILEY, 1973). This rearranged compound, hydroxylysino-δ-oxonorleucine (HyloxoN1e), gives Hy1OHN1e on reduction, as does the aldimine. However, although the compound may be stable to acid and heat treatment (BAILEY et al., 1969) it is possibly degraded on dissolution of the collagen in certain protein denaturants. Whether the denaturing agents have a direct chemical effect on the crosslinking bonds or an indirect effect via the disruption of the helical structure of the collagen, is unknown. It is known that all denaturing chemicals are not effective in solubilizing the collagen of chicken bone. For instance, 8M urea was little more effective than dilute acetic acid or 0.1M NaC1 (GLIMCHER and KATZ, 1965). The urea did denature the collagen, however, as observed by x-ray diffraction analysis (BONAR and GLIMCHER, 1972). It may be significant that the most effective extractants of chicken bone collagen are protein denaturants that are ionic salts.

There was no evidence that peptide bonds were broken in the chicken bone collagen by the denaturing solvents. Rather, it is concluded that, under certain conditions, the reducible crosslinking bonds break, thus allowing some of the collagen to be solubilized.

Fig. 1 a and b. Ion-exchange column chromatography of tritium-labeled crosslinking compounds from chicken bone collagen. The collagen samples were reduced with NaB^3H_4 (10 Ci/mole), hydrolyzed in 3N HC1 and eluted from the 60 cm column of an amino acid analyzer with a complex gradient in sodium citrate buffer. (a) Demineralized chicken bone 10 mg). (b) Soluble collagen (10 mg) that was extracted from chicken bone in 4M $CaCl_2$. Over 40% of the total collagen in the matrix was solubilized for this preparation.

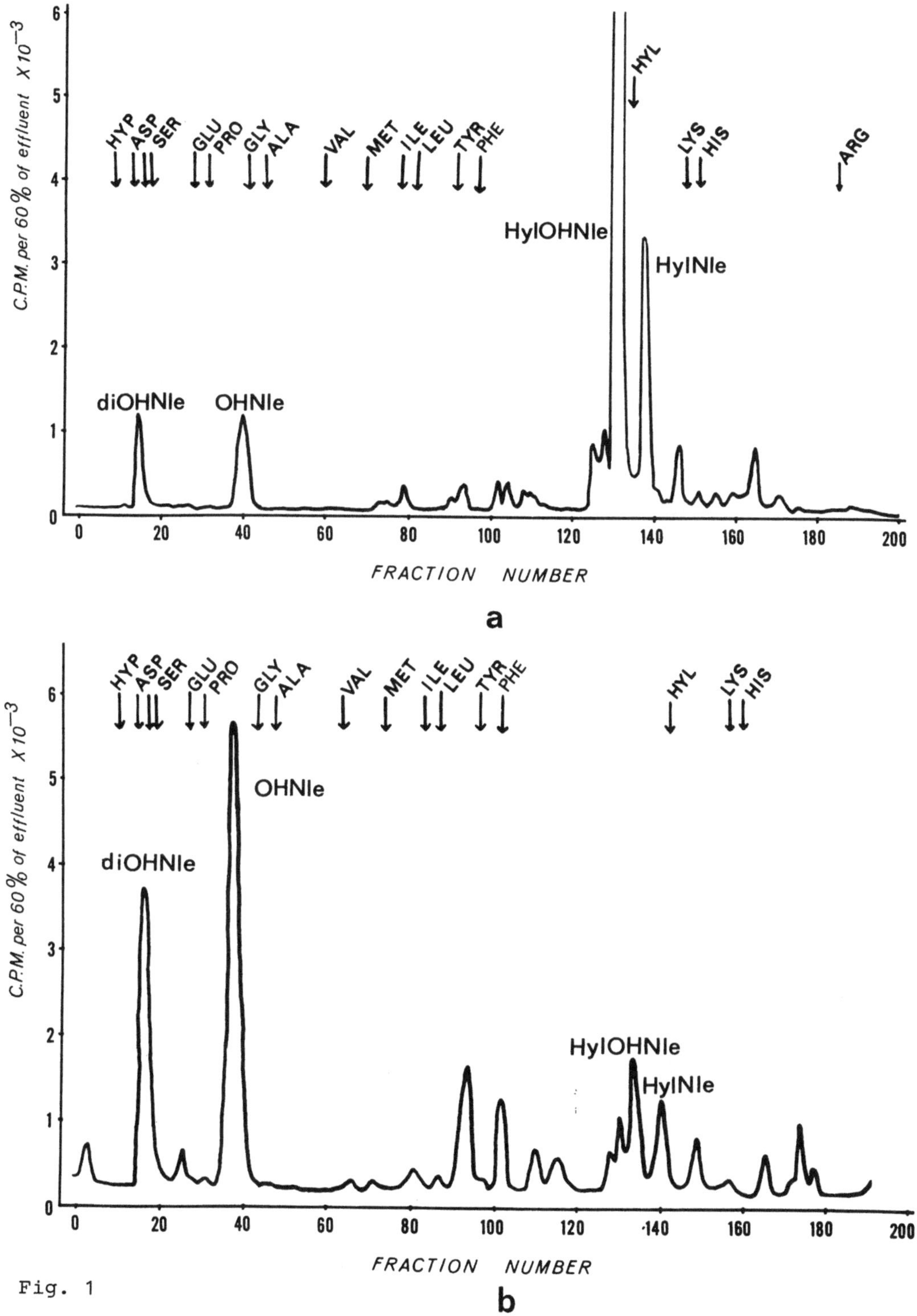

C.P.M. per 60% of effluent X 10⁻³
HYP ASP SER GLU PRO GLY ALA VAL MET ILE LEU TYR PHE HYL LYS HIS ARG
HylOHNle
HylNle
diOHNle
OHNle
FRACTION NUMBER
a
HYP ASP SER GLU PRO GLY ALA VAL MET ILE LEU TYR PHE HYL LYS HIS
OHNle
diOHNle
HylOHNle
HylNle
FRACTION NUMBER
b

Fig. 1

The extraction of large amounts of collagen from bone is a phenomenon that is apparently peculiar to chicken bone. Much less of the collagen in the matrices of mammalian bones can be solubilized by the protein denaturants (EYRE and GLIMCHER, unpublished).

References

BAILEY, A.J., FOWLER, L.J., PEACH, C.M.: Biochem. Biophys. Res. Comm. 35, 663 (1969).
BONAR, L.E., GLIMCHER, M.J.: J. Bone Joint Surg. 54A, 1129 (1972).
DAVIS, H.R., BAILEY, A.J.: Biochem. Biophys. Res. Comm. 45, 1416 (1971).
EYRE, D.R., GLIMCHER, M.J.: Biochim. Biophys. Acta 278, 206 (1972).
EYRE, D.R., GLIMCHER, M.J.: Biochem. Biophys. Res. Comm. 52, 663 (1973).
FURTHMAYR, H., TIMPL, R.: Anal. Biochem. 41, 510 (1971).
GALLOP, P.M., BLUMENFELD, O.O., SEIFTER, S.: Ann. Rev. Biochem. 41, 617 (1972).
GLIMCHER, M.J., KATZ, E.P.: J. Ultrastruct. Res. 12, 705 (1965).
MECHANIC, G., GALLOP, P.M., TANZER, M.L.: Biochem. Biophys. Res. Comm. 44, 1518 (1971).
MILLER, E.J., MARTIN, G.R., PIEZ, K.A., POWERS, M.J.: J. Biol. Chem. 242, 5481 (1967).
PIEZ, K.A.: Ann. Rev. Biochem. 37, 547 (1968).
ROBINS, S.P., BAILEY, A.J.: FEBS Letters 33, 167 (1973).
STEGEMANN, H.: Z. Physiol. Chem. 41, 331 (1972).
TANZER, M.L., MECHANIC, G., GALLOP, P.M.: Biochim. Biophys. Acta 207, 548 (1970).
TRAUB, W., PIEZ, K.A.: Adv. Prot. Chem 25, 243 (1971).

Comparison of Ca, P, and S Levels in Predentine and Tendon

W.A.P. NICHOLSON, B.A. ASHTON, H.J. HÖHLING AND A. BOYDE

Over the last few years our group has been involved in the determination of the elemental composition of mineralizing tissues by electron microprobe techniques. Recently we have concentrated on the mineralization process in the predentine layer of the continuously growing rat incisor.

Several methods of tissue preparation have been used. Initially the tissue was freeze dried and then embedded in vacuo in methacrylate, or alcohol fixed and then embedded. Ultra-thin sections of the tissue were then prepared. Our recent work, however, has been carried out on 5 μm-thick freeze dried cryostat sections.

In the continuously growing rat incisor the odontoblasts are separated from the mineralized dentine by a band of uncalcified predentine approximately 20 μm thick. As it is not possible to differentiate between the cells and predentine in the back scattered electron image formed in the microprobe our measurements have been restricted to the predentine up to 12 μm away from the mineralized border.

The influence of preparation technique on the mass fractions of Ca, P and S in predentine is shown in Table 1. It can be seen from the mass ratios that, compared with apatite, there is a great excess of phosphate in the cryostat sections, which is reduced by the short exposure to water involved in the sectioning of the vacuum embedded preparations. Following alcohol fixation the ratio is greatly decreased and only 45% of the calcium at the most could be associated with P in apatite.

The Ca concentration appears to be independent of the treatment, which may suggest that this element is fairly tightly bound to matrix components. It is quite clear from these figures that there is no real alternative than to work with cryostat sections.

To put these values of the mass fractions in perspective, the predentine values have been compared with those determined in freeze dried cryostat sections of rat tail tendon (Table 2). It can be seen that each element is present in higher amounts in predentine, although by different factors.

As microprobe analysis is still associated with some problems, for example the loss of material in the electron beam, the amounts of Ca, P and S in tendon obtained by microprobe analysis were compared with chemically determined values. The figures for Ca and P agreed within the experimental error of the methods, but the S values were widely dissimilar and further work is in progress to explain this difference.

It is interesting to speculate on the state of these elements within the tissue. For instance it can be calculated from the data of OWEN et al., (1973), that predentine has a water content of 66% (w/w). If the assumption is made that the measured amounts of Ca and P are in solution in the tissue water, the concentrations of 35 mM Ca; 175 mM P in predentine, and 5 mM Ca; 35 mM P in tendon would be far greater than the concentrations determined in rat interstitial fluid 2.2 mM Ca,

Table 1. Influence of preparation technique on the mass fraction of Ca, P and S in continuously growing rat incisor predentine

	Cryostat		Mass fraction (%w/w) in Vacuum[a] embedded		Alcohol[a] fixed	
	Range	Mean	Range	Mean	Range	Mean
Ca	0.1-1.3	0.28	0.2-0.6	0.3	0.2-0.6	0.3
P	0.3-3.4	1.1	0.2-0.7	0.3		⩽0.06
S	0.5-2.2	1.0		⩽0.6	-	-
Mass ratio Ca:P		1 : 4		1 : 1		1 : 0.2
Mass ratio Ca:P in apatite is				2.15: 1		

[a]Minimum weight fractions, i.e. embedding material included.

Table 2. Comparison of the Ca, P and S mass fraction (%w/w) in cryostat sections of rat incisor predentine and tail tendon

		Mass fraction (%w/W) Predentine	Tendon	Ratio Predentine:Tendon
Ca	microprobe	0.28	0.04	7 : 1
	chemical	-.-	0.05	
P	microprobe	1.1	0.22	5 : 1
	chemical	-.-	0.17	
S	microprobe	1.-	0.41	2.4 : 1
	chemical	-.-	0.07	

1.8 mM P (RASMUSSEN, 1972); rat plasma 2.5 mM Ca, 2.1 mM P (RASMUSSEN, 1972); bovine bone fluid 0.5 mM Ca, 1.8 mM P (NEUMAN, 1969). As the ion product (Ca^{2+}x P_i) of the predentine tissue fluid would be far higher than that required for the spontaneous formation of hydroxyapatite, we believe that precipitation is prevented by the Ca ions being bound to the organic matrix, and the P, which is largely removable by treatment with the aqueous solutions, being present not only as orthophosphate but also as pyro- and/or polyphosphate which could inhibit calcification until the stimulus for its onset is received.

So far only the mean values obtained from series consisting of several hundred point measurements have been discussed. However one of the great advantages of microprobe analysis is the high degree of spatial resolution possible, which has allowed us to monitor the variation of the calcium content within the predentine. In Fig. 1 the variation in the Ca content both along and perpendicular to the border is represented graphically. It is apparent that the Ca concentration is not uniform, and that close to the mineralizing front there are regions in which the Ca content is several times greater than in the regions

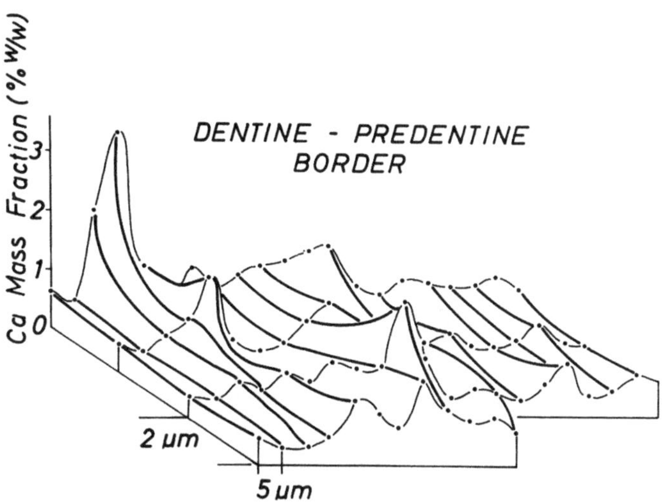

Fig. 1

neighboring them. There is also a general decrease in the Ca content towards the cells, although small zones of Ca enrichment can be seen. In the measurement series made so far it has not been possible to correlate the increases in Ca content with fluctuations in the levels of P and S.

It is proposed that the elevation of the Ca content in microregions to values several-fold above the average for the tissue is a necessary stage in the calcification process, and the results suggest that mineralization is not uniform along the front, and that some microregions can calcify while neighboring regions do not.

References

NEUMAN, W.F.: Fed. Proc. 28, 1846-1850 (1969).

OWEN, M., TRIFFITT, J.T., MELICK, R.A.: In: Hard Tissue; Growth, Repair and Mineralization, Ciba Foundation Symposium 11 (New Series), p. 263-293. Amsterdam: ASP (Elsevier-Excerpta Medica-North Holland) 1973.

RASMUSSEN, P.: Calc. Tiss. Res. 9, 200-206 (1972).

The Role of Lysyl Residue in Collagen Structure with Calcification and Aging

M. IGARASHI, Y. HAYASHI, M. MATSUURA AND S. YOSHINO

Collagen comprises about 90% of bone matrix. There are conflicting theories of calcification mechanism. Ultrastructural studies of cartilage and bone matrix revealed the cellular process of initial calcification (BONUCCI, 1967). On the other hand, from the epitactic theory by NEUMAN and NEUMAN (1953) to the charge neutralization theory by URRY (1971), the major role of collagen remains that of inducing mineralization. Crosslinks of collagen occur following the enzymatic conversion of the lysyl or hydroxylysyl residue in peptide linkage to aldehyde. The degree of crosslinks may affect the number of residues of lysine or hydroxylysine obtained from collagen hydrolysate.

We had previously reported the increase of the number of lysyl redidues in osteoporotic bone collagen not only in experimental rats but also in human beings. However, no change of hydroxylysyl residues was noted, (HAYASHI et al., 1973a; HAYASHI et al., 1973b).

Our present investigation is the change of amino acid composition of insoluble collagen from various rat bones with the purpose of obtaining further knowledge on collagen structure with calcification. Firstly, we report the change by osteolathyrism; secondly, we present the protective action of estriol on experimental osteoporosis as well as its normalizing action on the collagen. Finally, the change of lysyl and hydroxylysyl residues of insoluble collagen are shown depending on the varying ages of rats.

Exp. I. Eighteen male Sprague-Dawley rats were divided at random into three equal groups at 4 weeks of age. Group A were control rats. Group B were fed on the low calcium diet containing 0.01 % calcium. Group C were fed on the normal diet containing 0.4 % β-aminopropionitrile. After 4 weeks, the rats were killed. Exp. II. Twenty-four male Wistar rats were divided at random into four equal groups at 3 weeks of age. Group A were control rats. Group B were fed on low calcium diet. Group C were fed on low calciun diet and received daily oral administration of 40 μg of estriol. Group D were also fed low calcium diet and received daily oral administration of 300 μg of estriol. Estriol was suspended in distilled water and administered through the stomach tube. After four weeks, the rats were killed. Exp. III. Male Wistar rats of different ages: 30 rats of 3 weeks, 6 rats of 7 weeks, 3 rats of 10 weeks, 7 rats of 3 months, 3 rats of 6 months and 7 rats of about 2 years 5 months were killed for the experiment.

Both femoral and tibial shafts of each rat were removed and dissected free of muscles, visible connective tissues and periosteum. Then x-ray pictures of bones were taken. The bone marrows were washed away with water. A portion of bone was used for the determination of calcium concentration which was compared with the dry weight of the same bone. The other bones were decalcified. After extensive extraction of soluble collagen and fatty tissues, the remaining insoluble collagen was hydrolysed in 6 N HC1 at 110° for 24 hours under nitrogen in sealed tubes. The acid was removed in a rotary evaporator at about 50° under vacuum. The dried sample was dissolved in 0.2 M sodium citrate buffer, pH 2.20, and analyzed on a two-column automatic JEOL JLC-6AH analyzer with internal standards of homoarginine and norleucine. Every six runs included one run of standard mixture. No corrections were made for the possible partial destruction or incomplete release of individual amino acids due to hydrolysis conditions. Hydroxylysine was also determined by the method of BLUMENKRANTZ (BLUMENKRANTZ and PROCKOP, 1971) and lysine was determined by the microdiffusion method using purified lysine decarboxylase from Bacterium cadaveris.

In the first experiment the x-ray picture shows the marked difference between the control and the osteolathyritic bones. Table 1 shows remarkable decreases of calcium contents in group B and C. On the other hand, lysyl residues increase from group A to B and to C. No difference is noted on hydroxylysyl residues. In this case hydroylysine was determined by the Blumenkrantz method. This value of hydroxylysine is slightly lower than that which is later shown by amino acid analyzer.

Now we proceed to the second experiment in the effects of estriol. An x-ray picture of groups A, B, C and D was taken. The bone of group C indicates a slight increase of cortical thickness compared with that of group B. Amino acid composition of each group is compiled with the values of means and standard errors. Calculation was done whether any amino acid possessed the statistical difference between the groups. It was found that lysine and hydroxylysine possessed this difference. The summary of statistical calculation is shown in Table 2. The calcium content of group A is higher than others. However, estriol recovers its loss by low calcium diet, which is evident when comparison is made between groups B and C. Lysyl residues of groups B and C increase when

Table 1. The amount of lysyl and hydroxylysyl residues in insoluble collagen from bone[a]

		Ca/Dry weight of right Tibia (%)	Lysyl Residues (moles/10^5 g Collagen)	Hydroxylysyl Residues (moles/10^5 g Collagen)
Group A	normal calcium diet	28.1 ± 0.5	24.7 ± 0.5	8.62 ± 0.22
Group B	0.01 % low calcium diet	23.9 ± 0.4[b]	26.9 ± 0.5[d]	8.29 ± 0.22[c]
Group C	β-aminopropionitrile diet	22.4 ± 0.3[b]	31.3 ± 0.6[b]	8.33 ± 0.18[c]

[a] The values presented in this table represent means with standard errors.

[b] Statistically significant to group A. ($p < 0.001$).

[c] Not statistically significant to group A.

[d] Statistically significant to group A. ($p < 0.02$).

Table 2. The amount of lysyl and hydroxylysyl residues in insoluble collagen from rat bones[a]

Group	Number	Ca/Dry weight of Bone (%)	Lysine	Hydroxylysine	Sum of Lys. and Hyl.
			(Residues/1.000 Total Residues)		
A Normal Calcium Diet	6	27.3 ± 0.21	22.8 ± 0.16[2,3,]	12.3 ± 0.18 4,5,	35.0 ± 0.32[2,] 5,
B 0.01% Low Calcium Diet	6	22.9 ± 0.41[1,4,]	24.4 ± 0.34[2,] 3',	13.5 ± 0.37[2,4,]	37.9 ± 0.64[2,4,] 5',
C Low Calcium Diet + 40 μg Estriol	6	24.4 ± 0.27[1,4,]	24.3 ± 0.43 3,	11.7 ± 0.12[2,] 5,	36.2 ± 0.41 5,5',
D Low Calcium Diet + 300 μg Estriol	6	24.0 ± 0.45[1,]	23.1 ± 0.28 3',	12.6 ± 0.29	35.8 ± 0.31 4,

a The values presented in this table represent means with standard errors.
Statistically significant difference;

1,--P<0.001 to group A.
2,--P<0.005
3,3',--P<0.01
4, --P<0.025
5,5',--P<0.05

compared to group A. Hydroxylysyl residues have no consistent change as noted on lysyl residues. However, when sums of lysyl and hydroxylysyl residues are calculated, it is evident that the value increases from group A to B and then decreases through group C to D. No statistical difference is revealed between groups A and D, though definite differences are noted between groups. Estriol restores the sum from the high value of group B to the normal value of group D.

Finally we mention age-related changes of amino acid compositions of insoluble collagen of rat bones. Summarized results show that calcium contents are lowest in 3-week old rats and highest in aged rats. On the contrary, lysyl residues are highest in 3-week old rats and lowest in aged rats. Hydroxylysyl residues do not have a constant tendency. Maturation accompanies the increase of hydroxylation of lysine, and then it almost reaches a plateau. A comparison of the change of hydroxylation of proline and lysine shows a remarkable inverse relationship in the ratios of the two amino acid residues.

Although the biophysical structure of collagen is amply understood, how far the minute alteration of collagen structure affects the calcification of bone is scarcely revealed. The past few years have shown rapid progress in determining the structure of crosslinks in various collagens. We previously reported that the amount of lysyl residues increase in senile osteoporosis. This led us to investigate further the amino acid composition in several experimental conditions in relation with calcium content of bone. An inverse relation of calcium contents and lysyl residues was shown in the first experiment. The highest value of lysine was disclosed in the osteolathyrism group. β-aminopropionitrile was a potent inhibitor of lysyl oxidase according to Pinnel and Martin. Low calcium diet also invokes a high sum of lysyl and hydroxylysyl residues and this is reduced to the normal value by estriol. Our present results clearly demonstrate the protective effect of estriol against the loss of bone mineral in low calcium diet in the rat. The mechanism of estriol on bone is yet not elucidated in this experiment. Amino acid compositions of insoluble collagen from various ages of rats present the same tendency in calcium content and lysyl residue, but it is not so evident as in the preceding experiments. In any case, aging implicates several changes in amino acid composition of insoluble collagen. DAVIS (1973) recently discovered that every two tropocollagens are joined by one aldimine crosslink in various collagens, and indicated that the maturation of collagen does not involve a simple *in vivo* reduction of the aldimine bond. Our results of the main change of lysyl residues indicate a further necessary elucidation on the nature of crosslinks regulating the degree of calcification of bone.

Summary: we have illustrated the change of amino acid composition, particularly lysine and hydroxylysine, intimately related to the degree of calcification of bone.

References

BLUMENKRANTZ, N., PROCKOP, J.: Quantitative assay for hydroxylysine in protein hydrolyzates. Anal. Biochem. 39, 59-64 (1971).

BONUCCI, E.: Fine structure of early cartilage calcification. J. Ultrastruct. Res. 20, 33-45 (1967).

DAVIS, N.R.: A method for the determination of lysine-derived collagen crosslinks. Biochem. Biophys. Res. Commun. 52, 877-883 (1973).

HAYASHI, Y., IGARASHI, M., YOSHINO. S.: The amount of lysine from bone collagen in experimental osteoporosis. Bone Metabolism. 6, 81-84 (1973a).

HAYASHI, Y., TAKIZAWA, H., IGARASHI, M.: Aging of bone $\frac{I}{M}$/Collagen in the senile osteoporotic bone. Jap. J. Geriatrics. 10, 185-190 (1973b).

NEUMAN, W.F., NEUMAN, M.W.: Nature of the mineral phase in bone. Chem. Rev. 53, 1-36 (1953).

URRY, D.W.: Neutral sites for calcium ion binding to elastin and collagen: A charge neutralization theory for calcification and its relationship to atherosclerosis. Proc. Nat. Acad. Sci. U.S.A. 68, 810-814 (1971).

Evidence for a Role of Lysozyme in Endochondral Calcification

D. S. Howell, J. C. Pita and K. Kuettner

It was previously shown that a proteoglycan aggregate is involved as the principal inhibitor of mineral growth in the micropuncture fluids collected from endochondral growth plates of untreated phosphate-deficient, vitamin D-deficient rachitic rats. Furthermore, our data indicate degradation of proteoglycan at the onset of the calcification in such cartilages, with disappearance of the inhibitory aggregate form of proteoglycans in micropuncture fluids from the calcifying zone of healing rachitic rats.

This report concerns the agent which appears to cause the degradation of this proteoglycan inhibitor in cartilage undergoing calcification after 48 h of healing of the rickets with phosphate and vitamin D. First evidence of an enzymic degradation of the inhibitor was shown by the spontaneous progressive disappearance of proteoglycan aggregates from the puncture fluid aspirated at 48 h and incubated in vitro up to 36 h. During this peroid the average sedimentation coefficient of the proteoglycans is reduced to about one third of initial values. This effect could not have resulted from cathepsin D since degradation occurs at pH 7.6. In confirmation of previous data on micropuncture fluids of untreated rachitic rats (KUETTNER, K., et al., in press) assay for lysozyme with the micrococcus lysodeikticus method in agar following electrophoresis indicated the presence of lysozyme at levels of circa 80 mg %. Similar concentrations are found in the untreated rachitic rat cartilage, but the enzyme was in an inactive form and failed to degrade proteoglycan aggregates until electrophoresis was performed. The n-acetylglucosamine trimer inhibitor specific for lysozyme (at 10^{-3}M) totally halted degradation which spontaneously occured in all control 48-h micropuncture fluids. The data offer strong evidence that one of the principal enzymes degrading proteoglycans in the distal hypertrophic cell cartilage is lysozyme.

Reference

KUETTNER, K., SORGENTE, N., GROXEN, R., PITA, J.C., HOWELL, D.S.: Biochem. Biophys. Res. Coun. (in press).

The Formation of Bone Mineral*

A.S. POSNER, N.C. BLUMENTHAL, A.L. BOSKEY AND P.J. TANNENBAUM

I. Introduction

Synthetic calcium phosphate systems have long been used to study the mineralization process in hard tissue. It has been demonstrated that a hydrated, amorphous tricalcium phosphate (ACP) is an unstable precursor in the synthetic precipitation of hydroxyapatite (HA) (EANES and POSNER, 1965; EANES et al., 1965). By analogy with these in vitro systems and from independent x-ray diffraction and infrared studies on bone, it has been suggested that in tissue mineralization an amorphous calcium phosphate is formed first before any bone apatite is precipitated (EANES et al., 1967). As tissue mineralization proceeds some portion of the amorphous phase is stabilized (in some yet unknown manner) while the largest portion is transformed to the finely-divided apatite of bone. An example may be seen in the femur of the developing rat: at 3 days this bone contains only 33% apatite with the balance amorphous; at 80 days the femur is 63% crystalline and only 37% amorphous (TERMINE and POSNER, 1966).

The importance of the two-phase nature of bone mineral has been noted in a recent bone resorption study by TANNENBAUM et al. (1974). A study was made of the changes in the nature of bone mineral during active resorption of the medullary portion of female pigeon femurs during the egg-laying cycle. It was shown that the bone pool which contains more amorphous mineral (i.e., the most recently formed bone) is resorbed preferentially leaving the apatite-rich bone intact in this bone loss cycle. This corroborates the work of LOPEZ et al. (1970) on the spine of the eel.

This paper will describe experiments on the kinetics of conversion of ACP to HA under varying conditions. A better understanding of this process and the stabilization of ACP will add to our knowledge of bone mineral ultastructure and the process of tissue mineralization.

II. Synthetic Systems

Amorphous calcium phosphate transforms in aqueous medium to hydroxyapatite by first dissolving and then by the process of renucleation of hydroxyapatite (BOSKEY and POSNER, 1973). Fig. 1 shows the effect of changing pH on this transformation. It is clear that this reaction can be divided into three parts: (a) an induction period during which part of the ACP dissolves and the first HA nucleation takes place, (b) a proliferation period during which newly formed HA is nucleated on already present crystals and (c) a termination period during

+ This is Publication number 89 of the Laboratory of Ultrastructural Biochemistry. Work reported in this paper was supported, in part, by grants DE-01945 and AM-05414 from the National Institutes of Health.

which the rate of the appearance of new crystals decreases markedly. The pH 10 curve in Fig. 1 illustrates these divisions; the induction period (a) ends at about 3 hours, the proliferation period (b) proceeds then to about 7 hours and the remainder is the termination period (c). The rate of conversion (a) during the proliferation period can be described by: $dx / dt = k_o + k_1 a$; thus, the conversion is in this range a first order, autocatalytic reaction. By performing the ACP to HA transformation at pH 8 and at 10° C, and 48° C respectively, the activation energy for the (a) and (b) regions was calculated (BOSKEY and POSNER, 1973). It required 33 kilocalories per mole to dissolve ACP and nucleate the first HA, i.e., to complete the induction period. On the other hand, the activation energy for the proliferation period (b) was only 16.4 kilocalories per mole. Thus, energetically, once the first HA crystals are formed, the nucleation of new crystals is not as difficult as the formation of the initial HA.

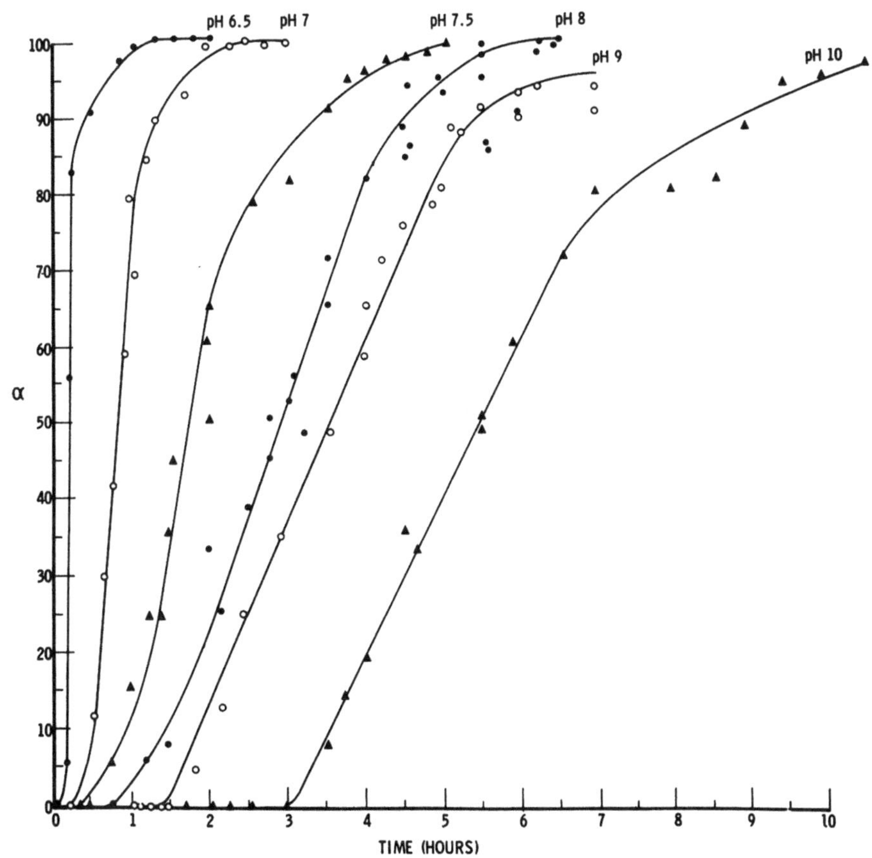

Fig. 1. Effect of pH on the transformation of ACP to HA as represented by a, extent of conversion reaction, vs. time; a was determined by the x-ray diffraction method described earlier (EANES and POSNER, 1965). The curves represent a visual best fit of the data points

III. Stabilization of ACP

It can be seen by Fig. 1 that the ACP to HA transformation rate is increased at lower pH values. It has been shown that ACP must dissolve first before it can transform to HA (BOSKEY and POSNER, 1973). It seems reasonable to assume that the increased ACP solubility accompanying the lower pH results in an increased transformation rate. The question remains, can we stabilize ACP,i.e., prevent its transformation to HA in solution, by reducing its solubility?

If present in sufficient amount, magnesium has been reported (EANES and POSNER, 1968) to stabilize ACP. A study was made on the kinetics of conversion of ACP to HA (pH 8; 26.0° C and 48.0° C) in the presence of increasing amounts of Mg. Two sets of experiments were performed where, (a) Mg-free ACP was added to solutions containing different amounts of Mg and (b) ACP precipitated in the presence of Mg was left in contact with the precipitating solution. If the Mg/Ca molar ratio in the system exceeded 0.2 no conversion was observed. Lower amounts of Mg increased the induction period bud did not affect the proliferation and termination periods. As an example, the induction period was practically zero at pH 8, 26° C, for the transformation. However, at the same temperature and pH, for a Mg/Ca molar ratio of 0.004 the induction period ended at about 30 minutes, while at a Mg/Ca of 0.04 the induction period ended at about 80 minutes. As noted, above Mg/Ca = 0.2 the induction period never ended, the ACP was stable.

The explanation for the Mg effect can be seen in the ACP solubility under varying Mg/Ca conditions. Table 1 illustrates this effect. The moles of Ca per liter of slurry present during the induction period is reduced by the presence of Mg. In the 0.20 Mg/Ca ratio (where ACP is not transformed) the Ca in solution is about the same value as that observed for HA in contact with water for 24 hours. Thus, we reduce the ACP solubility by (a) precipitating it in the presence of Mg and/or (b) by adding Mg in solution to Mg-free ACP. The exact mechanism of this reaction is unknown . It should be noted that Mg concentrations in the physiological range (Mg/Ca ratios of 0.004 to 0.04) slowed down the transformation time but did not prevent the reaction from taking place.

Table 1. Effects of Mg on the ACP solubility as measured by calcium concentration in transformation slurry

Mg/Ca molar ratio	Moles Ca/liter[a]
0.00	11.5×10^{-4}
0.008	10.9×10^{-4}
0.004	10.7×10^{-4}
0.040	9.6×10^{-4}
0.200	4.55×10^{-4}

[a] This figure represents the Ca concentration when it levels off during the induction period.

IV. HA Crystals

Magnesium did not appear to affect the growth of the final crystallites of HA. However, the presence of another ion, carbonate, although it did not seem to affect the transformation kinetics, had a marked effect on the final HA crystals. If ACP transformed in a series of carbonate solutions the final HA crystal size is diminished in high carbonate systems. Table 2 illustrates this point with an experiment where ACP is transformed in 72 hours in varying amounts of carbonate (2g ACP in 200cc H_2O with increasing amounts of Na_2CO_3, pH 7.4, Tris-HCl buffer). It is clear from Table 2 that a higher incorporation of carbonate decreases the HA crystal size. The serum level of carbonate ion is 0.26 molar. Thus, it seems reasonable to assume that bone apatite crystals are affected in some way by the presence of carbonate. In fact, bone mineral contains in the order of 4 weight percent carbonate. This is strikingly similar to the amount found in this experiment as given in Table 2.

V. Summary

The ACP-HA transformation in water is here used as a prototype system to understand bone mineral formation. The transformation can be prevented by the presence of Mg in greater quantity than a Mg/Ca molar ratio of 0.2; lower ratios can slow up the conversion. The conversion is slowed or prevented by reducing the ACP solubility. Finally, the presence of carbonate ion decreases the final HA crystal size but does not slow up conversion.

Table 2. Effect of carbonate on final HA crystal size when it is formed from ACP in solution. The crystal size is inversely proportional to the width at half maximum in degrees 2θ (Copper-Kα radiation) of the 002 x-ray diffraction peak. Thus, high carbonate resulted in smaller crystals of HA

Moles Na_2CO_3	Wt.% CO_3 in HA	X-Ray line width[a]
0.00	1.1	0.45° 2θ
0.0026	1.2	0.43
0.026	3.7	0.59
0.26	15.6	0.73

[a] Uncorrected for instrument broadening.

References

BOSKEY, A.L., POSNER, A.S.: Conversion of amorphous calcium phosphate to microcrystalline hydroxyapatite. A pH-dependent, solution-mediated, solid-solid conversion. J. Phys. Chem. 9, 2313-2317 (1973).

EANES, E.D., GILLESSEN, I.H., POSNER, A.S.: Intermediate stages in the precipitation of hydroxyapatite. Nature 208, 365-367 (1965).

EANES, E.D., POSNER, A.S.: Kinetics of conversion of non-crystalline calcium phosphate to crystalline hydroxyapatite. Trans. N.Y. Acad. Sci. 28, 233-241 (1965).

EANES, E.D., POSNER, A.S.: Intermediate phases in the basic solution preparation of alkaline earth phosphate. Calc. Tiss. Res. 2, 38-48 (1968).

EANES, E.D., TERMINE, J.D., POSNER, A.S.: Amorphous calcium phosphate in skeletal tissue. Clin. Orthoped. 53, 223-235 (1967).

LOPEZ, E., LEE, H.S., BAUD, C.A.: Etude histophysique de l'os d'un anguilla L. au cours d'une hypercalcémie provoquée per la maturation expérimentale. C.R. Acad. Sci. Paris 270, 2015-2017 (1970).

TANNENBAUM, P.J., SCHRAER, H., POSNER, A.S.: Crystalline changes in avian bone related to the reproductive cycle. II. Percent crystallinity changes. Calc. Tiss. Res. 14, 83-86 (1974).

TERMINE, J.D., POSNER, A.S.: Infrared analysis of rat bone: age dependency of amorphous and crystalline mineral fractions. Science 153, 1523-1525 (1966).

V. Biochemistry and Histochemistry of Bone Disease

Chairmen: P. J. GAILLARD and R. V. TALMAGE

Biochemistry of Bone Diseases

H. J. Dulce

I would like to give you a view of the known biochemical results concerning the most frequent bone diseases and, in conclusion to describe some biochemical kinetics which may lead to alterations in bone.

I. Osteoporosis

The osteoporosis that occurs locally or generalized is morphologically mainly considered as an atrophy, that means a disproportion of bone substance and medullar space. Age, hormones, inactivity and nutritional diseases play a part at the origin. Fig 1. demonstrates the decreased mineral concentration. The x-ray density is reduced 20 - 60% in comparison to normal bone, because the medullar space and the haversian spaces increase and cortical thickness or bone diameter decreases (NORDIN et al., 1968; OESER and KROKOWSKI, 1963; KROKOWSKI, 1964; HURXTHAL and VOSE, 1969; DOYLE, 1961). These alterations are seen in many bones (Fig. 1). In females the corticalis width is decreased more than in males and the medullar space is comparatively increased (NORDIN et al., 1968; DEQUECKER et al., 1971). The analytical density of osteoporotic bones amounts to 1,97 kg/l those of healthy bones to 1,93 kg/l (DEQUECKER et al., 1971).

Chemical analyses of bones show that in elderly patients the content of fat increases but bone tissue and water content decreases (FRERCKS, 1966). In 1966 Birkenhäger-Frenkel compared bone biopsies from the

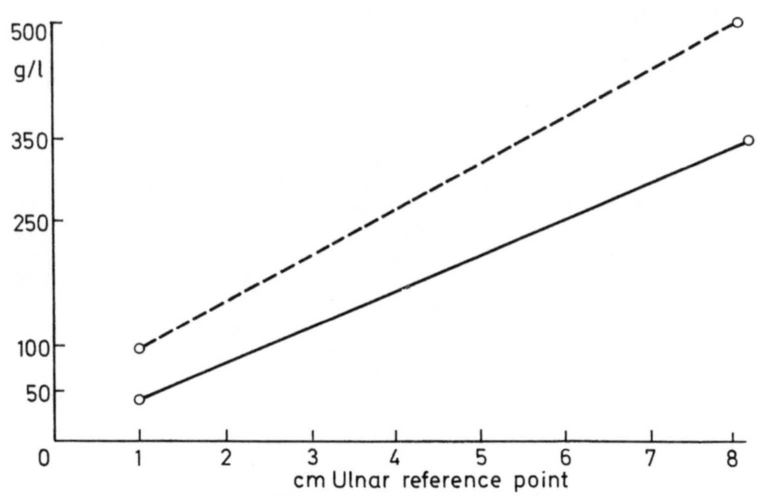

Fig. 1. Ulna mineral concentration in normal (---) and osteoporotic (——) man (DOYLE, 1961)

Table 1. % Extractable collagen from trabecular bone of 69 males and females (DEQUECKER et al., 1971)

Age	NACL	EDTA	LICL	KSCN	Residue	Collagen fat free dry weight
	1,0 M	0,5 M	5,0 M	5,0 M	%	g/100 g
20 - 39	0,37	0,57	1,62	7,48	90	22,6
50 - 59	0,33	0,38	2,06	5,48	92	23,0
70 - 89	0,21	0,41	1,19	3,65	95	24,5

iliac crest of healthy and osteoporotic people. It is possible to prove the reduction of bone mass as a result of lower mineralisation and matrix decrease, showing by the OH-proline and Ca- content. Earlier analyses of rib bones from SEIDEL et al. (1960) agree with these findings. In osteoporosis qualitative alterations apparently occur within the collagen fractions of bone. KELLY et al.(1959) found a decrease of the citrate extractable collagen of bone in osteoporotic patients. The citrate extractable collagen showed a Debye-Scherrer diagram like that of insoluble collagen. DEQUECKER et al.(1971) pointed out a small increase of bone collagen content parallel to the loss of bone mass. This behavior speaks for a demineralisation of osteoporotic bones. DEQUECKER further demonstrates that the soluble collagen fractions decrease when osteoporosis occurs(Table 1). We see the greatest reduction of soluble collagen during extraction with LiCl and KCNS solutions which split inter-and intra-molecular collagen linking. Therefore one may conclude that osteoporotic bones have more insoluble collagen with covalent linking than healthy bones. Concentrations of Ca, Mg, P, alk. phosphatase, protein and citrate in the blood plasma do not alter in most cases of osteoporosis (RIGGS, 1973). Only in cases of postmenopausal osteoporosis and acute Sudeck atrophy are small P-increases in the plasma noted. Furthermore, in Cushing osteoporosis, postmenopausal osteoporosis, and disuse osteoporosis, small increases of alk. phosphatase in the plasma occur (HAAS, 1966; MUSSGNUG, 1960).

While given a standard diet containing 700 - 800 mg Ca/d, older patients with osteoporosis excrete less Ca and hydroxyproline than normals (SALEH and COENEGRACHT, 1968). On the other hand we know single cases of osteoporosis with hypercalcuria (HAAS, 1966). Higher Ca-intake leads to a lesser hydroxyproline excretion in urine (NORDIN et al., 1965). These results speak for an altered turnover of calcium and collagen tissue in osteoporosis. For this reason many investigators have studied the turnover of calcium in patients with osteoporosis. NORDIN (1961) found out by an extensive study that the calcium intake with food was less in osteoporotic than in normal patients. The urine Ca-excretion in osteoporosis was independent of the calcium intake, the urine P-excretion was dependent. For this reason a calcium-depleted diet does not lower the calcium excretion in urine in osteoporotic patients (BHANDARKAR and NORDIN, 1962). Numerous examinations prove that in osteoporosis the calcium absorption through the intestine is defect. Osteoporosis has a high fecal and a low urinary ^{45}Ca-excretion. Normally the Ca-absorption decreases slowly after the 70th year of life (CANIGGIA et al., 1965). In senile osteoporosis this deficiency can be seen earlier and more intensively (CANIGGIA et al., 1963). If one increases the Ca-intake to 0,5 - 1,0 mMol/kg/d the calcium absorption is raised and the balance turns positive (HARRISON et al., 1961; SPENCER et al., 1964). But the calcium absorption does not increase in proportion to the intake as in healthy people. Only vitamin D supply increases the Ca-absorption in osteoporosis noticeably (SCHWARTZ et al.,

Table 2. Mean values of Ca-dynamics in different osteoporoses (CANIGGIA et al., 1964; HEANEY, 1962; BRONNER et al., 1963).

	Ca Intake mg/d	Ca Urine mg/d	Ca FAEC mg/d	BAL mg/d	Intest Ca-Absorpt. % Intake	Ca Secr. mg/d	Ca Pool mg/kg	Accret. Rate mg/kg/d	Resorpt. Rate mg/kg/d	Pool Turnover-R. mg/h
Postgastr. (7) Osteoporosis	1064	1029	99	<u>-64</u>	<u>14</u>	115	221	<u>40</u>	<u>41</u>	91
Normals (3)	940	725	172	+43	43	110	91 g/m^2	12,5 mg/m^2/d	12 mg/m^2/d	47 mg/m^2/d
Disuse (5) Osteoporosis	505	319	578	<u>-392</u>	<u>17</u>	157	3,62	<u>478</u>	<u>742</u>	794
Normals (4)	708	130	584	-6	38	142	3,28 g	310 mg/d	311 mg/d	471 mg/d
Postmenop. (5) Osteoporosis	680	156	678	<u>-154</u>	<u>16</u>	106	2,6	<u>375</u>	<u>528</u>	636
Normals (2)	669	189	527	-47	31	64	4,19	675	721	927

1968). Calcium balances are found mostly negative in untreated osteoporotic patients (HEANEY et al., 1958; HEANEY, 1962; BRONNER et al., 1963; CANIGGIA et al., 1964; TORO et al., 1958) (Table 2). In proportion to more negative Ca-balances the hydroxyproline excretion increases (NORDIN et al., 1965). That means that together with the demineralisation, precollagen is more metabolised. Calcitonin lowers the Ca-balance in osteoporosis because a hypercalcuria occurs (SLUYS-VEER et al, 1970). Measurements of all single steps of Ca-dynamics in osteoporotic patients show corresponding results. The Ca-balance is negative as a result of high urine or fecal excretion, the intestinal Ca-absorption is lowered, the bone resorption is higher than the bone accretion. The turnover of the Ca-pool seems to be low in postmenopausal osteoporosis (BRONNER et al., 1963) and high in postgastrectomy and disuse osteoporosis (CANIGGIA et al., 1964, HEANEY, 1962)(Table 2). By morphometry RIGGS et al. (1969) found in osteoporosis a bone resorption of 15,3% of total surface in comparison to a bone accretion of 4,4%. ^{45}Ca studies of LAFFERTY et al. (1964) agree with these results. With rising bone resorption rate the hydroxyproline excretion in urine increases. Probably there is no connection in osteoporosis between increased bone resorption and higher secretion of parathyroid hormone (RIGGS et al., 1973). The bone resorption might be based on an increased sensibility of bone cells to parathyroid hormone. The bone formation rate amounts to 0,03%/d in osteoporosis and to 0,09%/d in normal patients (BAUER et al., 1957). After infusion of Ca-gluconate solutions to osteoporotic patients the Ca-retention is partly higher and partly lower than normal (HAAS, 1966; v. BUCHEM, 1959). P-balances are not systematically changed (LAFFERTY et al., 1964; SPENCER et al., 1964). Mg-balances seem to be more positive than in healthy patients (HARRISON et al., 1961). The demonstrated results show that the osteoporosis of human beings is marked by four characteristics:

1. Deficient calcium absorption through the intestine.
2. Increased calcium resorption in the bone.
3. Increased and continuous calcium excretion in urine.
4. Deficient formation of soluble collagen in bone due to increased catabolism of hydroxyprolinepeptidase.

The first three points cause the negative Ca-balance.

A calcium-deficient diet effects an osteoporosis in rats, mice, dogs and cats which may serve as metabolic models (SALOMON and VOLPIN, 1970; GERSHON et al., 1964; CAMPBELL and DOUGLAS, 1965).

The results of the animal experiments correspond for the most part with observations in osteoporotic patients. The main difference is the secondary hyperparathyroidism in animals which is not yet proved in human beings.

II. Osteogenesis Imperfecta

The osteogenesis imperfecta is often called the inborn form of osteoporosis. According to chemical results this assumption is only partly correct. Histologically we found a typical osteoporosis with a lack of osteoblasts similar to osteoporosis in C-avitaminosis and copper deficiency. Contrary to osteoporosis the formation of insoluble matured collagen is defect. A so-called reticular collagen is formed (BETHKE, 1962). The hydroxyproline-protein of the plasma amounts to 21,7 mg/l in osteogenesis imperfecta and to only 9,3 mg/l in normals (LANGNESS and BEHNKE, 1971). At the same time we see an increased urinary hydroxyproline excretion (CANIGGIA and GENNARI, 1972).

There is a clear increase of neutral sugars and a_2-macroglobulines in the plasma (BETHKE, 1962). Neutral sugars are an essential part of reticular fibrils; a_2-macroglobulines are signs of a defect connective tissue degradation and the cause of the high blood sedimentation rate found in osteogenesis imperfecta. The Ca, P and Mg-content of the plasma is mostly normal. In some cases increased plasma-Ca is observed (CANIGGIA and GENNARI, 1972). Plasma alkaline phosphatase is more often found increased (HAAS, 1966; KEATS and ANAST, 1960; ZEITOUN, 1963; BETHKE, 1962; KUNZE, 1960; CANIGGIA and GENNARI, 1972). In the periost of bones from osteogenesis imperfecta no alkaline phosphatase was found (HANSEN, 1934). Succinodehydrogenase is missing also in these bones, similar to C-avitaminosis and copper deficiency. Calcitonin treatment lowers the phosphatase activity and the Ca-content of the plasma and the urinary Ca- and hydroxyproline excretion in these cases of osteogenesis imperfecta (CANIGGIA and GENNARI, 1972).

III. Rickets and Osteomalacia

Rickets and osteomalacia both have a demineralized bone matrix. The diseases might be vitamin D-sensitive or vitamin D-resistant (WALKER and ARWIDSSON, 1954; WILLENBOCKEL, 1969). There remains a matrix which is to a great extent able to store parenterally applied calcium as mineral. All investigators found in vitamin D-sensitive osteomalacia a calcium retention of 63 - 95% after calcium infusion in vitamin D-sensitive osteomalacia in comparison to normal 50 - 60% (v. BUCHEM, 1959; CLERKIN et al., 1964; FOURMAN and HAAPANEN, 1964; HAAS, 1966). This behavior is typical of osteomalacia in contrast to osteoporosis which often shows a normal or decreased calcium retention after Ca-infusion. The strong affinity of bones from osteomalacia to calcium is further proven by the high incorporation of ^{90}Sr (HEUCK, 1963).

In vitamin D-resistant osteomalacia or rickets the Ca-retention is mostly normal or only slightly varied (HAAS, 1966). Bone accretion and bone resorption rates are proportionally strongly increased in vitamin D-resistant rickets. In vitamin D-sensitive rickets and osteomalacia the calcium and phosphorus absorption in the intestine is lowered, and the Ca- and P-balances becomes negative (WILLENBOCKEL, 1969; CANIGGIA et al., 1968; STANBURY, 1962). After oral administration of $^{32}PO_4$ one finds low plasma activities compared to normals, proving the lack of absorption. Vitamin D compensates this absorption defect (JACKSON, 1962). In vitamin D-resistant rickets the calcium absorption in the intestine seems to be normal and the balances are mostly compensated (LAFFERTY et al., 1963). In inherited rickets there exists a defect of the phosphate transport through the intestine (SHORT et al., 1973). In all different forms of rickets and osteomalacia we find in the plasma high alkaline phosphatase activities and low levels of phosphorus and partly of calcium (GOSSMANN, 1957; HENNEMANN et al., 1962; LAFFERTY et al., 1963; FOURMAN and HAAPANEN, 1964; HAUGE, 1956; NITSCHKE and GIEGLER, 1961; BAIRD and OLESKY, 1957; HOSOKAWA et al., 1964; BORDIER et al., 1969; ROSE, 1956; HAAS, 1966; v. BUCHEM, 1959; JACKSON et al., 1958; JACKSON, 1962; GOUGH et al., 1964).

In many cases there results a low Ca x P product with the tendency to a physico-chemical dissolution of bone mineral or to inhibition of mineralization. The increase of plasma alkaline phosphatase is most likely based on bone phosphatase (WILHELM, 1958), as a lot of this phosphatase is to be found in rachitic matrix. Plasmaprotein content is usually normal. In rickets and osteomalacia the urinary calcium excretion during Ca-deficient diet is normal or slightly decreased (HAAS, 1966). The urinary Ca-excretion increases only in renal tubular defects. The citrate excretion is lowered (WILLENBOCKEL, 1969). The tubular phosphate reabsorption is distinctly decreased and increases when vitamin D is given. In osteomalacia the tubular phosphate reabsorption amounts to about 77,7% in contrast to normals with about 87,8% (BORDIER et al., 1969). In vitamin D-resistant rickets JACKSON et al. (1958) found the T_mP of 23 mg/l in contrast to normal 41 mg/l. In rickets and osteomalacia we see the high hydroxyproline excretion that speaks for a high collagen turnover. Rickets of rats and chicken developing by a vitamin D-deficient diet is like a defect of bone mineralization and a defect of collagen maturation with high collagen turnover and without loss of matrix in human beings. Therefore these experiments represent models for clearing up the mode of action of vitamin D.

IV. Primary Hyperparathyroidism

In hyperparathyroidism morphologically we have an osteoclastosis with demineralization, matrix catabolism and bone formation processes at the same time (HAAS, 1966). Ca-retention after infusion of Ca-salts is mostly increased only in strong hypercalcuria; low values will also be observed. The Ca-infusion effects a decrease of the parathyroid hormone content of the plasma (COE et al., 1973) and the high urinary hydroxyproline excretion (NORDIN et al., 1965a). The calcium absorption through the intestine, the bone accretion and bone resorption are very high in hyperparathyroidism (LAFFERTY and PEARSON, 1963a; REINER et al., 1970) (Table 3). There is a clear trend towards a lower Ca-balance. The renal phosphate clearance is increased. Investigators observed values between 16 and 47 ml/min (ROCKNEY et al., 1959; REYNOLDS et al., 1960; HORN et al., 1961; LAFFERTY and PEARSON, 1963a). The increased renal phosphate excretion is caused by a low phosphate reabsorption in the tubules (HORN et al., 1964; BORM, 1961). The high phosphate clearance is mostly accompanied by a hypercalcuria (HAAS, 1966; LAFFERTY and PEARSON, 1963a). A high urinary hydroxyproline excretion is a reference to the increased turnover of the collagen matrix. In hyperparathyroidism values between 55 and 236 mg/d were observed. Normal values are between 20 and 30 mg/d (CERDA et al., 1970; WOESSNER, 1968; GOIDANACH et al., 1965). The plasma calcium increases in hyperparathyroidism especially the ionized Ca (LLOYD and ROSE, 1958; ROCKNEY et al., 1959; TEMPLETON et al., 1962; GRÜNDIG et al., 1970; REYNOLDS et al., 1960; BORM, 1961; BELLABARBA et al., 1963; KUHLENCORDT and LOZANO-TONKIN, 1964; LAFFERTY and PEARSON, 1963a). Combined with the increase of Ca one finds, as a rule, a decrease of anorganic phosphorus and the Ca x P

Table 3. Ca-dynamics in hyperparathyroidism (LAFFERTY and PEARSON, 1963)

	Ca	Endog	% Ca	Ca	P	Bone	Bone
	Intake mg/d	Fecal Ca mg/d	Absorpt.	Bal. mg/d	Bal. mg/d	Accret. Rate mg/d/kg	Resorpt. Rate mg/d/kg
Hyperparathyroidism	789	121	51	+ 38	+ 88	16,5	16,8
Normals	822	160	38	+ 59	+ 30	8,5	8,2

product. The plasma magnesium level is mostly below 0,5 mMol/l (ELIEL et al., 1969). The alkaline phosphatase of the plasma increases in correlation to the urinary hydroxyproline excretion(CERDA et al., 1970; DENT, et al., 1961; BELLABARBA et al., 1963; HAAS, 1966; GRUENDIG, 1970; NAEGELE, 1955; LLOYD and ROSE, 1958). The alkaline phosphatase probably comes from bone cells. Increases of acid phosphatase are partly observed (GRUENDIG et al., 1970; NAEGELE, 1955). This enzyme augmentation probably originates from the hypertrophic parathyroid tissue. Furthermore in hyperparathyroidism plasma citrate contents of about 51 mg/l are to be seen three times as high as in healthy people (KOMARKOVA et al., 1959). This citrate originates from mobilized bone tissue. All alterations of bones and calcium dynamics in hyperparathyroidism may be caused by parathyroid hormone.

V. Renal Osteodystrophy

Renal osteodystrophy appears in cases of chronic nephrophathy with renal acidosis. The bones show signs of osteomalacia, osteoclasia and osteosclerosis. Renal osteodystrophy is only influenced by high doses of vitamin D. Bones are demineralized, their ash content is lowered from about 30% in normals to about 25% (HEUCK, 1963). Bone mineral of uraemic patients is altered (KAYE et al., 1970). We see a high F-content and a lower CO_3-content. The solubility of bone mineral is diminished as seen by a lower Ca-release. The ^{90}Sr-incorporation into bones of renal osteodystrophy amounts to about 2 $\mu C\,^{90}Sr/g$ Ca. Normal bones have an incorporation of about only 0,8 $\mu C\,^{90}Sr/g$ Ca (HEUCK, 1963). As in osteomalacia we find an increased Ca-retention after i.v. Ca-infusion (HAAS, 1966). The chronic acidosis causes a quick sodium loss from bones of about 10 - 25%. Ca-balances are mostly negative. As in rickets and osteomalacia an increase of plasma alkaline phosphatase was observed and a decrease of plasma Ca, but in contrast to osteomalacia an increase of plasma phosphorus so that the Ca x P product remained normal (KAYE et al., 1970; WILLENBOCKEL, 1969; HAAS, 1966). The phosphorus increase is caused by the renal excretion defect, the Ca-decrease is caused by the impaired intestinal Ca-absorption (HAAS, 1966). A direct correlation between creatinine clearance and intestinal Ca-absorption is not proved. Apparently, in renal osteodystrophy the influence of vitamin D and its effective metabolites on the Ca-absorption through the intestine is partly excluded so that parathyroid hormone is ineffective, too. With alkalisation, the acidosis and the demineralization get better, but the intestinal Ca-absorption is not raised. This is only effected by high vitamin D doses following increases of plasma Ca and Ca x P product so that in some cases heterotopic mineralizations occur. For the development of renal osteodystrophy the parathyroids are essential (HAAS, 1966). Acidosis or the so-called pseudohypoparathyroidism leads to a secondary hyperparathyroidism with hyperplasy of parathyroids, increase of plasma parathyroid hormone level, and bone resorption without plasma Ca increase.

VI. Pseudohypoparathyroidism

Pseudohypoparathyroidism is a hereditary disease but is as well to be found as a symptom of other bone diseases. Pseudohypoparathyroidism is mostly accompanied by a secondary hyperparathyroidism as discussed in osteomalacia, osteodystrophy and osteoporosis. Pseudohypoparathyroidism in contrast to hypoparathyroidism after parathyroidectomy does not react upon parathyroid hormone. Morphologically the bones often show an osteoclasia. Intestinal Ca-absorption and urinary Ca-excretion are low, renal phosphate reabsorption is high (BELL et al., 1963; SCHWARZ, 1964). Ca-balance is negative, plasma Ca is decreased, plasma phosphate, the Ca x P product, and the alkaline phosphatase activity are increased. In some cases heterotrophic mineralizations will be observed.

VII. Paget's Disease

Paget's disease is morphologically characterized by an extensive bone resorption combined with an unarranged secondary bone formation. In total we have a quicker turnover of bone tissue (HAAS, 1966; HEANEY and WHEDON, 1958). Bone formation rate amounts to 100 - 150 mg/kg/d, as opposed to 9 mg/kg/d in healthy people. There exists a negative Ca-balance, an increased Ca-retention, and a three times increased exchangeable Ca-pool (HAAS, 1966). Bone alkaline phosphatase is high in Paget's disease, acid phosphatase is unchanged (WOODWARD, 1959; HAAS, 1966; MARTIN et al., 1964; HIRSCH, 1960; JESSERER, 1959; CERDA et al., 1970; SÜSSMANN, 1970). Plasma alkaline phosphatase is also increased . (WOODARD, 1959). The plasma enzyme originates to an extent of 90% from the liver as SÜSSMANN (1970) was able to prove with the specific antibody test. In many patients plasma Ca level is high (MARTIN et al., 1964; V.D. BIJVOET et al., 1968; HAAS, 1966; HADDA and CALDWELL, 1972). Plasma phosphorus is mostly normal or slightly decreased. By the augmented urinary Ca-excretion, the Ca-balances become more negative (HAAS, 1966; MARTIN et al.,1964). The high hydroxyproline content of plasma and urine reflects a high collagen turnover V.D. BIJVOET et al., 1968; GOIDANACH et al., 1965; WOESSNER, 1968). In urine values up to about 350 mg/d are found (CERDA et al., 1970; KRANE et al., 1973). Calcitonin treatment of Paget's disease is able to lower plasma and urinary hydroxyproline values and the plasma Ca and phosphatase activity (HADDAD and CALDWELL, 1972; BIJVOET et al., 1968). Especially the urinary hydroxyproline polypeptide fraction is concerned which is a proportion of bone collagen synthesis (KRANE 1973). On the strength of biochemical results m. Paget belongs to the group of hyperparathyroidism.

VIII. Hypoparathyroidism and Osteopetrosis

Hypoparathyroidism and osteopetrosis are today partly interpreted as hyperthyrocalcitonism. In the medullar space of bone there occurs an intensive mineralization that is uninfluenceable by parathyroid hormone in cases of osteopetrosis and may be redeveloped in cases of hypoparathyroidism (FRASER et al., 1968). Bone formation dominates over bone resorption. The medullar bone mineralization is accompanied by a matrix alteration. The collagen part decreases at the cost of the m u coproteid part (VEJLENS, 1972) as shown by the movements of the hexosamine and OH-proline contents. The Ca-retention after infusion of Ca-solutions is very high in hypoparathyroidism. HAAS (1966) found amounts of 81 - 94%. The bone formation rate was about 31 mg Ca/kg/d compared to about 9 mg Ca/kg/d in normals (HEANEY and WHEDON, 1958). Ca-balance is positive, intestinal Ca-absorption is raised to 73%, urinary Ca-excretion and bone turnover are strongly decreased (DENT et al., 1965; HASSENHÜTTL, 1962; HAAS, 1966; FRASER et al., 1968).

Plasma Ca and phosphate levels are normal or slightly decreased. Only in hypoparathyroidism plasma is phosphate often increased (DENT et al., 1965; 1966). In both diseases plasma alkaline phosphatase is mostly high. In mice we find an inborn osteopetrosis with a defect bone resorption and plasma alterations as in human beings (MURPHY, 1969, 1970; WALKER, 1966).

The most important biochemical findings in bone diseases distinguish four groups of bone diseases

1. bone resorption with low turnover, negative Ca-balance, matrix defect and deficient intestinal Ca-absorption (osteoporosis, osteogenesis imperfecta).
2. demineralization with high turnover, negative Ca-balance, collagen maturation defect and deficient intestinal Ca-absorption (osteomalacia, rickets).
3. bone resorption with high turnover, negative Ca-balance, matrix resorption and different intestinal Ca-absorption (hyperparathyroidism, renal osteodystrophy, m. Paget).
4. hypermineralisation with low turnover, positive Ca-balance, matrix defect and high intestinal Ca-absorption (hypoparathyroidism, osteopetrosis).

In the third group you would further find osteoclastic tumors. The main metabolic mechanisms that may lead to bone diseases are:

a) mineral resorption
b) matrix resorption
c) mineral formation
d) matrix formation

A mineral resorption may result when the Ca x P product of the extracellular fluid decreases by about one third or a local or generalized acidosis occurs (DULCE, 1970). An acidosis transfers trivalent phosphate ions to bivalent phosphate ions. By this way acidosis dissolves the apatite lattice. Osteoclasts and osteocytes may probably produce H^+ions and may concentrate Ca and PO_4 cellularly to about 200 mMol/kg (NICHOLS, 1970). It is supposed that membrane bound alkaline phosphatase and adenylate cyclase participates in the Ca and PO_4 influx (HEKKELMAN, 1970; NICHOLS, 1970). Parathyroid hormone activates the H^+ion formation of osteoclasts and osteocytes and the Ca and PO_4influx. The aerobic glycolysis with lactate concentration, the H^+ion secretion and lysosomal acid proteases increase (VAES, 1970). Calcitonin inhibits these reactions. Probably carboanhydrase is engaged in the H^+ion secretion (DULCE, 1970)(Fig. 2). We suppose that a parathyroid hormone dependent carboanhydrase partly eliminates the buffer effects of carbonic acid on lactic acid. Therefore a H^+ion efflux appears that inhibits the mineralization and dissolves bone mineral. Under normal cellular conditions H^+ions will be buffered and appear in the cell water. This hypothesis will be supported by therapeutical investigations with carboanhydrase inhibitors (Table 4). HARRISON et al. (1961) demonstrates in patients that the bone Ca-resorption was inhibited by hydrochlorothiazid combined with an increased Ca-balance. A Ca-mobilisation by citrate will be declined today on quantitative grounds. Other investigations of HOLKE (1973) show that parathyroid hormone activates the membrane-bound adenylate cyclase followed by an increase of the intracellular content of 3,5 cyclic AMP and an activation of a protease that phosphorylized carboanhydrase to an active enzyme. An increase of endogenous 3,5 AMP and an osteoclastic bone resorption is observed by dibuturyl 3,5 AMP application in rat calvaria and mice radius (VAES, 1968; HEERSCHE et al., 1971; HERRMANN-ERLEE, 1970). In contrast to in vitro cyclic 3,5 AMP did not bring about mineral resorption (TASHIJAN et al., 1972) but parathyroid hormone and prostaglandine E_2 had strongest effects. Perhaps parathyroid hormone effects the morphological differentiation from immature bone cells to osteoclasts by alterating

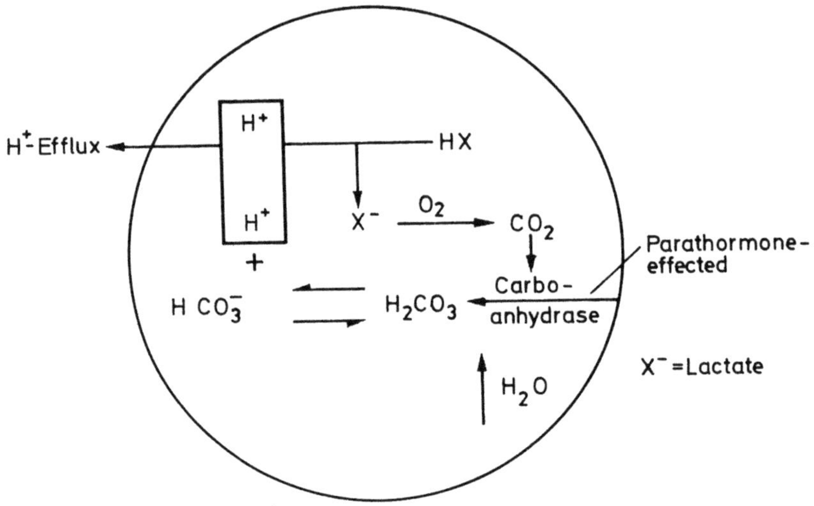

Fig. 2. H^+-Secretion in osteoclasts

Table 4. Inhibition of bone-Ca-resorption by hydrochlorothiazid in osteoporosis (HARRISON and HITCHMAN, 1971)

	Ca Accretion mg/d	Ca Resorption mg/d	Ca Balance mg/d
I Control O	760	1.178	- 418
9m 50 g Thiazid/d	255	258	- 3
27m Thiazid/d	380	220	+ 160
II Control O	470	443	+ 27
7,5m 50 g Thiazid/d	300	0	+ 300

the cell metabolism to acid secretion and proteolysis. The matrix resorption follows together with or after a demineralization probably by a parathyroid hormone-dependent bone collagenase and lysosomal proteases which split collagen and release hydroxyproline peptides (VAES, 1966, 1968; WALKER et al., 1964; WOOD and NICHOLS, 1965; REYNOLDS, 1970; HERRMANN-ERLEE, 1970).

MEARS (1969) and MARTIN et al. (1965) have results which speak for a stimulation of the RNA synthesis in osteoclasts by parathyroid hormone followed by a synthesis of osteolytic enzymes. Osteoclastic bone diseases have a strong matrix resorption. The turnover of soluble immature bone collagen amounts to about four days, that of insoluble mature collagen to about 40 days (GERBER, 1960). The kinetics of mineral formation and matrix formation were discussed yesterday by Dr. POSNER, Dr. BACHRA, Dr. MECHANIC and others. We believe that an increasing cellular CO_2-formation combined with an inhibition of the carboanhydrase or an increasing alkalinity might promote a bone mineralization. Calcitonin increases the calcium secretion of bone cells (VAES, 1970; REYNOLDS, 1970; NICHOLS, 1970) probably calcitonin activates mineralization by inhibition of the cell metabolism leading to osteoclasia.

Bone collagen possesses very good nucleation abilities. In bone diseases with hypermineralization these reactions are assumed to be very active. Matrix formation in bone cells is a special protein synthesis within the ribosomes. Parathyroid hormone prostaglandine E_1 and cyclic AMP which stimulate the adenylcyclase of bone cells activate the influx of glyzine, proline and hydroxyproline into bone cells. An increased collagen synthesis might be thereby possible (PHANG and DOWNING, 1973) like the increase caused by somatotrope hormone (VAES and NICHOLS 1962a). In contrast to earlier investigations it shows that parathyroid hormone like glucocorticoids inhibit collagen synthesis (VAES and NICHOLS, 1962; JOHNSTON et al., 1962; SMITH and ARMSTRONG, 1961; FLANAGAN, 1964a). Calcitonin inhibits the degradation of insoluble mature collagen by parathyroid hormone (HEERSCHE, 1968). A bone matrix with a high content of acid mucopolysaccharides is difficult to mineralize (DULCE, 1960). We know some defects of matrix formation which lead to bone diseases. In vitamin D-avitaminosis, defect collagen is formed because proline is not hydroxylated. In osteogenesis imperfecta, experimental lathyrism and copper deficiency, more defect tropocollagen with altered numbers of cross-linkings and less insoluble mature collagen is formed (SIEGEL and MARTIN, 1970) because of the inactive lysyloxydase.

Today biochemical results are able to discriminate bone diseases.
Today biochemical results are able to recognize the pathogenesis of bone diseases.
Today biochemical results are able to develop new views in the therapy of bone diseases.

References

BAIRD, I.M., OLEESKY,S.: Gastroenterology 33, 284 (1957).
BAUER, G.C., CARLSSON, A., LINDQUIST, B.: Acta med. scand. 158, 139 (1957).
BELL, N.H., GERARD, E.S., BARTTER, F.C.: J. clin. Endocr. 23, 759 (1963).
BELLABARBA, U., KANIG, K., KOPTAGEL, G.: Klin. Wschr. 41, 789 (1963).
BETHKE, J.: Klin. Wschr. 40, 558-567 (1962).
BHANDARKAR, S.D., NORDIN,B.: Brit. med. J. 145 (1962).
BIJVOET, O.v.d., SLUYS VEER, J., JANSEN,A.: Calc. tiss. Res. 2, 22 (1968).
BIRKENHAEGER-FRENKEL, D.H.: 4th Europ. Symp. on Calcified Tissues 1966. Leiden.
BORDIER, Th., HIOCO, D., ROUQUIER, M., HEPNER, G., THOMPSON, G.: Calc. tiss. Res. 4, 78 (1969).
BORM, D.: Dtsch. med. Wschr. 86, 1541 (1961).
BRONNER, F., RICHELLE, L., SAVILLE, P., NICHOLAS. J.: J. clin. Invest. 42, 898 (1963).
BRONNER, F.: Amer. J. Physiol. 198, 605 (1960).
BUCHEM, F.S. van: Brit. med. J. 933, (1959).
CAMPBELL, J.R., DOUGLAS, T.: Brit. J. Nutr. 19, 339 (1965).
CANIGGIA, A., GENNARI, C.: Calc. tiss. res. 9, 243 (1972).
CANIGGIA, A., GENNARI, C., BENCINI, M., PALAZZUOLI, V.: Calc. tiss. res. 2, 299 (1968).
CANIGGIA, A., GENNARI, C., CESARI, L., ROMANO, S.: Gerontologia 10, 193 (1965).
CANIGGIA, A., GENNARI, C., BIANCHI, V., GUIDERI, R.: Acta med. Scand. 173, 613 (1963).
CANIGGIA, A., GENNARI, C., CESARI, L.: Acta. med. Scand. 176, 599 (1964).
CERDA, J., TOSKES, P., SHOPA, N.: Clin. chim. acta 27, 437 (1970).
CLERKIN, E.P., HAAS, H., MINTZ, D., MELONI, R., CANARY, J.: Metabolism 13, 161 (1964).
COE, F.L., CANTERBURY, J., FIRPO, J., REISS, E.: J. clin. Invest. 52, 222 (1973).

DEQUECKER, J., MERLEVEDE, W.: Bioch. biophys. acta 244, 410 (1971a).
DEQUECKER, J., REMANS, J., FRANSSEN, R., VAES, J.: Calc. tiss. Res. 7, 23 (1971).
DENT, C.E., FRIEDMAN, M.: Brit. med. J. 1676 (1964).
DENT, C.E., SMELLIE, J., WATSON, L.: Arch. Dis. Childhood 40, 7 (1965).
DENT, C.E., HARTLAND, B.V., HICKS, J., SYKES, E.D.: Lancet II, 336 (1961).
DOYLE, F.H.: Brit. J. Radiol. 34, 698 (1961).
DULCE, H.-J.: Hdb. der medizin. Radiologie, Bd. IV/1. Berlin, Heidelberg, New York: Springer 1970.
DULCE, H.-J.: Z. physiol. Chem. 319, 257 (1960).
ELIEL, L., SMITH, W., CHANES, R., HAWRYLKA, J.: Am. N.Y. Acad. Sci. 162, 810 (1969).
FLANAGAN, B., NICHOLS, G.: Endocrinology 74, 180 (1964a).
FOURMAN, P., HAAPANEN, E.: Schw. med. Wschr. 94, 886 (1964).
FRASER, D., KOCH, S., CHAN, A., CHERIAN, A.: Calc. tiss. Res. 252 (1968).
FRERCKS, E.: Dissertation Kiel, 1966.
GERBER, G.: J. biol. Chem. 235, 2653 (1960).
GERSHON-COHEN, J., JOWSEY, J.: Metabolism 13, 221 (1964).
GOIDANACH, I., LENZI, I., SILVA, E.: Clin. chim. acta. 11, 35 (1965).
GOSSMANN, H.H.: Die Medizinische 443 (1957).
GOUGH, K.R., LLOYD, D.C., WILLS, M.: Lancet. 1261 (1964).
GRUENDIG, E., CZITOBER, H., MARUNA, R.: Clin. chim. acta. 30, 33 (1970).
HAAS, H.: Knochenstoffwechsel und Parathyreoideaerkrankungen. Stuttgart: Thieme 1966.
HAAS, H., DAMBACHER, M.: Calc. tiss. Res. 2, 21 (1968) Suppl.
HADDAD, J., CALDWELL, J.: J. clin. Invest. 51, 3133 (1972).
HANSEN, A.: Proc. soc. Exp. Biol. 31, 1023 (1934).
HARRISON, M., FRASER, R., MÜLLAN, B.: Lancet I, 1015 (1961).
HASSENHÜTTL, K.: J. bone Joint Surg. 44A, 359 (1962).
HAUGE, B.N.: Acta. med. Scand. 153, 271 (1956).
HEANEY, R.P., WHEDON, G.D.: J. clin. Endocr. Metab. 18, 1246 (1958).
HEANEY, R.P.: Am. J. Med. 33, 188 (1962).
HEERSCHE, J., FEDAK, S.A., AURBACH, G.D.: J. biol. Chem. 246, 6770 (1971).
HEERSCHE, J.: Calc. tiss. Res. 2, 80 (1968) Suppl.
HEIKKINNEN, E., KÜLONEN, E.: Experientia 20, 310 (1964).
HEKKELMAN, J.: Calc. tiss. Res. 4, 73 (1970.
HENNEMANN, P.H., DEMPSEY, E., CARROLL, E., HENNEMANN, D.: Metabolism 11, 103 (1962).
HARRISON, J., HITCHMAN, A.J., FINLAY, J.M., FRASER, D., YENDT, E., BAYLEY, T.A., MCNEILL, K.G.: Metabolism 20, 1107 (1971).
HERRMANN-ERLEE, M.: Calc. tiss. Res. 4, 70 (1970).
HEUCK, F.: Radiologia Austria XIV 1, 29 (1963).
HIRSCH, W.: Med. Welt 2456 (1960).
HOLKE, M.: Dissertation. Berlin: 1973.
HORN, H.D., GROTE, G., NIETH, H.: Klin. Wschr. 42, 991 (1964).
HOSOKAWA, O., YOSHIKAWA, S., KAWABATA, M., HATSUYAMA, Y., FUJISTA, T.: J. bone Jout. Surg. 46A, 998 (1964).
HURXTHAL, L., VOSE, G.P.: Calc. tiss. Res. 4, 245 (1969).
JACKSON, W.P., DOWDLE, E., LINDER, G.C.: Brit. med. J. 1269 (1958).
JACKSON, W.P.: J. clin. Endocr. Met. 22, 195 (1962).
JESSERER, H.: Med. Klinik 54, 2151 (1959).
JOHNSTON, C.C., DEISS, W.P., MINER, E.B.: J. biol. Chem. 237, 3560 (1962).
KAYE, M., FRUEH, A., SILVERMAN, M.: J. clin. Invest. 49, 442 (1970).
KEATS, T., ANAST, C.: Am. J. Röntgenol. 84, 492 (1960).
KELLY, M., LITTLE, K., COURTS, A.: Lancet II, 1125 (1959).
KOMARKOVA, A., PACOVSKY, V., VOSTAL, J., BLEHA, O., VITKOVA, E.: Z. ges. Inn. Med. 14, 778 (1959).

KRANE, St., HARRIS, M., SINGER, F.: Metabolism 22, 51 (1973).
KROKOWSKI, E.: Fortschr. Röntgenstr. 100, 359 (1964).
KUHLENCORDT, F., LOZANO-TONKIN, C.: Internist 5, 197 (1964).
KUNZE, D.: Ztrbt. Chirurgie 85, 1541 (1960).
LAFFERTY, F.W., PEARSON, O.H.: J. clin. Endoc. Met. 23, 891 (1963a).
LAFFERTY, F.W., HERNDON, C., PEARSON, O.H.: J. clin. Endocr. Met. 23, 903 (1963).
LAFFERTY, F.W., SPENZER, G.E., PEARSON, O.H.: Am. J. Med. 36, 514 (1964).
LANGNESS, U., BEHNKE, H.: Metabolism 20, 456 (1971).
LLOYD, H.M., ROSE, G.A.: Lancet II, 1258 (1958).
MARTIN, M.M., BARR, A., HOWE, J.: Arch. int. Med. 114, 482 (1964).
MARTIN, G.R., MECCA, C.H., SCHIFFMANN, E., GOLDHABER, P.: in Parathyroid Glands. P.J. Gailland, R. Talmage, A.M. Budy. Chikago: University Press 1965.
MEARS, D.: Exp. all. Res. 58, 427 (1969).
MURPHY, H.: Calc. tiss. Res. 3, 176 (1969).
MURPHY, H.: Calc. tiss. Res. 4, 152 (1970).
MUSSGNUG, G.: Medizin. Welt 1468 (1960).
NAEGELE, E.: Dtsch. med. Wschr. 80, 1400 (1955).
NICHOLS, G.: Calc. tiss. Res. 4, 61 (1970).
NITSCHKE, U., GIEGLER, I.: Klin. Wschr. 39, 733 (1961).
NORDIN, B.E., YOUNG, M.M., BENTLEY, B., SYKES, J.: Calc. Tiss. Res. 2, 56 (1968) Supplement.
NORDIN, B.E., SMITH, D.A., MCGREGOR, J.: in Parathyroid Glands. P.J. Gailland, R. Talmage, A.M. Budy. Chikago: University Press 1965a.
NORDIN, B.E., SMITH, D.A., MCFADYEN, I., JOHNSTON, S.: 2nd European Symposium on calcified tissues Liège (1965).
NORDIN, B.E.C.: Lancet I, 1011 (1961).
OESER, H., KROKOWSKI, E.: Brti. J. Radiol. 36, 274 (1963).
PHANG, J.M., DOWNING, S.J.: Am. J. Physil. 224, 191 (1973).
REINER, M., NADARAJAH, A., LEESE, B., JOPLIN, G.: Calc. tiss. Res. 4, 95 (1970) Suppl.
REYNOLDS, J.: Calc. tiss. Res. 4, 52 (1970).
REYNOLDS, T.B., LAUMAN, H., TUPIKOVA, N.: J. clin. Endocrin. 20, 1136 (1960).
RIGGS, L., ARNAUD, C.D., JOWSEY, J., GOLDSMITH, R.S., KELLY, P.: J. clin. Invest. 52, 181 (1973).
RIGGS, L., JOWSEY, J., KELLY, P., JONES, J., MAHER, F.: J. clin. Invest 48, 1065 (1969).
ROCKNEY, R.E., KLEEMANN, C.R., MAXWELL, M.: Arch. int. Med. 104, 797 (1959).
ROSE, G.A.: Brit. med. J. 805 (1956).
SALEH, A., COENEGRACHT, J.: Clin. chim. acta. 21, 445 (1968).
SALOMON, C.D., VOLPIN, G.: Calc. tiss. Res. 4, 80 (1970) Suppl.
SCHWARTZ, E., PANARILLO, V., SAELI, J.: Calc. tiss. res. 2, 55 (1968) Suppl.
SCHWARZ, G.: Pseudohypoparathyreoidismus und Bendopseudohypoparathyreoidismus. Berlin, Heidelberg, New York: Springer 1964.
SHORT, E., BINDER, H.J., ROSENBERG, L.E.: Science 179, 700 (1973).
SIEGEL, R., MARTIN, G.: Calc. tiss. res. 4, 42 (1970).
SLUYS-VEER, J.v.d., BIJVOET, O., SMEENK, D.: Calc. tiss. Res. 4, 88 (1970).
SMITH, Q.T., ARMSTRONG, W.D.: Am. J. Physiol. 200, 1330 (1961).
SPENCER, H., MENCZEL, J., LEWIN, J., SAMACHSON, J.: Am. J. Med. 37, 223 (1964).
STANBURY, S.W.: Schw. med. Wschr. 92, 883 (1962).
SÜSSMANN, H.: Clin. chim. Acta 27, 121 (1970).
TASHIJAN, A.H., VOELKEL, E.F., LEVINE, L., GOLDHABER, P.: J. exp. Med. 136, 1329 (1972).
TEMPLETON, A.W., GANCONETTE, J., ORMOND, R.S.: Radiology 78, 955 (1962).

FOLLIS, H.R.: J. Paediatrics 41, 713 (1952).
TORO, G., ACKERMANN, P.G., KOUNTZ, W.B.: Proc. soc. exp. Biol. Med. 97, 819 (1958).
VAES, G.: Nature, 219, 939 (1968).
VAES, G.: Calc. tiss. Res. 4, 57 (1970).
VAES, G.: La résorption osseuse et l'hormone parathyroidienne p. 79. Louvain: E. Warny 1966.
VAES, G., NICHOLS, G.: Endocrinology 70, 546 (1962).
VAES, G., NICHOLS, G.: Endocrinology 70, 890 (1962a).
VEJLENS, L.: Calc. tiss. Res. 9, 54 (1972).
WALKER, D.: Endocrinology 79, 836 (1966).
WALKER, D.G., LAPIERE, C.M., GROSS, J.: Bioch. biophys. Res. 15, 397 (1964).
WALKER, A.R., ARWIDSSON, U.B.: Metabolism 3, 385 (1954).
WILHELM, G.: Dtsch. med. Wschr. 83, 1428 (1958).
WILLENBOCKEL, U.: Zur Physiologie und Pathologie des Phosphatstoffwechsels. Stuttgart: Enke, 1969.
WOESSNER, J.: in Treatise on Collagen Vol. 2, B. B.S. Gould. London: Academic Press 1968.
WOOD, J.F., NICHOLS, G.: J. cell. Biol. 26, 747 (1965).
WOODARD, H.: Cancer 12, 1226 (1959).
ZEITOUN, M.: Arch. Dis. Childhood 38, 289 (1963).

Tetracycline Staining of Bone in Normal and Pathological States

J. E. Aaron and F. G. E. Pautard

I. Introduction

Tetracycline (TC) is usually associated in bone physiology with labeling procedures, where calculated doses of the antibiotic are administered at one time, or at intervals, and the resultant fixation of the molecule in the bone substance is assessed by the disposition and thickness of the fluorescent complex in ultraviolet light of specific wavelength.

While earlier literature described the transitory association of TC with soft tissues, recent reports have been largely concerned with the nature of the TC binding to the mineral complement of bone (KAITILA, 1971) and the use of this fact in examining the metabolic state of the tissue. The characteristic fluorescent regions have been regarded as wholly extracellular and cells appear to take no part in the process. There is some evidence, however, that isolated bone cells can bind tetracycline; and that certain osteocytes become fluorescent in pathological conditions such as renal failure (AARON, in preparation).

When tetracycline is used as a stain for fresh or fixed tissue, the molecule is then found within the cell in regions and in amounts which depend on the specimen and its activity. In developing bone (AARON and PAUTARD, 1973), the pattern of binding within cells appears to be associated closely with the cell cycle of maturation, commencing in the region of the nucleus and extending throughout the cytoplasm. The following preliminary report suggests that the pattern of cell staining with TC reflects phases of the synthesis of mineralized structures in the juxtanuclear region, particularly of the osteocyte. An analysis of this pattern may be useful in the diagnosis of abnormal states.

II. Materials and Methods

The preparation of the calvaria of 5-day-old mice and the selection of sites has been described by AARON (1973). Femora and soft tissues from the same, or from mature animals were excised, fixed in neutral formol saline or glutaraldehyde/formaldehyde mixtures, embedded in wax and sectioned. Specimens of human iliac crest were similarly fixed but embedded in methyl/butyl polymethacrylate mixtures before sectioning. All tissues stained with TC were treated in bulk, fixed and unfixed, at a concentration of 8 mg% (AARON and PAUTARD, 1973). Cell contents, vacuoles and nuclei were identified by staining with gentian violet (BAST, 1921), neutral red (BARNICOT, 1947), osmic acid (2% in cacodylate buffer) and by the Feulgen reaction (HUMASON, 1967). The methods for "calcium" and "phosphorus" have been examined by AARON (1973).

III. Results

After oral administration of TC at intervals of several days, the newly-formed regions of bone clearly labeled as distinct zones (Fig. 1a). The cells are not fluorescent, appearing as dark spaces. When sections of similar tissue are stained with TC for 3 hours, the bone is fluorescent over the entire surface, although local differences in brightness reflect the position of the lamellae as well as an occasional osteocyte (Fig. 1b).

When biopsies of the iliac crest are stained with TC before embedding and sectioning, only certain regions of the bone fluoresce. In some cases, areas of extracellular mineral are stained (in renal failure, Fig. 2a): in others, cells alone and their immediate surroundings may be involved (in Cushing's Syndrome, Fig. 2b). Osteoclasts usually contain small particles which stain with TC.

Some soft tissues, such as the liver (Fig. 3a) and the mammary gland (Fig. 3b), show a pattern of intracellular staining with TC. A cell, or group of cells, may be densely stained while adjacent cells may not fluoresce at all. In bone such as the diaphysis of the developing femur (Fig. 4a) a similar pattern may be present; or the intensity of staining may increase in all cells from the least, to the most, mature region, as in the developing calvarium (Fig. 4b). In all preparations where the tissues were stained before sectioning, the gross bone mineral showed only a faint autofluorescence; TC fluorescence was confined to the periphery of resorbing edges or in the periosteocytic spaces.

Fig. 1. (a) TC labeling of human iliac crest. The two bands arrowed 1 and 2 represent typical intervals of time after oral administration of antibiotic, x 70. (b) TC staining of sections of human iliac crest. The entire bony area shows varying degrees of fluorescence. Here and there a brighter region (arrowed) suggests that some cells might be heavily stained, x 170

Fig. 2a and b. Tetracycline staining of whole tissue before sectioning. Human iliac crest from pathological cases. (a) Patches of diffuse stain in certain regions only, in contrast to Fig. 1b where the fluorescence is general. Renal failure, x 170. (b) Staining of cells alone with some local associated fluorescence. Note that the cell processes are also stained; an adjacent cell (arrowed) does not fluoresce. Cushing's Syndrome, x 430

Fig. 3a and b. TC staining of whole soft tissue. (a) Liver, 5-day-old mouse, where numerous pale cells are sharply contrasted with small clusters of strongly fluorescent cells, x 290. (b) Mammary gland, lactating mouse, where bright cells (arrowed) are the dominant feature and there is some added fluorescence from cell products, x 290

Fig. 4a and b. TC staining of whole bone in the 5-day-old mouse. (a) In the cortex of the femur in L.S., densely-stained cells (arrowed) are contrasted with pale cells elsewhere. The edge of the bone tends to show some scatter because of the intensity of fluorescence, x 225. (b) In the calvarium in thin section, the gradation of stain is clearly shown from the early osteoprogenitor areas (1) which are hardly visible, through the developing osteoblasts to the intensely-stained osteocytes (3) enveloped in bone. Other osteoblasts (labeled B) and the mature bone show only a faint autofluorescence, x 170

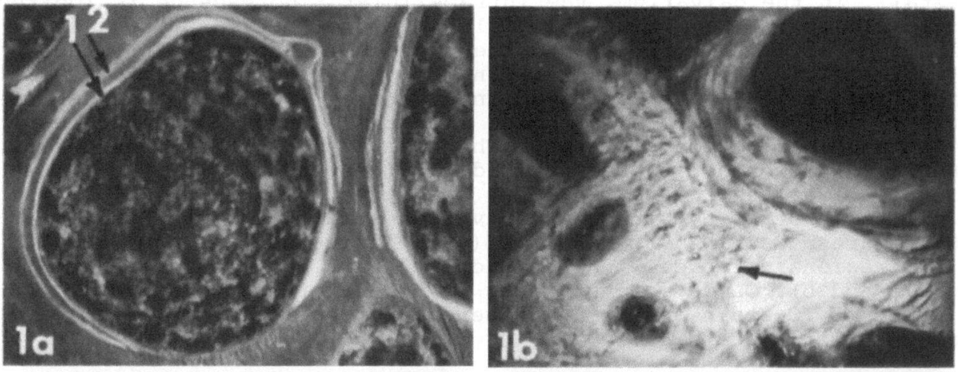
1
2
1a
1b

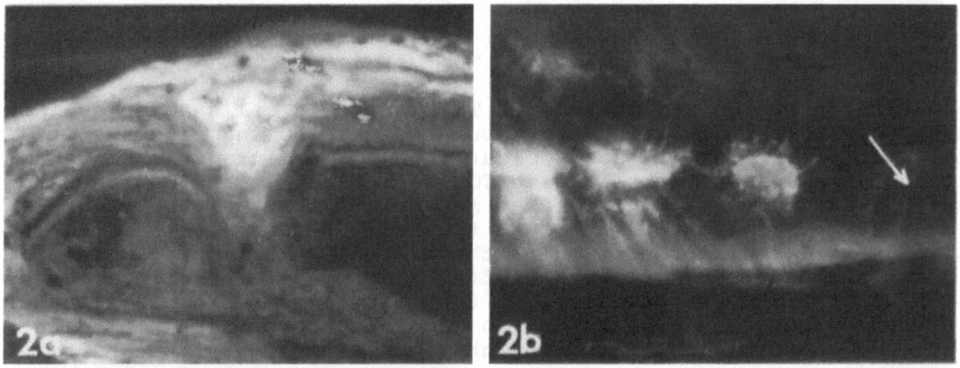
2a
2b

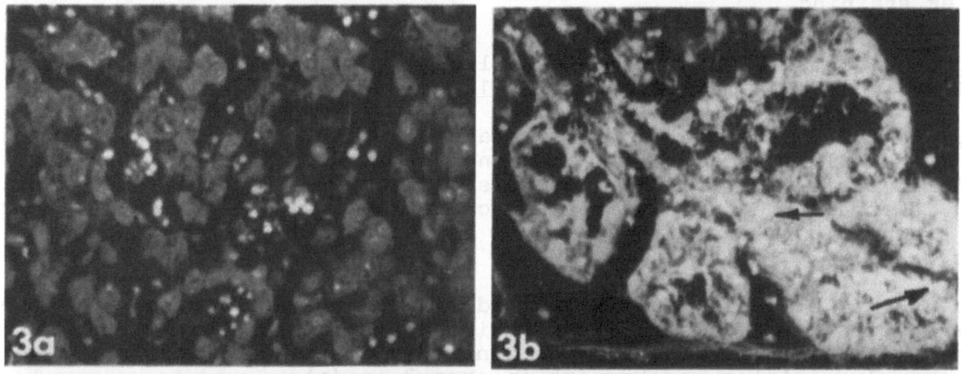
3a
3b

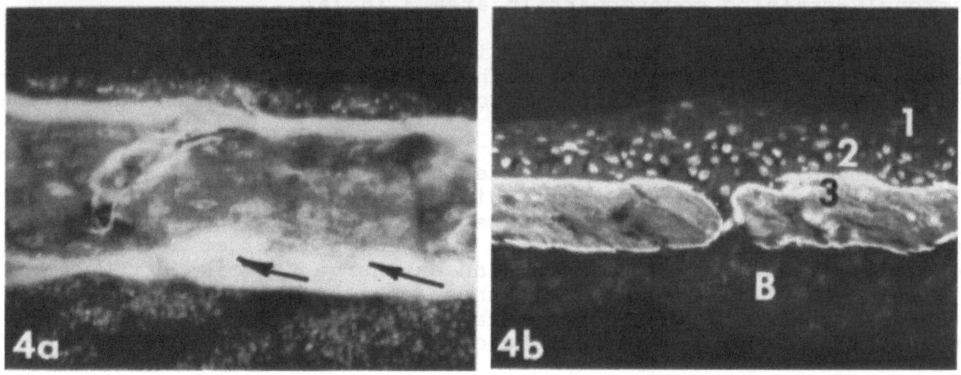
1
2
3
B
4a
4b

In detail, in the calvarium, the intracellular fluorescence is located principally in the region of the nucleus in the "early" phase and in a more generalized network around the nucleus in the "late" phase. When the intracellular fluorescent pattern is compared with the bright field optical appearance under polarized interference contrast, the features of the TC pattern (Fig. 5a) coincide with the vacuoles or particles in the juxtanuclear region (Fig. 5b). When compared with features described by BAST (1921) for osteocytes stained with gentian violet, the TC regions coincide with the juxtanuclear apparatus (Fig. 6a and b), although not all cells staining with gentian violet stain with TC. When compared with features described by BARNICOT (1947) using neutral red, the TC stained regions coincide with the vacuoles in the juxtanuclear region (Fig. 7a and b). Osmic acid staining of similar regions indicates that the areas of TC stain are in areas generally accepted for the golgi apparatus (Fig. 8a and b).

IV. Discussion

The nature, position and pattern of TC binding in cells after staining immediately raise many questions about the origins and pathways of calcification in bone. It seems clear that the failure to observe cell patterns in specimens labeled with TC (as in Fig. 1a) in contrast to specimens stained with TC (Fig. 2a) may be a matter only of concentration and time. If transport through the cell is rapid, or limited metabolically to a small proportion of the mineral, the fluorescence will be too weak to detect, or absent, after a few days. There is evidence (AARON and PAUTARD, in preparation) that after periods of labeling of less than an hour, bone cells do, in fact, show appreciable binding of TC in a pattern similar to, though less intense than, the pattern after staining.

Fig. 5-8. Detailed appearance of cells in the intact, 5-day-old mouse calvarium at high magnification, x 1400

Fig. 5a and b. The cell strongly stained with TC in (a) shows a number of intracellular features which coincide with the image (b) as seen under polarized interference contrast optics. The general area of staining is around the nucleus, with discrete regions (an example is arrowed) of strong fluorescence appearing in the same position as vacuoles or particles

Fig. 6a and b. The two cells labeled 1 and 2 in (a) correspond to the juxtanuclear area as seen after staining with gentian violet in (b). The cell 3 staining densely with gentian violet in (b) does not have a strongly-staining counterpart in area 3 in (a)

Fig. 7a and b. The TC-stained area in (a) corresponds to the area stained with neutral red in (b) although the cell A stained with neutral red in (b) does not have a fluorescent counterpart in (a) in the same region

Fig. 8a and b. Osmic acid illustrates the disposition and appearance of the golgi apparatus. The osmiophilic region may appear as a "crescent" (arrowed C in (a)) at one edge of the nucleus, or it may surround the nucleus entirely (as at O arrowed in (a)); at a later stage, the stained region may spread into an extensive network (arrowed N in (b)) which may fill the cell. All these counterparts are observed in TC-stained specimens, but counterstaining is usually obscured by the dense general precipitation of the osmic acid ►

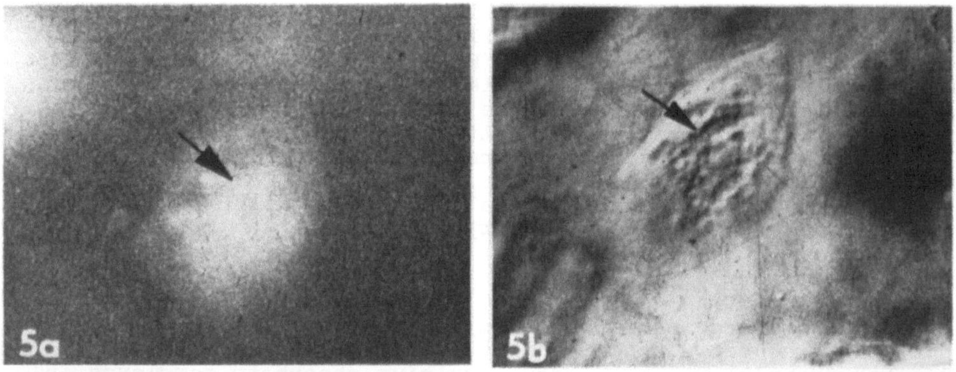
5a
5b

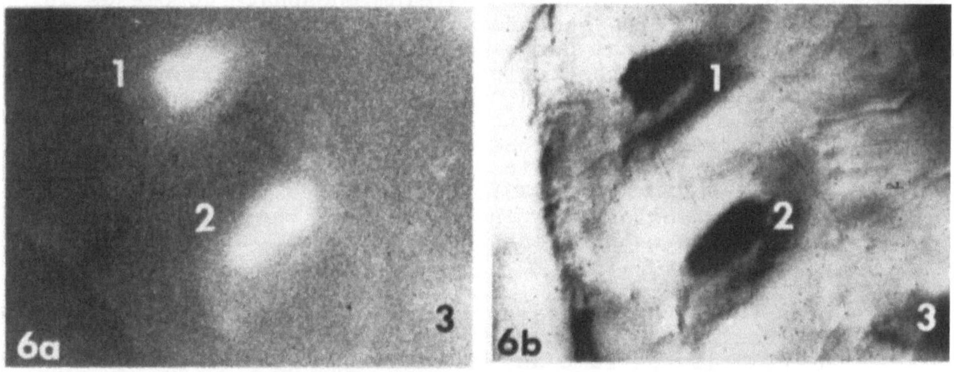
1
2
3
6a
1
2
3
6b

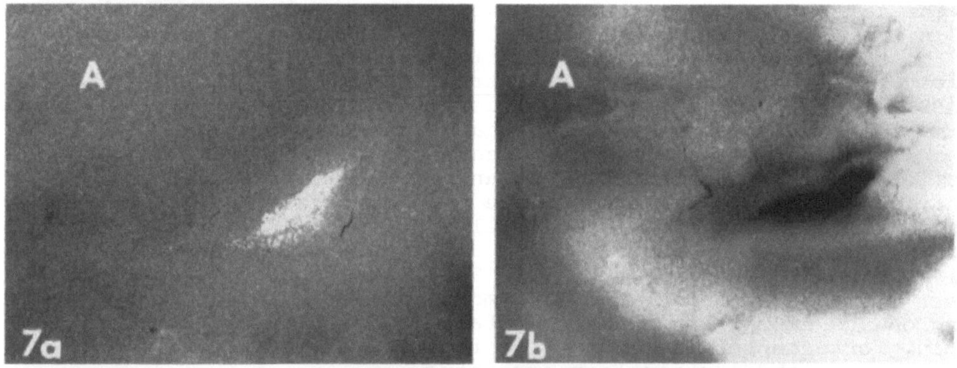
A
7a
A
7b

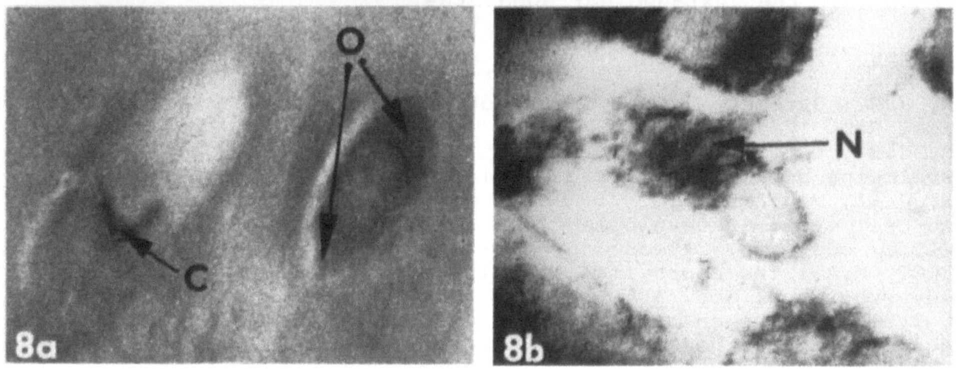
O
C
8a
N
8b

The absence of fluoresence in extracellular bone after staining suggests that the TC-binding site is not normally accessible, although it may become so after damage by sectioning (Fig. 1b). The possibility that the TC-binding site is normally occupied by a biological molecule of similar character is attractive, since it explains the "packaging" of the mineral in the cell and the stability of inorganic matter outside the cell. Moreover, the staining of gross bone with TC before sectioning suggests that in these examples (renal failure, Fig. 2a) some regions of extracellular "mature" mineral may be incompletely covered with the biological molecules necessary to maintain the integrity of the area.

The staining of soft, as well as hard, tissues may indicate no more than two separate phenomena. But it may be that the difference is only a matter of degree between bone and other cells. The liver (as in Fig. 3a, for example) may manufacture calcium phosphate complexes in the same way as bone, but the end product may be recycled immediately for metabolic and cytoskeletal purposes and not translocated to the extracellular space. Again, the regional and discontinuous nature of the cell-staining suggests a periodic mechanism of some kind, where cells "switch on" and "switch off" the fabrication of the TC-binding complex. In bone, particularly in the calvarium (Fig. 4b), the binding increases in the order osteoprogenitor - osteoblast - osteocyte (Fig. 4b, see also AARON and PAUTARD, 1973). The staining, often intense, of chondrocytes and osteoclasts is in accord with the evidence that the former is associated with mineral deposition and the latter with its removal.

Finally, it seems to us that the association of TC with the juxtanuclear apparatus of the cell is a matter of great importance. If the TC complex within the cell is the same as, or similar to, the TC complex outside the cell, it is impossible to escape the conclusion that the mineral in the matrix may have commenced as such within the cell. Coupled with the fact that there are intracellular features which closely resemble those attributed to mineral structures in the matrix (AARON and PAUTARD, 1972), but which are sufficiently soluble to escape detection in the electron microscope after preparation with conventional fixatives and stains, then the packaging and export of mineralized structures is not easy to reconcile with an extracellular epitaxy. If the synthesis of bone mineral takes place in the golgi apparatus, it follows the same path as many other known cell products, particularly coccoliths, which appear to be formed and completely calcified within the golgi vesicles (OUTKA and WILLIAMS, 1971). Bone manufacture may then be part of cell differentiation and most of the factors now supposedly acting on the extracellular bone (vitamin D, parathyroid hormone, calcitonin, for example) may need to be regarded as somehow affecting the performance of the juxtanuclear apparatus.

References

AARON, J.E.: Osteocyte types in the developing mouse calvarium. Calc. Tiss. Res. 12, 259-279 (1973).

AARON, J.E., PAUTARD, F.G.E.: Ultrastructural features of phosphate in developing bone cells. Israel J. med. Sci. 8, 625-629 (1972).

AARON, J.E., PAUTARD, F.G.E.: A cell cycle in bone mineralization. In: The cell cycle in development and differentiation (M. BALLS and BILLET, eds.), p. 325-331. London: Cambridge University Press 1973.

BARNICOT, N.A.: The supravital staining of osteoclasts with neutral red. Proc. roy. Soc. 134, 467-485 (1947).

BAST, T.H.: Studies on the structure and multiplication of bone cells facilitated by a new technique. Amer. J. Anat. 29, 139-157 (1921).

HUMASON, G.L.: Animal tissue techniques. London: W.H. Freeman and Co. (1967).
KAITILA, I.: Effect of tetracycline on mineralization in cycloheximide-treated bones in vitro. Calc. Tiss. Res. 7, 46-57 (1971).
OUTKA, D.E., WILLIAMS, D.C.: Sequential coccolith morphogenesis in Hymenomonas carterae. J. Protozool. 18, 285-297 (1971).

Biochemical Determinations in Serum and Bone Homogenates from Patients with Femurhead Necroses

R.F.L. Maruna and E. Trojan

The following determinations were made in serum and femurhead homogenates of patients with necroses, as an incident of fractures of the upper thigh some time previously: total activities and isoenzyme activities of alkaline phosphatases, adenosinetriphosphatases, (AT-) and pyrophosphatases.

I. Method

The total activities of the alkaline phosphatases were determined according to the prescription of King & Delory. To determine AT- and pyrophosphatases we modified the prescription of Taussky & Shorr. Isoenzyme activities are separated by thin layer starch gel electrophoresis and are developed on special filter paper with substratum buffer solution. After a short incubation time the fraction can be estimated densitometrically by a MPS-Photometer from Vitatron (The Netherlands). We made the homogenates by soft pressing with a hydraulic press and mixed it in a homogenising-mill with five times the amount of trisbuffer-acetate (0,75 mol. pH 8,6). Then we centrifugalised. We used the supernatant fluid for our analyses. Total activities are expressed as mU per gramme protein in 100 ml homogenate, isoenzyme activities in relative percentage. In serum the results are specified in mU/ml, as usual.

Chemicals were provided by Calbiochem (USA), Austro-Anal GMBH (Austria) and Merck (G.F.R.)

II. Results

In Table 1 the total activities of three phosphatases are given: alkaline, AT- and pyrophosphatases. On the left are results from femurheads, on the right results of femurneck determinations. Sometimes femurnecks and femurheads of the same patient were not available, so there were no results. Two of the results, nos. 80 and 63a are from parts of femurheads well supplied with blood and therefore well nourished. The following results, beginning with number 60 are from patients with more or less heavy necroses. An activity of 140 mU/g% protein is below 1% of the femurheads well supplied with blood. Comparing results of femurnecks, we can say that these parts of bone must sometimes also be diseased.

In Table 2, results of isoenzyme activities (in relative percentage) in femurhead homogenates of 17 patients with femurhead necroses are described. Ø-fraction is in accordance with IgM-fraction of protein electrophoresis. Fraction A_6 is equivalent to albumin-fraction.

Table 1. Total activities of alkaline AT- and pyrophosphatases in caput and collum femoris homogenates from 17 patients (mU/g% protein)

Pat.nos.	Alk.phts. in caput homogen.	AT-phts.	Pyrophts.	Alk.phts. in collum homogen.	AT-phts.	Pyrophts.
T80	16.500	1.270	540			
63a	15.400	2.400	2.680	6.700	1.470	1.490
23a	5.720	3.120	3.780			
60	4.020	1.720	230			
65	1.700	520	860			
55	1.330	120	350			
63b	1.150	300	160	6.700	1.470	1.490
23b	900	410	310			
54	660	110	170	2.760	1.260	210
91	430	130	150			
66	290	390	240	1.540	1.330	240
38	290	250	380	4.800	1.620	1.860
88	290	110	140	2.160	2.040	1.120
97	270	360	300	430	670	350
98	230	340	280	2.520	1.870	1.600
53	140	160	80	6.490	630	780
99	140	30	120	460	570	730
58	110	40	110			

Table 2. Alkaline phosphatases isoenzyme fractions from 17 patients with caput femoris necroses and Ø-fractions of AT- and pyrophosphatases of the same patients (in rel. percentages). In homogenates.

Fractions:	A_7	A_6	A_5	A_4	A_3	A_2	A_1	Ø	K_1
$\bar{x}$	7,4	10,7	9,5	9,8	8,8	9,4	10,9	25,8	7,4
SD	0,9	0,9	0,5	1,2	1,0	1,3	1,1	4,0	0,6
AT-phosphatases $\bar{x}$								7,4	...
SD								1,3	
Pyrophosphatases $\bar{x}$ (4 pat.)..................								29,6	...
(13 pat.)..................								9,8	...
SD								1,06	

A is a symbol of anodic and K of cathodic electrophoretic run. It is surprising to see fraction A_6 so low. All the other fractions are similar to the fractions in serum. The Ø-fraction has a mean value comparable to the upper range of serum Ø-fraction. The activity of osteoblasts is included in this fraction, and is therefore very important to us. The Table is completed with an appendix of the activities of Ø-fraction from the AT-phosphatases as well as from pyrophosphatases. The value of AT-phosphatases is 7,4%, a fractional part of the alkaline isoenzyme. The activities of pyrophosphatases in this fraction are found at 23,6% in 4 patients but in 13 cases at 9,8% only!

Table 3 registers a few results of isoenzyme of alkaline, AT- and pyrophosphatases Ø-fractions in mU/g% protein, not in relative percentage. From patient no. 63a we had a part of femurhead, well supplied with blood and 63b is a necrotic place of the same bone. The activities of the following are less than 10% of the well nourished ones. In this series we can see fraction activities with less than 5% of the non necrotic piece.

Table 3. Isoenzymes Ø-fractions of alkaline, AT- and pyrophosphatases in femur homogenates of a few patients (in mU/g% protein, n = 17)

Pat.nos.	Alk.phosts.	AT-phosts.	Pyrophosts.
63 a	7.400	240	350+)
b	715	40	17++)
55	570	30	74
54	175	30	36
38	80	18	25
53	62	35	17
66	56	36	9
88	40	4	18
58	18	7	3

Table 4. Serum alkaline, AT- and pyrophosphatases isoenzyme fractions from patients with caput femoris necroses (in rel. percentages, n = 17)

Fractions:	A_7	A_6	A_5	A_4	A_3	A_2	A_1	Ø	K_1	
Alk.phsts.	8,2	13,3	9,3	7,4	7,2	10,0	8,2	21,6	7,2	$\bar{x}$
	1,8	2,6	2,4	1,6	2,4	2,8	1,8	3,6	2,6	±2 SD
Normal	-,-	20,8	9,7	8,2	12,3	15,0	10,2	21,0	2,8	$\bar{x}$
		2,4	2,4	2,0	2,9	3,0	1,9	3,7	1,0	±2 SD
AT-phsts.	8,7	17,0	9,0	10,3	6,4	9,2	7,6	23,1	7,2	$\bar{x}$
	1,6	3,2	1,8	2,4	1,4	2,4	2,0	2,6	2,0	±2 SD
Pyrophsts.	6,0	17,0	7,9	4,4	7,6	10,9	10,9	21,6	5,1	$\bar{x}$
	1,2	2,8	1,6	1,4	1,0	3,0	2,4	3,6	1,8	±2 SD

In Table 4 the mean values of alkaline, AT- and pyrophosphatases-isoenzyme activities in serum are summarized. It is interesting to notice here: the alkaline phosphatase fraction A_6 with 13,3% compared with 20,8% in normal serum. The A_3-fraction, the one of the osteoclast-activity, is approximately less than 40% of the normal value. For this reason we must understand that the necrosis is not caused by the action of osteoclasts. Ø-fraction is within normal range. This fraction is reduced in the AT- and pyrophosphatases. The fraction A_4 of pyrophosphatases is less than half of the normal value. Activities of Ø-fraction of AT- and pyrophosphatases are within normal range. There is thus no difference in the osteoblast activities.

III. Discussion

In homogenates of femurhead necrosis the A_6-fraction of alkaline and the Ø-fraction of AT- and pyrophosphatases are reduced. This reduction means an accumulation of pyrophosphate, in other words: an inhibition of forming hydroxyapatite in bone. The decrease of A_6-fraction of alkaline phosphatase-activities in serum shows a diminution of liver activity. No signs of a higher activity of osteoclasts are to be found. So we see: the problem of femurhead necrosis is not localized in the place of action only, it must rather be a reaction of a higher metabolic disorder.

A Clinical and Biochemical Survey of Osteogenesis Imperfecta with Evidence for a Generalised Collagen Defect*

M.J.O. Francis, R. Smith and R.J. Bauze

In osteogenesis imperfecta (O.I.) the clinical features of excessive bone fragility and blue sclerotics suggest an abnormality of those tissues which contain collagen, but previous biochemical evidence of this has not been convincing.

We report here the results of a combined clinical and biochemical study of a group of patients with this disorder and present evidence that, in severe O.I., the polymeric structural collagen of the skin is unstable.

I. Patients and Methods

Forty-two patients were studied. Severe bone disease was characterised by long bone deformity, many fractures and inability to walk unaided. The color of the sclerae was estimated clinically and by photography, using a standard color chart.

Skin biopsies were taken from over the lateral aspect of the greater trochanter in 16 patients. Two were taken at operation and two *post mortem* from patients dying within a week of birth.

Total, soluble and polymeric collagen of skin was estimated according to the methods of FRANCIS and MACMILLAN (1971). The stability to depolymerisation of the major polymeric collagen fraction (65-85% of total skin collagen in control subjects) was determined. This gives an indirect measure of the numbers of covalent cross links between individual collagen molecules (see FRANCIS et al., 1973, for control data).

Urinary total hydroxyproline (THP) and the hydroxyproline fractions of plasma were measured in 28 patients (PROCKOP and UDENFRIEND, 1960; LEROY et al., 1964).

+ We are grateful to the physicians and surgeons who have kindly allowed us to study these patients and to the patients themselves for their willing collaboration. This work was done during the tenure of a Girdlestone Scholarship by R.J.B. and was supported by grants to M.J.O.F. from the Medical Research Council and to R.S. from the Wellcome Trust.

Table 1. Clinical and Biochemical details of Patients with Osteogenesis Imperfecta

Clinical						Biochemical	
No.	Age (years)	Sex	Family History	Fractures (Age of onset	(Approx. no.	Total skin collagen (μg/mg ffdw skin)	Major polymeric Collagen fraction of skin (μg/mg ffdw skin)
A. Blue sclerae							
1.[b]	8 days	M	-ve	Birth	Many	405[a]	0[a]
1A[b]	3 mths	M	-ve	Birth	Many	335[a]	0[a]
2.[c]	5	F	-ve	13m	~5	515	80[a]
3.	17	F	-ve	5y	5	400[a]	80[a]
4.	33	F	+ve	~2y	12	445	130[a]
5.	35	F	+ve	18m	11	400[a]	115[a]
6.	39	F	+ve	2-3y	25	595	270
7.	40	M	+ve	6y	3	500	280
8.	40	F	+ve	12m	28	570	185[a]
9.	56	F	+ve	2y	11	470	255
10.	64	M	+ve	14y	3	425[a]	215[a]
B. Intermediate (pale blue) sclerae							
11.[c]	11	F	+ve	18m	30	630	195
12.	17	M	+ve	4m	45	600	440
13.	25	M	-ve	Birth	70	485	225[a]
C. White sclerae							
14.[c]	12	M	-ve	Birth	50	405[a]	55[a]
15.	19	M	+ve	< 3y	60	525	410
16.	24	M	-ve	Birth	30	575	250
17.	25	M	-ve	Birth	Many	660	450
18.	39	M	-ve	13m	>100	610	490
19.[d]	56	F	+ve	7y	12	490	145[a]

Table 1. (Continuation)

Controls (15-70 years)	620±195(39)[e]	450±210(30)[e]

[a] Abnormal values (more than 2xS.D. from appropriate control means)

[b] Post-mortem sample

[c] Leg sample taken at operation

[d] Upper-arm sample

[e] Mean ± 2S.D. (number)

II. Results and Discussion

Clinical. Twenty-two out of 42 patients had mild bone disease, which was closely associated with blue sclerae; those with severe bone disease (17), tended to have sclerae of normal color. Details are given for those on whom the skin collagen was examined (Table 1).

Biochemical. In 28 subjects the ethanol insoluble "hypro-protein" of plasma was within our normal range (4-10 mg/l), except for one patient with severe bone disease (19.5 mg/l). Similarly the ethanol soluble hydroxyproline in the plasma was normal (0.4-2.5 mg/l) except in one rapidly growing adolescent. Allowing for the wide variation with age the urine THP expressed as mg/24 h, was also normal. In most patients, especially those with severe bone disease, attempted correction for surface area based on height and weight was unrealistic. We were not able to confirm the abnormalities described by some previous workers

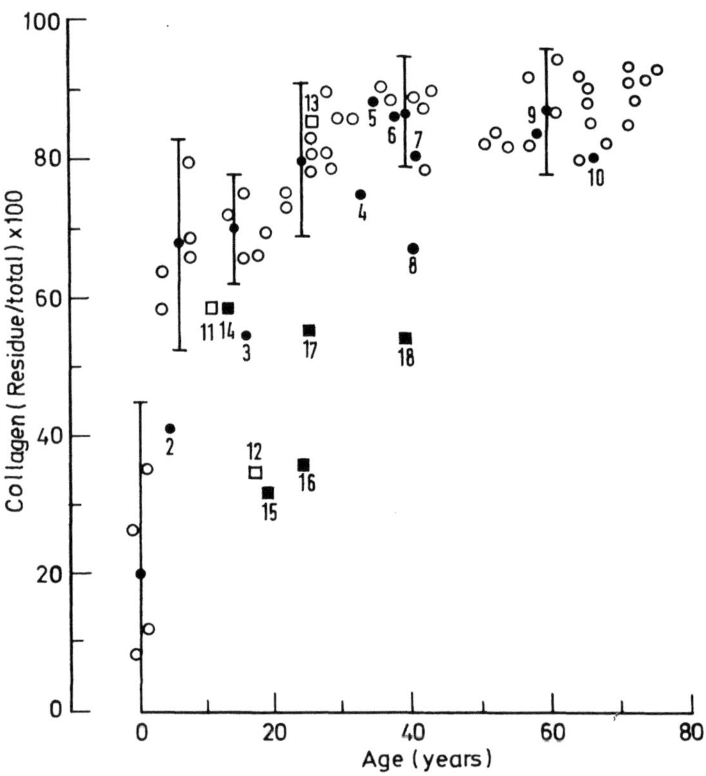

Fig. 1. Changes with age in the stability of polymeric collagen extracted from human skin of controls (o) and of patients with osteogenesis imperfecta with blue sclerae (●), intermediate (pale blue) sclerae (□) and white sclerae (■). Patients are numbered as in the Table. The collagen in the residue after treatment with cold alkali for 6 days at 4° C is expressed as a percentage of total collagen. Mean ± 2 S.D. are given for controls in the age ranges 0-3, 4-9, 10-19, 20-29, 30-49, 50-75 years

(see KIVIRIKKO, 1970, for review; and LANGNESS and BEHNKE, 1971). The total and polymeric collagen of the skin in the group with blue sclerae over 15 years old (Table 1) was significantly ($p < 0.001$) lower than controls. In those with white sclerae the difference from controls was not significant. The proportion of collagen soluble in 5% NaC1 was within the normal range (less than 0.8% of total collagen) except in patients no. 3 (2.0%) and no. 6 (0.9%). The stability of the major polymeric collagen fraction to depolymerisation by cold alkali (2.0M NaOH at 4° C for 6 days) is shown (Fig. 1). In those with white sclerae and severe bone disease, the stability was less than normal for age (no. 14-18). Individual patients with blue sclerae had polymeric collagen of normal stability. Only patients no. 7 and 9 had no significant collagen abnormality; these both had mild bone disease.

These results suggest that the clinical heterogeneity of O.I. patients is associated with different collagen abnormalities which may result from alterations in amino-acid sequence. In one there is a reduction of normally cross-linked collagen, leading to mild bone disease of an osteoporotic sort and to thin sclerae (which allow the blue choroid to show through). In the other abnormality the severe bone disease is due to the production of unstable collagen; since it is present in normal amounts, the sclerae are often of normal color.

References

FRANCIS, M.J.O., MACMILLAN, D.C.: The extraction of polymeric collagen from biopsies of human skin. Biochim. biophys. Acta 251, 236-245 (1971).

FRANCIS, M.J.O., SMITH, R., MACMILLAN, D.C.: Polymeric collagen of skin in normal subjects and in patients with inherited disorders of connective tissue. Clin. Sci. 44, 429-438 (1973).

KIVIRIKKO, K.I.: Urinary excretion of hydroxyproline in health and disease. Intern. Rev. Connective Tiss. Res. 5, 93-163 (1970).

LANGNESS, V., BEHNKE, H.: Collagen metabolism in plasma and urine in osteogenesis imperfecta. Metabolism 20, 456-463 (1971).

LEROY, E.C., KAPLAN, A., UDENFRIEND, S., SJOERDSMA, A.: A hydroxyproline-containing collagen-like protein in plasma and a procedure for its assay. J. Biol. Chem. 239, 3350 (1964).

MCKUSICK, V.: Heritable disorders of connective tissue. 4th ed. St. Louis: C.V. Mosby Co. 1972.

Analysis of Tissue PO_2 and PCO_2 in Healing Bone*

J. KIVISAARI AND J. NIINIKOSKI

At the Ninth European Symposium on Calcified Tissues in Baden near Vienna, we showed results of the measurements of tissue oxygen tensions in healing rabbit tibias (NIINIKOSKI and HUNT, 1972). These studies have been continued to include the analysis of tissue carbon dioxide tensions in the same model.

Oxygen and carbon dioxide tensions were assessed in healing rabbit tibias over a period of 55 days by using a loop-shaped Silastic tube 17 cm long implanted into the marrow cavity. The endosteum was injured during the implantation to ensure a circumferential trauma around the Silastic tube. For the measurement of tissue gases the tube ends were exposed through a small incision. The Silastic tube was then filled with hypoxic saline solution which equilibrated to the average PO_2 and PCO_2 of the surrounding regenerating bone within two minutes (KIVISAARI and NIINIKOSKI, 1973). The equilibrated fluid sample was then collected in an Astrup glass capillary tube.

During each experiment several Astrup capillaries were filled. They were inserted one at a time into a Radiometer microsample injector and the sample was injected into a thermostated cuvette containing either O_2 or CO_2 electrode calibrated prior to the measurement.

Each tibia was tested once by measuring baseline gas tensions and the response to breathing of a gas mixture containing 95% O_2 and 5% CO_2- Response of bone tissue gases to occlusion of local circulation was tested and bone pCO_2 was recorded after a single intravenous injection of acetazolamide, a potent inhibitor of carbonic anhydrase (MINKIN and JENNINGS, 1972).

A rapid decline in baseline pO_2 from 32 to 25 mmHg occurred within the first three days postimplantation (Fig. 1). During the following 52 days the PO_2 rose gradually to 40 mmHg. The maximum PO_2 during breathing of 95% O_2 and 5% CO_2 rose progressively from 35 mmHg at 3 days to 140 mmHg by the 30th day.

The baseline PCO_2 rose from 64 to 85 mmHg within the first three days and then gradually declined to 57 mmHg by the 30th day (Fig. 2). Correspondingly, the maximum PCO_2 during breathing of 95% O_2 and 5% CO_2 increased from 65 to 101 mmHg during the first three days and then decreased to a level of 65 mmHg by the 30th day. The increase in PO_2 and decrease in PCO_2 during the course of healing are probably due to increasing vascularity and washout in the repair tissue. The production of carbon dioxide is probably at its highest during the early phases of healing.

Occlusion of the circulation to the leg by means of a tourniquet decreased the bone PO_2 from the normal level to a minimum of 2-5 mmHg within a few minutes. At the same time bone pCO_2 increased to a maximum

+ This work was supported in part by Contract No. DAJA-37-73-C-2681 for the U.S. Army through its European Research Office.

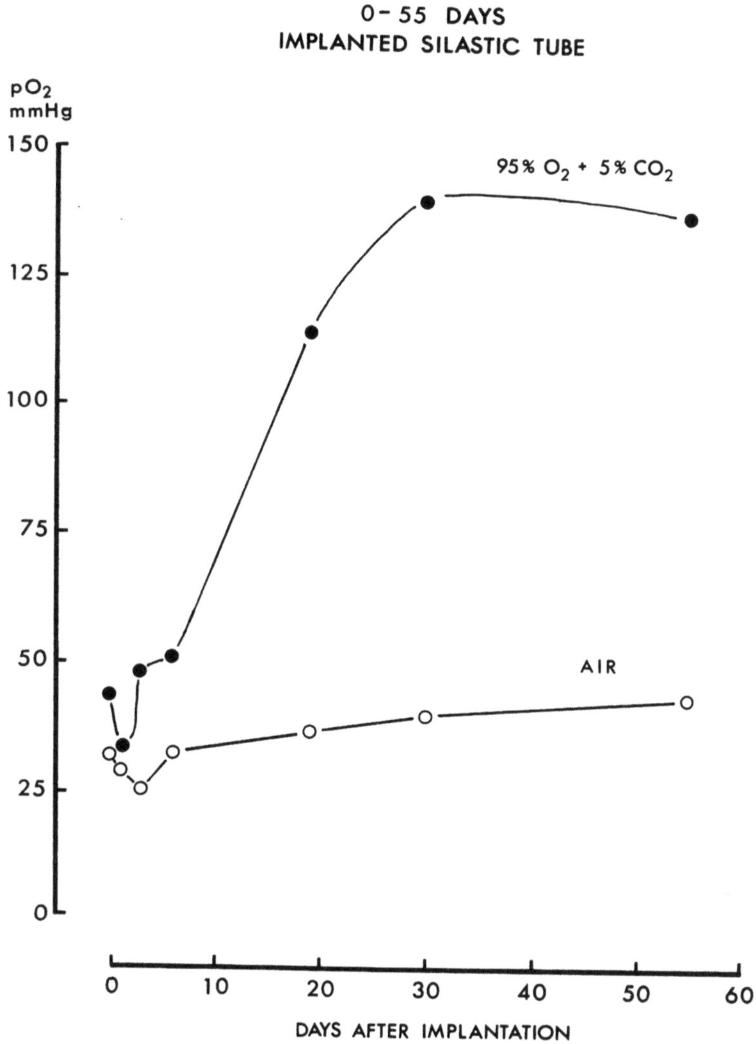

Fig. 1 Oxygen tension in healing bone

of approximately 120 mmHg. After the release of circulation the gas tensions returned to normal levels in 10 minutes.

In repeated experiments, intravenous administration of acetazolamide, 100 mg/kg of body weight, increased the bone pCO_2 at early phases of healing (19 days post-implantation) and decreased the pCO_2 at the later period (55 days). Arterial blood pCO_2 remained unchanged which suggests that the changes in bone pCO_2 were due to a local effect of carbonic anhydrase inhibitor. It can be speculated that from the reactions catalyzed by carbonic anhydrase, the change of carbon dioxide to bicarbonate is dominating during calcification process whereas at the later, resorptive phase the main direction of the reaction is from bicarbonate

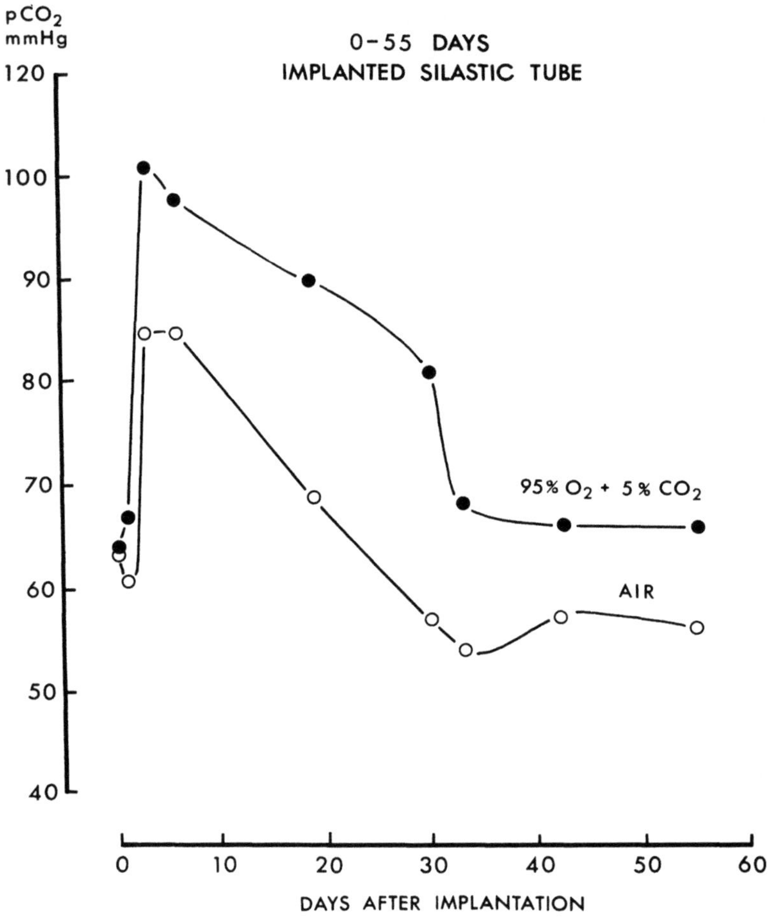

Fig. 2 Carbon dioxide tension in healing bone

to carbon dioxide. This would fit with the concept of DULCE et al. (1960) that carbonic anhydrase aids in mineralization and demineralization of bone by controlling the local secretion of hydrogen ions.

On histologic examination, the Silastic tube was surrounded by a trabeculated structure of rapidly calcifying bone on the 19th day. On the 30th day, trabeculation had greatly diminished and by the 55th day the marrow cavity was almost completely remodeled. The tonometer itself produced very little tissue reaction.

References

DULCE, H.J., SIEGMUND, P., KORBER, F., SCHUTTE, E.: Zur Biochemie der Knochenauflösung. III. Über das Vorkommen von Carboanhydratase im Knochen. Z. Physiol. Chem. 320, 163-167 (1960).

KIVISAARI, J., NIINIKOSKI, J.: Use of Silastic tube and capillary sampling technic in the measurement of tissue PO_2 and PCO_2. Amer. J. Surg. 125, 623-627 (1973).

MINKIN, C., JENNINGS, J.M.: Carbonic anhydrase and bone remodeling: sulfonamide inhibition of bone resorption in organ culture. Science 176, 1031-1033 (1972).

NIINIKOSKI, J., HUNT, T.K.: The measurement of oxygen tension in healing bone. Ninth European Symposium on Calcified Tissues. Baden near Vienna, Austria: October 1-6, 1972.

VI. Therapeutical Aspects of Bone Diseases

Chairmen: G.V. Foster and J.A. Parsons

Effect of Various Therapies on Bone Loss in Women

B.E.C. NORDIN, A. HORSMAN AND J.C. GALLAGHER

I. Introduction

The treatment of patients with the osteoporotic crush fracture syndrome will remain unsatisfactory until the pathogenesis of the condition is fully understood and precise methods are available to monitor the effect of therapy on bone loss.

As far as the first consideration is concerned, it is necessary to have a working hypothesis before attempting any kind of treatment, and we are continuing to explore the working hypothesis that osteopenia (normal bone loss) and osteoporosis (excessive bone loss) are analogous to the reduced bone mass that can be produced in animals by dietary calcium deficiency (NORDIN, 1960, 1961 and 1971). In post-menopausal osteopenia, it appears that the bone loss is secondary to increased sensitivity of the bone to parathyroid hormone in the absence of oestrogens (GALLAGHER and NORDIN, 1973). In the post-menopausal crush fracture syndrome, oestrogen deficiency seems to be associated with some other factor or factors which leads to accelerated bone loss. One such factor that we have identified is an impaired absorption of calcium (GALLAGHER er al., 1973). We have therefore been exploring the effect of calcium intake and oestrogens in post-menopausal osteopenia, and the effects of calcium supplements, vitamin D and oestrogens in post-menopausal women with the crush fracture syndrome.

As far as the monitoring of bone loss is concerned, we have developed a method of measuring sequential changes in the average thickness of the cortices of the metacarpals, which is our most sensitive procedure at the present time.

The present paper is therefore concerned with the effectiveness of various therapies in reducing the rate of bone loss assessed by this technique in normal post-menopausal women and in post-menopausal women with the crush fracture syndrome.

II. Clinical Material and Methods

Data are presented on 21 normal pre-menopausal women and 18 normal post-menopausal women observed for periods of 1 to 8 years, and on 39 post-menopausal women with crush fractures observed for 1 to 7 years.

Contact radiographs of both hands of each patient were obtained using a standard radiographic technique (HORSMAN and NORDIN, 1973). Serial measurements of total and medullary widths were made by the same observer, any one set of the patients' films being measured at the same sitting, the observer being unaware of the type of case or treatment given. The measurements were made at the mid-point of the second, third and fourth metacarpals with the needle calipers of British Indicators

Limited (NORDIN and SMITH, 1965). Each measurement was rounded to the nearest 0.05 mm. The precision of the calculated change in average cortical width has been shown to be ~0.04 mm (HORSMAN and SIMPSON, 1974).

Calcium absorption was monitored by means of an isotopic procedure involving kinetic analysis of the plasma radio-activity following an oral dose of radio-active calcium in 20 mg of calcium carrier as calcium chloride (BULLAMORE et al., 1970). The normal range is 0.3 - 1.4 of the dose absorbed per hour. We shall be reporting elsewhere a high correlation between absorption of calcium measured in this way and by conventional balance techniques.

The therapies discussed in this paper comprise calcium supplements, calcium citrate or calcium glycerophosphate, to yield 1 g of elemental calcium daily; ethinyl oestradiol, 0.05 mg daily for 3 weeks out of 4; and vitamin D, 1,000 to 40,000 i.u. daily, the dose being raised in a stepwise fashion until the radiocalcium absorption responded. (In four cases the radiocalcium absorption did not respond even to 40,000 i.u. daily and no further elevation of the dose was attempted.) Calcium supplements alone were given to 10 crush fracture cases with a normal initial absorption of calcium; the remainder were given vitamin D with or without calcium.

III. Results

1. Normal women

Sequential measurements on 21 pre-menopausal women followed for periods of up to 6.5 years showed no significant loss of bone in the group as a whole. Sequential measurements on 18 normal post-menopausal women followed for periods of up to 8 years revealed a significant loss of bone in nearly all of them and a highly significant loss in the group as a whole ($p<0.001$) with a mean rate of change of average cortical width of -0.044 mm/year (Fig. 1).

Sequential measurements were also obtained on 11 post-menopausal women followed for periods of up to 6 years on calcium supplements (Fig. 2). The mean rate of decrease of average cortical width (0.0081 mm/year) was lower in this group than in the untreated post-menopausal controls, but the difference of the mean rates of change was not significant. These results are compared in Fig. 2 with sequential measurements on 6 post-menopausal women treated with low calcium diets for idiopathic stone disease, one of whom had previously been on calcium supplements. In this group the mean rate of decrease of average cortical width (0.11 mm/year) was greater than in the group of untreated post-menopausal controls, but again the difference was not significant. However, the difference between low and high calcium groups was highly significant ($p < 0.01$).

A further group of 14 post-menopausal women were observed on oestrogen therapy for 2 to 6 years. The rate of bone loss in this group (0.010 mm/year) was significantly less than in the untreated controls ($p < 0.02$; Fig.2).

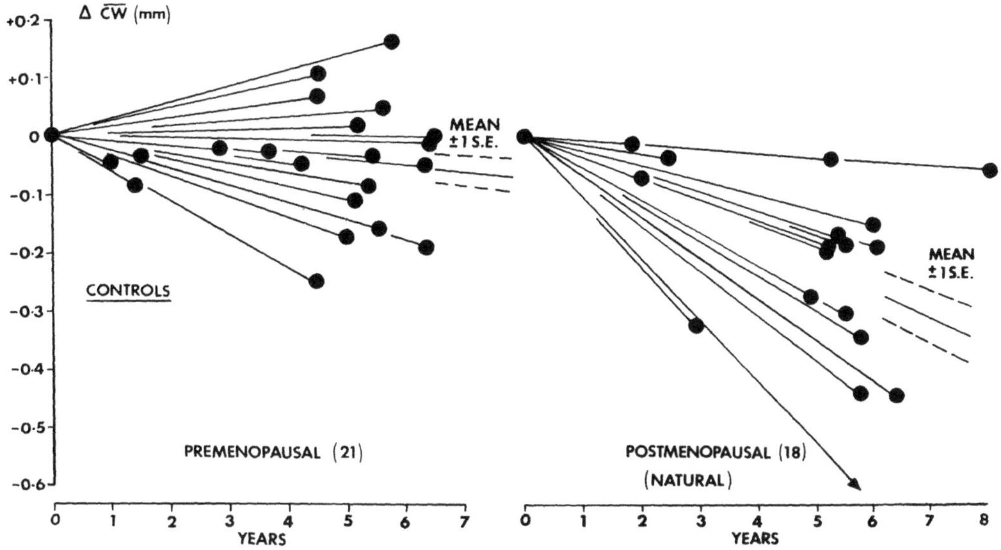

Fig. 1. Sequential changes in average cortical width, $\Delta\overline{CW}$ mm, in 21 pre-menopausal and 18 post-menopausal normal women

2. Crush fracture cases

Sixteen post-menopausal women with crush fractures were observed untreated for 3 months to 3 years. Sequential changes in average cortical width are shown in Fig. 3. It will be seen that in the majority of these cases bone loss proceeded more rapidly than normal, the mean rate of change of average cortical width being -0.12 mm/year which is significantly increased ($p<0.01$).

Ten post-menopausal crush fracture cases were observed both off and on treatment with vitamin D and/or calcium supplements. In two cases, one of which was apparently not losing bone, the other apparently gaining bone in the untreated state, there was bone loss during treatment. In one case no bone loss was observed either before or during treatment. In the remaining seven the rate of bone loss on treatment was less than it had been during the control period (Fig. 4). The effect of the treatments on the group as a whole is significant ($p<0.05$).

Altogether, 33 post-menopausal crush fracture cases have now been observed on various combinations of vitamin D and/or calcium supplements. In each treated group, the mean rate of bone loss was significantly less than in the untreated group (Fig. 4).

The effect of vitamin D treatment on bone loss appeared to be related to the mean dose during the period of observation and to the radiocalcium absorption. Fig. 5 shows the relationship between the mean daily dose of vitamin D and the rate of change of average cortical width, based on 27 observations on 19 subjects. Pairs of observations on the same subject are identified by symbols. It will be seen that at doses below 20,000 i.u. daily, 10 of the 16 cases were losing bone at a rate greater than the post-menopausal mean (-0.44 mm/yr). At doses of 20,000 i.u. or more, 10 of the 11 cases were losing bone at a rate less than the post-menopausal mean. Eight cases appear at both dose levels, and in 6 of these the rate of loss was reduced by raising the dose of vitamin D.

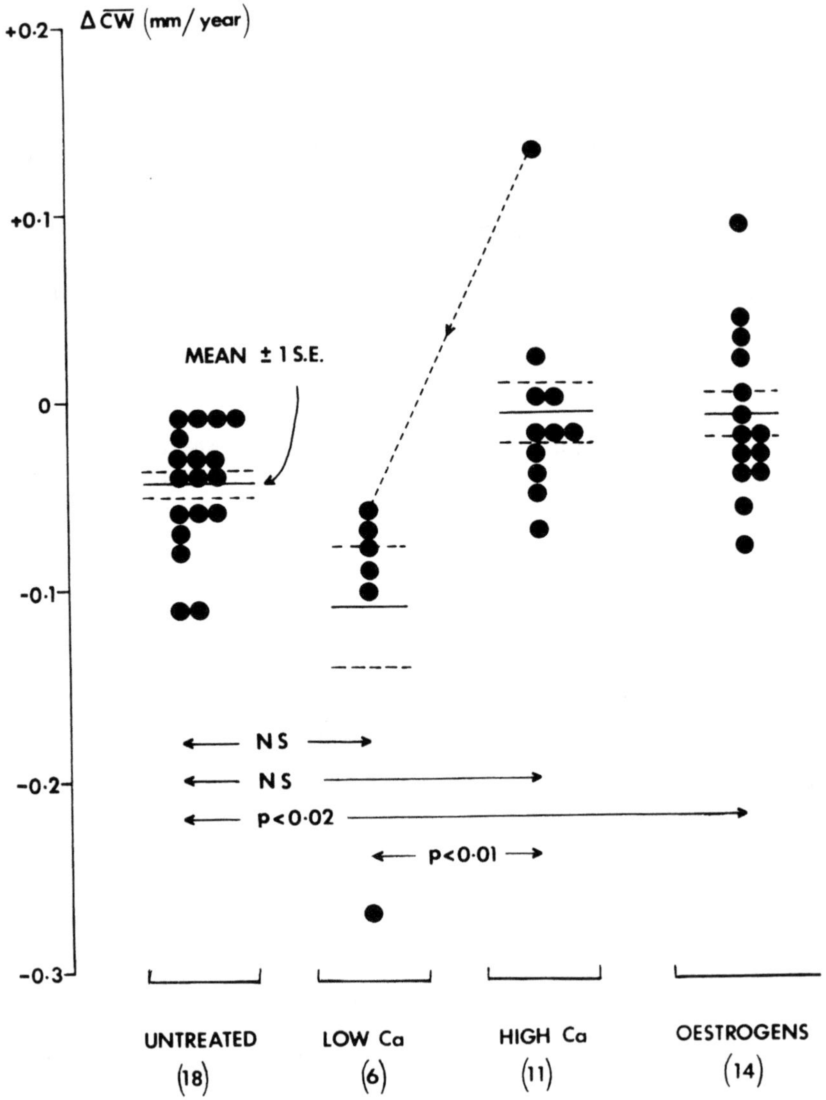

Fig. 2. The rate of change of average cortical width, $\Delta\overline{CW}$ mm/year, in groups of normal post-menopausal women treated with low and high calcium diets, and with oestrogens. The group of untreated post-menopausal controls is also included for comparison

Fig. 6 shows the relationship between the radiocalcium absorption rate (α) and the rate of change of average cortical width in 25 observations on 18 cases. The α value used was the mean value (if more than one was available) during the period within which the rate of bone loss was determined. It will be seen that the rate of loss of cortical width was greater than the post-menopausal mean in 8 of the 13 observation periods during which α was less than 0.6/h and less than the post-menopausal mean in 10 of the 12 observation periods during which α was 0.6/h or over.

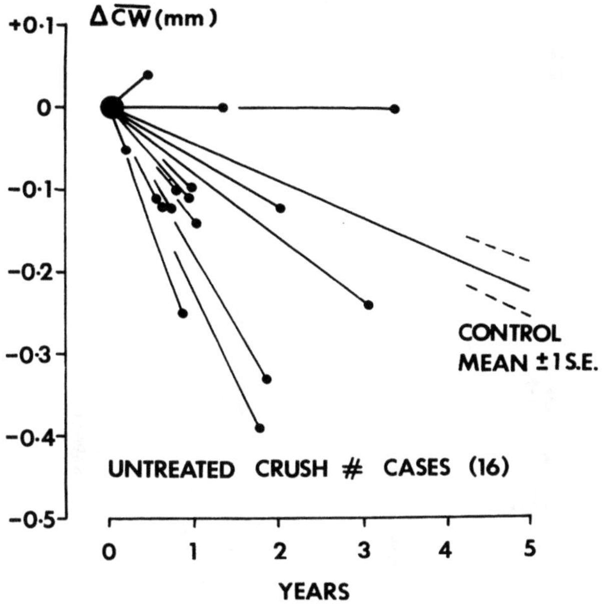

Fig. 3. Sequential changes in average cortical width, $\Delta\overline{CW}$ mm, in 16 untreated post-menopausal women with crush fractures

IV. Discussion

The calculation of the change in the average cortical width from duplicate measurements of the total and medullary widths of six metacarpals represents a significant advance in the detection of bone loss which enables the effect of therapy to be monitored in a relatively short time on quite small numbers of individuals. The rate of bone loss in the normal post-menopausal women is such that a significant change can be detected in the average woman in about 1.7 years, though naturally the length of time that must elapse before a significant loss is observed depends upon the rate of loss in the individual case. The procedure clearly shows that pre-menopausal women as a group do not lose bone and that the average cortical width of post-menopausal women decreases at a mean rate of 0.44 mm/year, representing an annual loss from the metacarpal cortex of about 1%, a figure which is very similar to the rate previously reported in cross-sectional studies (NORDIN, 1971).

That this loss of bone can be delayed or prevented by oestrogen therapy has previously been reported in two cross-sectional studies (DAVIS et al., 1970; MEEMA and MEEMA, 1968) and in one sequential study (AITKEN et al., 1973), and we have now confirmed this in a sequential study on a small series of 14 women. These observations are compatible with the biochemical evidence of increased bone resorption following oophorectomy previously reported by us (GALLAGHER et al., 1972) and with the reversal of these biochemical changes which can be achieved by oestrogen therapy (GALLAGHER and NORDIN, 1973). It should be emphasized that the observed decrease in cortical width in post-menopausal women is entirely attributable to endosteal bone loss. Similarly, the reduced rate of loss

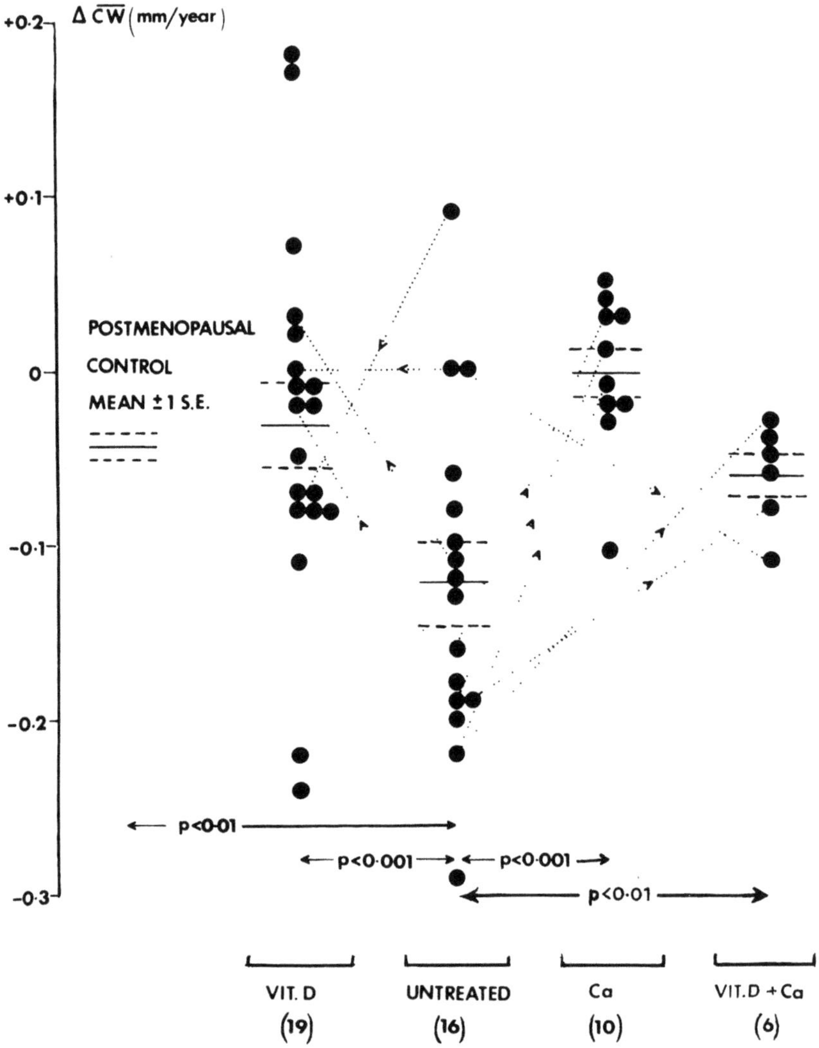

Fig. 4. The rate of change of average cortical width, $\Delta\overline{cw}$ mm/year, in post-menopausal women with crush fractures. The untreated fracture group is compared with the post-menopausal control group, and with other groups of fracture cases on vitamin D, calcium and vitamin D + calcium therapies

in oestrogen-treated women is entirely attributable to the reduced rate of endosteal bone loss since there is on average no significant change in total width during the period of observation in either the control group or the oestrogen-treated group.

Our data suggest that calcium intake also has an effect upon post-menopausal bone loss but perhaps only when the intake is exceptionally low or exceptionally high. Within the normal range of dietary calcium intake, we have not been able to establish any relationship between the rate of post-menopausal bone loss and dietary calcium as judged by

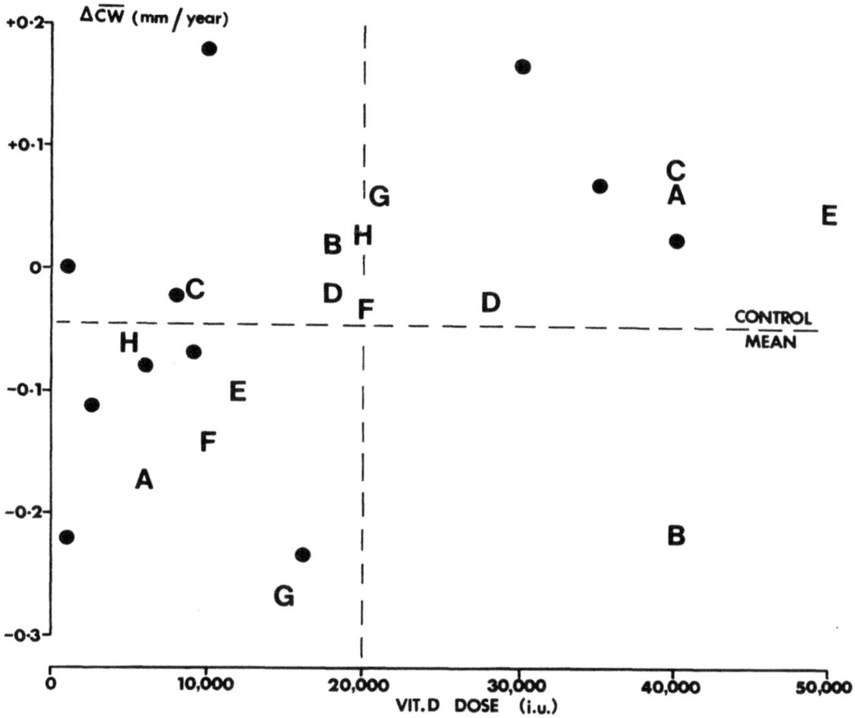

Fig. 5. The relationship between the rate of change of average cortical width, $\Delta\overline{cw}$ mm/year, and the dose of vitamin D given to post-menopausal women with crush fractures. Two observations on the same patient are identified by letters

diet history, and in this respect we agree with the observations of GARN et al. (1967) and WALKER (1972) who have failed to detect any relation between calcium intake and bone mass in cross-sectional studies. This may mean either that even post-menopausal women can adapt to a certain range of calcium intakes, or that the estimates of dietary calcium are too imprecise to yield correlations between bone mass or bone loss and calcium intake within the normal range. However, the supplementation of the diet with a gram of calcium, elevating the total calcium intake to 1,500-2,000 milligram daily, appears to have reduced the average rate of loss in the subjects we have studied, and in this respect our data bear out the findings of ALBANESE et al. (1973), who reported a significant reduction in the rate of bone loss in 19 elderly subjects given calcium and vitamin D supplements. Conversely, our data on a small group of idiopathic post-menopausal stone-formers suggest that the rate of loss is increased by a low calcium diet (estimated content 250 mg Ca daily). The rate of loss in this group was comparable with that in the group of untreated crush fracture cases, and we believe the risk of producing clinical bone disease to be so great that further observations on the effect of a low calcium diet in post-menopausal women will be impossible on ethical grounds. Although post-menopausal bone loss is hormonal in the sense that its immediate cause is increased bone resorption due to oestrogen deficiency, the rate of this bone loss can be modified if the calcium intake is varied sufficiently, either upwards or downwards, as would be expected if post-menopausal bone resorption were parathyroid hormone-mediated.

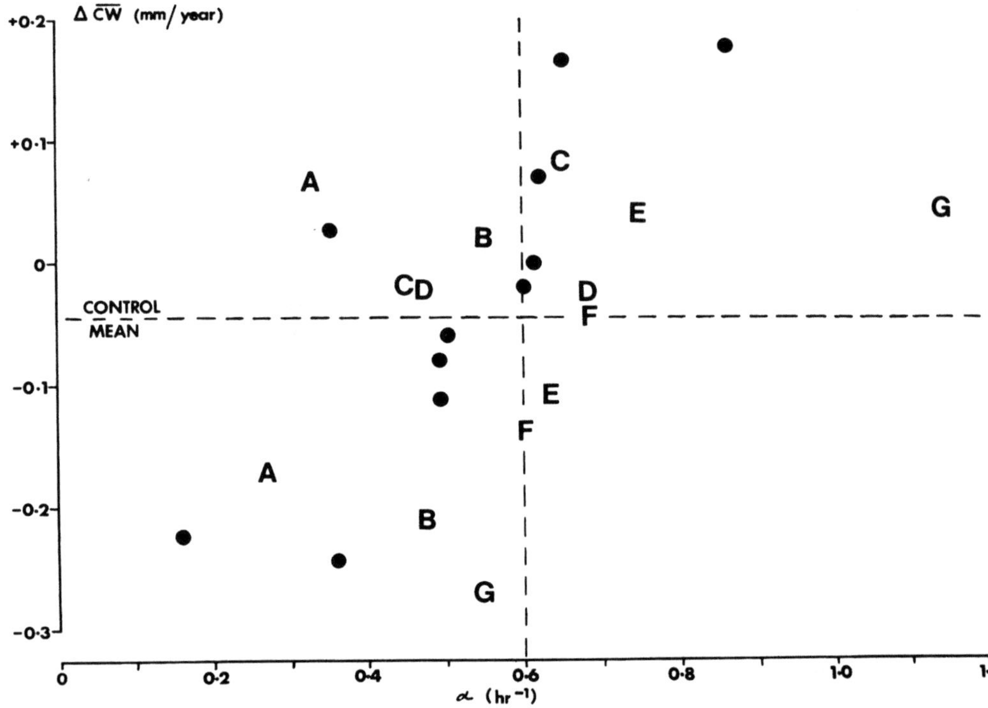

Fig. 6. The relationship between the rate of change of average cortical width, $\Delta\overline{cw}$ mm/year, and the radiocalcium absorption rate, α hr^{-1}, in post-menopausal women with crush fractures treated with vitamin D. Two observations on the same patient are identified by letters

The observed acceleration of bone loss in untreated crush fracture cases is of particular interest. Many workers have noted that, although patients with crush fractures as a group tend to have less peripheral bone than age-matched controls, many or most of their values fall within the normal range, albeit towards the lower end (GALLAGHER et al., 1973). Yet it is clear that crush fractures of the spine tend to occur in individuals with reduced amounts of vertebral trabecular bone (ARNOLD, 1973) and this reduction in trabecular bone mass is borne out by iliac crest biopsy data (MEUNIER et al., 1973). These observations are compatible with the concept that patients with crush fractures are suffering from some form of "calcium deficiency" as a result of which trabecular bone (possibly because of its increased surface/volume ratio) is preferentially mobilized in the same way that trabecular bone is preferentially mobilized when animals are placed on low calcium diets (NORDIN, 1960). Extrapolation of our sequential cortical width measurements back to the age at menopause does not support the suggestion that patients with crush fractures reached the menopause with a reduced amount of bone (MORGAN, 1973). Rather it appears that these patients' bone mass was on average normal at the time of the menopause and that there has been accelerated loss since then. The fact that many of the peripheral bone measurements still fall within the normal range when crush fractures develop is simply due to the large range of normal values at maturity; clearly sub-normal peripheral bone values have not yet been reached in all cases, but if the bone loss were to continue at its untreated rate the values would sooner or later fall below the normal lower limit.

In seeking to identify the cause of this accelerated bone loss we have found reduced absorption of calcium in balance studies on untreated crush fracture cases (GALLAGHER et al., 1973). We suggest that this malabsorption of calcium, combined with the increased sensitivity of bone to resorption which characterizes oestrogen deficiency, may be sufficient to explain spinal osteoporosis, although other contributory factors as yet unidentified cannot be excluded. On the assumption that malabsorption of calcium is at least a contributory factor, we have devoted our main therapeutic effort to correcting this malabsorption with vitamin D. Some patients respond to 1-2,000 units daily, but the majority require 10-20,000 units and a few do not respond even to 40,000 units daily. Such large doses might rule out true vitamin D deficiency in these cases but this must await a clearer definition of vitamin D requirement, particularly in the elderly, and the measurement of plasma vitamin D levels in this type of case. It is not impossible that the fundamental abnormality is not so much a deficiency of vitamin D as an alteration in its metabolism. In some preliminary studies with 1α-hydroxycholecalciferol we have found that small doses of this compound completely correct malabsorption of calcium even in crush fracture cases previously resistant to very large doses of vitamin D2. Whether this indicates an abnormality in vitamin D metabolism it is too early to say, but at least it indicates that the malabsorption of calcium in the crush fracture syndrome can be corrected and is not apparently due to any intrinsic abnormality in the gut.

In the vitamin D-treated group as a whole, a highly significant reduction in the rate of bone loss has been observed. Our data suggest that the response to vitamin D may be related to the dose used and to the calcium absorption rate that can be achieved; where vitamin D therapy has been effective in correcting malabsorption of calcium it has usually been effective in reducing the rate of bone loss. In this connection it should be noted that the beneficial effects of fluoride therapy reported by RIGGS et al. (1973) could possibly be attributable to the relatively large doses of vitamin D which they gave to their cases at the same time. Our data also indicate that the administration of calcium supplements to crush fracture patients with normal calcium absorption may reduce the rate of bone loss, just as it does in the normal controls.

In conclusion, we would suggest that there is a strong case for oestrogen replacement and/or calcium supplements in post-menopausal women. In view of the other possible benefits of oestrogen therapy it is presumably the treatment of choice, but when for any reason oestrogen therapy is contra-indicated or declined, calcium supplementation should be recommended. Conversely, it seems unwise to place post-menopausal women on low calcium diets and in this connection it should perhaps be noted that the elimination of milk from reducing diets because of its high calorie content may yield a diet that is possibly dangerously low in calcium.

As far as the treatment of post-menopausal crush fracture cases is concerned, we would suggest that the routine investigation of such cases should include the measurement of calcium absorption and the estimation of oestrogenic activity from a vaginal smear or by the measurement of total oestrogen excretion. Patients with malabsorption of calcium should be given vitamin D in a dose of 10-20,000 units daily; those with evidence of severe oestrogen insufficiency should receive oestrogen replacement therapy, and all of them should receive calcium supplements. Such therapy at least reduces the rate of bone loss, even if it does not result in the replacement of the bone that has been lost. From the purely clinical point of view, the results are excellent.

References

AITKEN, J.M., HART, D.M., LINDSAY, R.: Oestrogen replacement therapy for prevention of osteoporosis after oophorectomy. British Medical Journal 2, 515-519 (1973).

ALBANESE, A.A., EDELSON, A.H., WOODHULL, M.W., LORENZE, E.J., WEIN, E.H., ORTO, L.A.: Effect of a calcium supplement on serum cholesterol, calcium, phosphorus and bone density of "normal, healthy" elderly females. Nutrition Reports International 8, 119-130 (1973).

ARNOLD, J.S.: Amount and quality of trabecular bone in osteoporotic vertebral fractures. Clinics in Endocrinology and Metabolism 2, 221-238 (1973).

BULLAMORE, J.R., GALLAGHER, J.C., WILKINSON, R., NORDIN, B.E.C., MARSHALL, D.H.: The effect of age on calcium absorption. Lancet II, 535-537 (1970).

DAVIS, M.E., LANZL, L.H., COX, A.B.: Detection, prevention and retardation of menopausal osteoporosis. Obstetrics and Gynecology 36, 187-198 (1970).

GALLAGHER, J.C., AARON, J., HORSMAN, A., MARSHALL, D.H., WILKINSON, R., NORDIN, B.E.C.: The crush fracture syndrome in post-menopausal women. Clinics in Endocrinology and Metabolism 2, 293-315 (1973).

GALLAGHER, J.C., NORDIN, B.E.C.: Oestrogens and calcium metabolism. In: Aging and Estrogens. Frontiers of Hormone Research 2, pp. 98-117. Basel: Karger 1973.

GALLAGHER, J.C., YOUNG, M.M., NORDIN, B.E.C.: Effects of artificial menopause on plasma and urine calcium and phosphate. Clinical Endocrinology 1, 57-64 (1972).

GARN, S.M., ROHMANN, C.G., WAGNER, B.: Bone loss as a general phenomenon in man. Federation Proceedings 26, 1729-1736 (1967).

HORSMAN, A., NORDIN, B.E.C.: The quantitative assessment of sequential changes in cortical bone geometry. In: Proceedings of the 9th European Symposium of Calcified Tissues (ed. H. Czitober and J. Eschberger) pp. 185-188. Vienna: Facta-Publication 1973.

HORSMAN, A., SIMPSON, M.: A morphometric method for the measurement of sequential changes in cortical bone geometry. To be published (1974).

MEEMA, H.E., MEEMA, S.: Prevention of post-menopausal osteoporosis by hormone treatment of the menopause. Canadian Medical Association Journal 99, 248-251 (1968).

MEUNIER, P., COUPRON, P., EDOUARD, C., BERNARD, J., BRINGUIER, J., VIGNON, G.: Physiological senile involution and pathological rarefaction of bone. Clinics in Endocrinology and Metabolism 2, 239-256 (1973).

MORGAN, D.B.: Aging and Osteoporosis, in particular spinal osteoporosis. Clinics in Endocrinology and Metabolism 2, 187-201 (1973).

NORDIN, B.E.C.: Osteoporosis, osteomalacia and calcium deficiency. Clinical Orthopaedics 17, 253-258 (1960).

NORDIN, B.E.C.: The pathogenesis of osteoporosis. Lancet I, 1011-1014 (1961).

NORDIN, B.E.C.: Clinical significance and pathogenesis of osteoporosis. British Medical Journal 1, 571-576 (1971).

NORDIN, B.E.C., SMITH, D.A.: Diagnostic procedures in disorders of calcium metabolism. London: Churchill 1965.

RIGGS, B.L., JOWSEY, J., KELLY, P.J., HOFFMAN, D.L., ARNAUD, C.D.: Studies on pathogenesis and treatment in post-menopausal and senile osteoporosis. Clinics in Endocrinology and Metabolism 2, 317-332 (1973).

WALKER, A.R.P.: The requirement of calcium: should low intakes be supplemented? American Journal of Clinical Nutrition 25, 518-530 (1972).

Effect of Calcitonin in Paget's Disease

D. Hioco

Previous studies have indicated the beneficial effects of porcine, salmon and human calcitonin on the clinical and biochemical disturbances in Paget's disease (BELL et al., 1970; BIJVOET et al., 1968; HADDAD et al., 1970; SHAI et al., 1970; WOODHOUSE et al., 1971).

However, an assessment of these reports to determine the therapeutic activity of calcitonin in controlling the disease is limited by the small number of patients studied and the short periods of observation.

The present paper reviews our experience in the treatment of Paget's disease with extracted porcine calcitonin and describes the clinical efficacy and biochemical response.

I. Subjects and Methods

1. Patients

Thirty-nine patients with symptomatic Paget's disease were selected and treated for periods of 4 to 30 months; 34 patients were treated with Armour extractive porcine calcitonin, and 5 patients with Sandoz synthetic salmon calcitonin.

The patients received calcitonin in doses ranging from 160 MRC Units per day, to 100 MRC Units 3 times a week, administered in one intramuscular injection with gelatine diluent for porcine calcitonin, and subcutaneously in one injection with buffered isotonic saline solution for salmon calcitonin.

Side-effects, i.e. mild nausea and transient erythema, occurred in 15% patients.

2. Methods

Measurements were made during the basal period and at 4-months intervals during treatment, for as long as 20 months in some patients, of serum and urinary calcium, phosphate and urinary hydroxyproline, determined by standard procedures. Calcium-balance (REIFENSTEIN et al., 1945) and ^{45}Ca kinetic studies (AUBERT and MILHAUD, 1960) were performed during 10-day periods. The Ca-balance studies were carried out in a metabolic ward while the patients were ingesting a constant calcium diet approximating the content of that eaten at home.

II. Results

1. Clinical Response

Bone pain was an outstanding feature in this group and moderate to dramatic relief was observed in 70% of patients. In almost every patient, maximum clinical benefits were achieved during the first 4 to 10 months of treatment. No discernable changes in radiographs of the skeleton were noted in the majority of patients, despite the prolonged period of treatment. In two cases only was an apparent radiologic improvement observed.

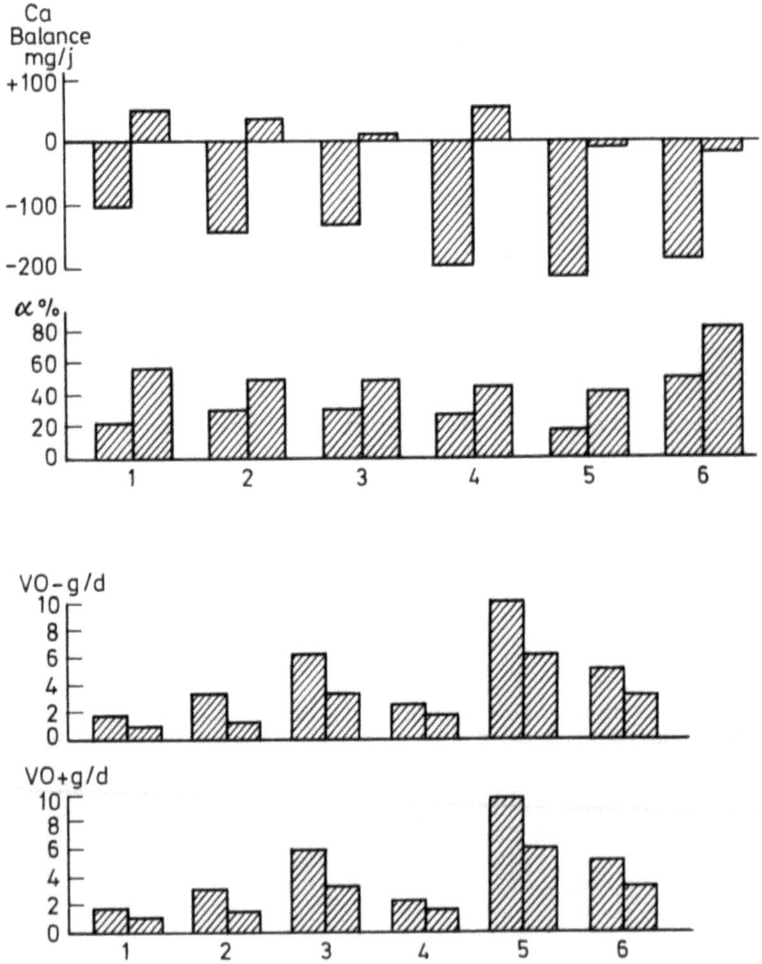

Fig. 1. Calcium balance and kinetic data: biochemical and kinetic data in 6 patients, before treatment and after 4 months of treatment

2. Biochemical Data

a) Biochemical Parameters

The fasting serum calcium and serum inorganic phosphate did not change after 4 to 8 months of treatment. Urinary calcium decreased significantly in a great number of patients who had a high urinary calcium level before treatment.

b) Calcium Balance and Kinetic Data

The calcium balance, which was markedly negative at the baseline, became positive or equilibrated and at the same time the true absorption index, measured by the technique of AUBERT and MILHAUD (1960) with ^{45}Ca, increased in every case.

Furthermore, there was a highly significant reduction in the bone turnover in all cases, as reflected in a fall of Vo^+ and Vo^- (Fig. 1). If Vo^+ can be assumed to represent an approximate measure of bone formation, and Vo^- an approximate measure of bone destruction, the study of the combined radiocalcium turnover and the calcium balance data indicates that calcitonin was effective in inhibiting bone resorption, and the consequent effect on bone formation was usually inhibitory as well.

c) Urinary Hydroxyproline and Serum Alkaline Phosphatase

The sequential changes in these two parameters were noted at 4, 8, 12 and 20 months. As shown in Fig. 2, after 4 months of treatment a decrease in serum alkaline phosphatase and urinary hydroxyproline was recorded, as compared with the initial values taken as 100% (Fig. 2). The initial absolute values were 30 to 170 K.A. Units for alkaline phosphatase, and 80 to 900 mg per day, for urinary hydroxyproline. After 8 months of treatment a plateau effect was observed in these two parameters. This plateau was noted after 12 months, despite wide variations due to a return towards the pretreatment values in about 20% of the patients. In 5 individual patients treated for 20 months, after a fall in the hydroxyproline and serum alkaline phosphatase at 4 months, the response pattern became variable when treatment was prolonged, and all but one seemed to relapse.

III. Conclusions

Considerable clinical and biochemical improvement was recorded in 39 patients with Paget's disease, treated with porcine calcitonin or salmon calcitonin. After an initial fall in urinary hydroxyproline and serum alkaline phosphatase, a plateau phenomenon, or a relapse, was observed in 12 to 16 patients treated for a period of 12 months, at a level of about 60% of the baseline value.

Five patients were treated for 20 months, but the parameters showed great variations, and no conclusions could be drawn because of the reduced number of cases studied.

This is at variance with the good clinical response observed in these patients, as manifested in persisting relief of bone pain and improved physical mobility. There is no satisfactory explanation for the plateau phenomenon. Data collected from the literature have failed to show a

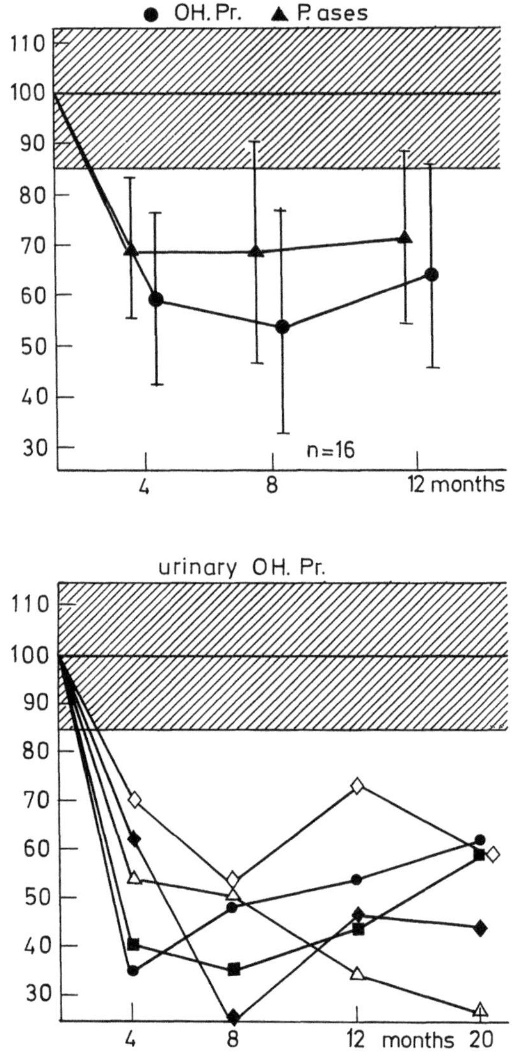

Fig. 2. Changes in serum alkaline phosphatase and urinary hydroxyproline during treatment

clear causal relationship between this phenomenon, antibody formation, or the possibility of a calcitonin-induced secondary hyperparathyroidism.

However, one may ask whether the 160 MRC Units of porcine calcitonin material administered daily were able to elicit maximum response in all cases.

References

AUBERT, J.P., MILHAUD, G.: Méthode de mesure des principales voies du métabolism calcique chez l'homme. Bioch. Bioph. Acta 39, 122-128 (1960).

BELL, N.H., AVERY, S., JOHNSTON, C.C. Jr.: Effects of Calcitonin in Paget's disease and polyostotic fibrous dysplasia. J. Clin. Endocr. 31, 283-90 (1970).
BIJVOET, O.L.M., VAN DER SLUYS VEER, J., JANSEN, A.P.: Effects of calcitonin on patients with Paget's disease, thyrotoxicosis or hypercalcemia. Lancet I, 876-881 (1968).
HADDAD, J.G. Jr., BIRGE, S.J., AVIOLI, L.V.: Effects of prolonged thyrocalcitonin administration on Paget's disease of bone. New Engl. J. Med. 283, 549-555 (1970).
REIFENSTEIN, E.C., ALBRIGHT, F., WELLS, S.L.: The accumulation, interpretation and presentation of data pertaining to metabolic balances, notably those of Calcium, Phosphorus and Nitrogen. J. Clin. Invest. 5, 367-372 (1945).
SHAI, F., BAKER, R.K., WALLACH, S.: The clinical and metabolic effects of porcine calcitonin on Paget's disease of bone. J. Clin. Invest. 50, 1927-40 (1970).
WOODHOUSE, N.J.Y., et al.: Human calcitonin in the treatment of Paget's disease of bone. Lancet I, 1139-1143 (1971).

Biochemical and Radiological Observations in Patients with Osteitis Deformans Treated with Synthetic Human Calcitonin*

P. B. Greenberg, F. H. Doyle, M. T. Fisher, C. J. Hillyard, G. F. Joplin, J. L. McLoughlin, I. MacIntyre, J. Pennock, M. K. Ward and N. J. Y. Woodhouse

I. Summary

Radiological and biochemical changes in 28 adults with Paget's disease of bone treated for 15 months to 4 years with synthetic human calcitonin are discussed. They indicate that human calcitonin is effective in long-term treatment.

II. Patients

Twenty-eight patients (age 47-74 years) were treated with synthetic human calcitonin for from 15 months to 4 years (mean 27 months). All had active Paget's disease as established by clinical, biochemical and radiological criteria. All had local symptoms considered due to Paget's disease and most were referred for treatment because of symptoms or complications. Mean serum alkaline phosphatase concentration was 86 King Armstrong units (range 12-363; normal 3-13) and mean urinary excretion of total hydroxyproline per 24 hours was 209 mg (range 28-1245; normal <45).

III. Methods

1. Treatment

Synthetic human calcitonin (each ampule containing 0.5 mg calcitonin M (Ciba 47 175-Ba) with 40 mg mannitol, in a lyophylised form) was administered by intramuscular injection dissolved in physiological saline. The initial dose was between 0.5 mg once weekly and 0.5 mg three times a day; the dose chosen in individuals depended only on calcitonin availability. Maintenance treatment was on doses between 0.5 mg twice daily and 0.5 mg once weekly. This dose was continued until biochemical remission occured and was maintained for at least 6 months, when the dose was reduced. If serial determinations indicated that the stage was reached in which no further improvement in biochemical parameters occured over a 3-6 month period, the dose was increased as soon as calcitonin supplies permitted. Treatment was stopped for brief periods in some of the first patients treated in order to determine if and at what rate relapse of the disease might occur.

+ We are indebted to Ciba-Geigy Ltd. for support and for supplies of synthetic human calcitonin. The patients were nursed by Sister K. Boon and Staff in the Metabolic Unit.

2. Biochemical Assessment

The patients were admitted to the Metabolic Unit for 4-5 day periods on a standard diet, before treatment and at intervals during treatment. Determinations of serum alkaline phosphatase concentrations and urinary total hydroxyproline excretion rates were measured each 3-6 months (WOODHOUSE et al., 1971) as was the 24-hour strontium space (FRASER et al., 1960). Plasma in dilutions as high as 1:12.5 were incubated with ^{125}I-labeled synthetic human calcitonin in order to detect antibodies formed during treatment; free and bound ^{125}I-labeled calcitonin were separated by means of dextran-coated charcoal. Immunologically reactive parathyroid hormone was measured in the fasting morning plasma of patients just before their next dose of calcitonin. Antibody was supplied by Burroughs Wellcome (211/32) and highly purified bovine parathyroid hormone was used as tracer and standard.

3. Radiological Assessment

Surveys included bones in which the patient had noted symptoms or over which the skin temperature appeared to be increased or which were judged clinically to be deformed.Irrespective of these features most patients had additional radiographs of skull, chest, pelvis and hands. Radiographs were taken when doses were changed and at intervals of 4-6 months during continuous treatment.

Serial radiographs were analyzed by one observer (FHD) who had no knowledge of individual treatment schedules. In classifying and measuring the changes that occured in several bones over different time periods with various doses calcitonin, the interval on a given dose was called a "treatment-period", and the changes occuring in the different bones during such peroids were summated (DOYLE et al, 1974). Radiological regression was defined as including reduction of external bone volume and a return to a more normal bone shape.

IV. Results

1. Biochemical

a) The mean reduction on current treatments of serum alkaline phosphatase is to 64% of pre-treatment values (range 5-105) and the mean reduction of urinary hydroxyproline excretion is to 53% of control values (range 12-100). Of the 5 patients who have reductions of less than 15% in either of these parameters, 4 were only receiving 0.5 mg weekly.
b) The rate of reduction of these parameters has been consistently greater in the first 3 months than in subsequent periods.
c) Five patients have been maintained clinically and biochemically normal on doses of at least 0.5 mg daily.
d) When treatment is ceased, prompt biochemical relapse occurs.
e) Changes in serial strontium space measurements have been similar to the changes in the other biochemical parameters.
f) Calcium balance in two patients reflecting treatment for 45 months (FA) and 39 months (IJ) was found to be + 0.3 mEq/day and + 3.4 mEq/day respectively. Both patients had been shown to be in negative balance before treatment and had an acute improvement in calcium balance on starting calcitonin.
g) Antibodies binding ^{125}I-labeled synthetic human calcitonin have not

been detected in the plasmas of any of the patients after intervals of treatment ranging from 4-36 months (mean 17.8 months).
h) Basal immunoreactive plasma parathyroid hormone concentrations have been within the normal range in all the 13 patients studied so far.

2. Radiological

a) 12 patients have shown radiographic changes.
b) 7 patients have shown radiological "regression".
c) The radiological responses were found to be related to calcitonin dosage.

V. Discussion

The results indicate that changes in biochemical and radiological parameters occur during treatment with synthetic human calcitonin, and that the direction of the changes is towards normalisation of these parameters. The natural history of Paget's disease, judged biochemically (WOODARD, 1959) or radiologically (GRAINGER and LAWS, 1957) is one of progression, suggesting that the changes seen by us reflect the action of calcitonin. The relapse seen after cessation of treatment and the dose/response relationship of the radiological findings further confirms that the effects are due to calcitonin. Human calcitonin is not immunogenic and in contrast to the results with salmon and pig calcitonins (SINGER et al., 1972; HADDAD and CALDWELL, 1972), acquired resistance associated with antibody formation does not occur.

The reason for the smaller rate of improvement in biochemical parameters during prolonged treatment is unknown. Possibilities include (a) inadequate dosage, (b) an acquired biological ineffectiveness, or (c) the skeletal remodeling which occurs during treatment with human calcitonin; biochemical parameters reflect this remodeling in addition to the activity of the disease process. Thus assessment of disease response must be based on both biochemical and morphological changes.

References

DOYLE, F.H., PENNOCK, J., GREENBERG, P.B., JOPLIN, G.F., MACINTYRE, I.: Radiological evidence of a dose-related response to long term treatment of Paget's disease with human calcitonin. Br. J. Radiol. 47, 1(1974).
FRASER, R., HARRISON, M., IBBERTSON, K.: The rate of calcium turnover in bone. Quart. J. M. 29, 85 (1960).
GRAINGER, R.G., LAWS, J.W.: Paget's disease - active or quiescent? Brit. J. Radiol. 30, 120 (1957).
HADDAD, J.G., CALDWELL, J.G.: Calcitonin resistance: clinical and immunologic studies in subjects with Paget's disease of bone treated with porcine and salmon calcitonins. J. clin. Invest. 51, 3133 (1972).
SINGER, F.R., ALDRED, J.P., MEER, R.M., KRANE, S.M., POTTS, J.T., BLOCH, K.J.: An evaluation of antibodies and clinical resistance to calcitonin. J. clin. Invest. 41, 2331 (1972).
WOODARD, H.Q.: Long term studies of blood chemistry in Paget's disease of bone. Cancer 12, 1226 (1959).
WOODHOUSE, N.J.Y., REINER, M., BORDIER, P., KALU, D.N., FISHER, M.T., FOSTER, G.V., JOPLIN, G.F., MACINTYRE, I.: Human calcitonin in the treatment of Paget's bone disease. Lancet 1, 1139 (1971).

Influence of a Diphosphonate and 1,25-Dihydroxycholecalciferol on Calcium Metabolism*

J.-P. Bonjour, U. Trechsel, U. Tröhler, H. Fleisch, L.A. Baxter and H.F. DeLuca

The diphosphonate, disodium ethane-1-hydroxy-1,1-diphosphonate (EHDP) has been shown to modify several variables of calcium metabolism in the rat (GASSER et al., 1972). In their kinetic and balance study the authors (GASSER et al., 1972) demonstrated that EHDP according to the dose administered could evoke opposite effects on some variables of calcium metabolism. At a daily dose of 1 mg P/kg s.c. EHDP brought about a decrease in the mobilization of calcium from bone, but did not affect significantly the rate of bone mineralization. At this dose also an augmentation in the intestinal absorption and retention of calcium was observed, and no alteration in the process of cartilage and bone mineralization could be detected morphologically (SCHENK et al., 1973). In the same studies (GASSER et al., 1972; SCHENK et al., 1973) it was observed that the effects of EHDP differ markedly at a dose ten times greater (10 mg P/kg s.c.). In this case the diphosphonate induced a decrease in the rate of bone mineralization accompanied by a significant reduction in intestinal absorption and retention of calcium (GASSER et al., 1972). Morphologically an impairment in the mineralization of growth cartilage and bone was described in rats treated with the same large doses of EHDP (SCHENK et al., 1973). The inhibitory action of large doses of EHDP on both bone mineralization and intestinal calcium absorption suggested (MORGAN et al., 1971) the existence of a link between the two processes, in accordance with the hypothesis conceived by NICOLAYSEN (1943). A more detailed study of the intestinal response to large doses of EHDP was then carried out. It revealed that the reduction in calcium absorption was associated with a biochemical depression in calcium binding protein, Ca^{++}-stimulated adenosine triphosphatase and alkaline phosphatase of the intestinal mucosa (BONJOUR et al., 1973a). All these findings suggested very strongly that EHDP given at high doses interfered with the action or metabolism of vitamin D. Such an interference was then disclosed at the level of the transformation of 25-hydroxycholecalciferol (25-OHD_3) into the active vitamin D_3 metabolite 1,25-dihydroxycholecalciferol (1,25-$(OH)_2D_3$. This was shown both by the demonstration of a preventive and corrective effect of 1,25-$(OH)_2D_3$ on the EHDP-induced depression of intestinal calcium absorption (BONJOUR et al., 1973b) and by the observation that rats (HILL et al., 1973) or chicks (BAXTER et al., submitted) treated with large doses of EHDP are not able to synthesize normally 1,25-$(OH)_2D_3$. The mechanism of this inhibition remains to be elucidated. The inhibition could be secondary to the blockage of bone mineralization, and the consecutive change in calcium or phosphorus metabolism. The production of 1,25-$(OH)_2D_3$ has been shown to be inversely related to the level of the calcemia (BOYLE et al., 1971) and phosphatemia (TANAKA et al., 1973). Therefore the hypercalcemia observed with the large

+ Some of the work presented has been supported by the US National Institutes of Health (Grant No. AM 07266), the Schweizerischer Nationalfonds für wissenschaftliche Forschung (Grant No. 3.326.70), by Procter and Gamble and a contract from the US Atomic Energy Commission No. AT-(11-1)-1668.

Table 1. Influence of dietary calcium and phosphorus on intestinal calcium absorption in EHDP-treated rats

Dietary Ca g %	1.1		0.1		1.1	
Dietary P g %	1.1		1.1		0.2	
	Control	EHDP	Control	EHDP	Control	EHDP
Mean calcium absorbed (%)	65.7	42.5++	80.1	74.0	86.3	46.1+++
SEM	5.1	2.6	4.6	2.8	2.8	3.2
Number of animals	8	8	4	4	4	4

Rats were fed the indicated diets for 14 days. Over the last seven days the animals received either EHDP (10 mg P/kg s.c. for 7 days) or equivalent volumes of 0.15 M NaCl vehicle. On the 15th day after fasting overnight, calcium absorption was measured by the in situ duodenal loop method. The duodenal segments were filled with 0.5 ml of 0.15 M NaCl solution containing 0.1 uCi ^{45}Ca. $CaCl_2$ was added at a concentration of $4 \cdot 10^{-4}$ M. Results represent the percentage of ^{45}Ca disappearance from the duodenal lumen during the 15 minutes incubation period.

++ $p < 0.01$; +++ $p < 0.001$, as compared to the corresponding control group. SEM = Standard error of the mean.

In control rats, lowering dietary calcium or phosphate led to a rise in calcium absorption capacity. EHDP-treated rats maintained on a low calcium diet have a calcium absorption close to that of control, whereas the effect of EHDP was not modified by lowering the phosphorus content of the diet.

dose of EHDP (GASSER et al., 1972; BONJOUR et al., 1973b) might play a role in the inhibition of 1,25-$(OH)_2D_3$ production. However, if so, it is unlikely to be due to a decrease in parathyroid hormone secretion, since the decreased calcium absorption was neither corrected nor prevented by supplementing EHDP-treated rats with various doses of parathyroid hormone. Furthermore in thyroparathyroidectomized animals, large doses of EHDP also induced a decrease in calcium absorption.

To date there is no experimental fact which suggests that inorganic phosphorus should play a role in the inhibition of 1,25-$(OH)_2D_3$ mediated by EHDP. Indeed, in rats contrary to what has been observed in man (RECKER et al., 1973), EHDP administered at various doses did not raise serum inorganic phosphorus (GASSER et al., 1972). Furthermore in contrast with the effect of a low calcium diet, decreasing the phosphate intake did not influence the inhibitory effect of EHDP on calcium absorption (Table 1). Finally the inorganic phosphorus content of the renal cortex was not modified in rats treated for 7 days with EHDP at a daily dose of 10 mg P/kg s.c.

The possibility that EHDP has a direct action on the activity or content of the 25-hydroxycholecalciferol-1-hydroxylase has to be considered, since EHDP can penetrate the renal cell (TRÖHLER et al., 1973).

Since EHDP given in large doses modifies not only intestinal calcium absorption, but also other variables of calcium metabolism, especially bone mineralization (GASSER et al., 1972), the problem arose, whether these latter changes were also due to a deficiency in 1,25-$(OH)_2D_3$.

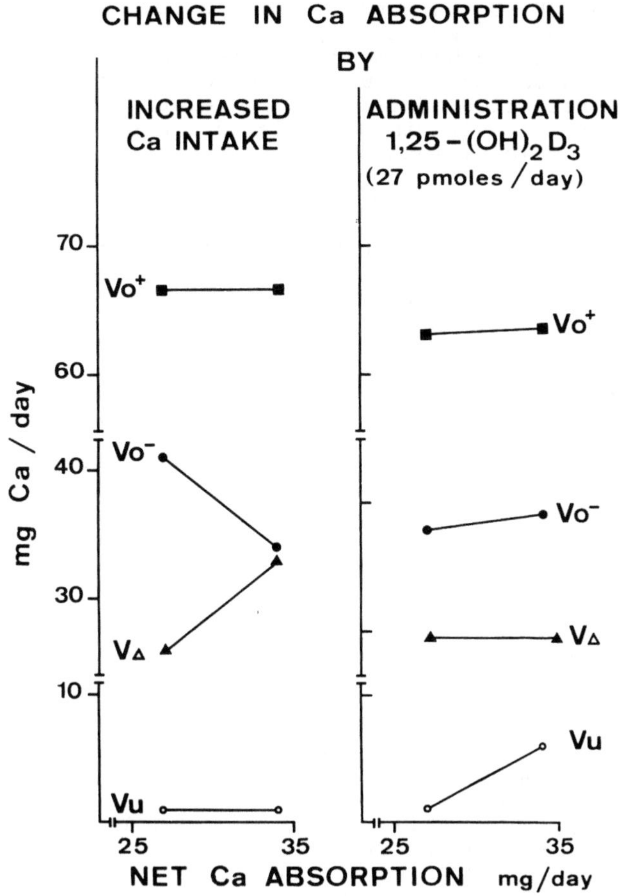

Fig. 1. The variation of several variables of calcium metabolism in vitamin D-repleted normal rats are represented in relation to a change in the net calcium absorption induced either by increasing dietary calcium or by i.p. administration of 1,25-dihydroxycholecalciferol. In both conditions the estimate of bone formation (V_{o+}) was not significantly influenced. When intestinal absorption was enhanced by greater calcium intake, bone resorption (Vo^-) decreased and a rise in calcium retention (VΔ) was observed. On dosing with 1,25-$(OH)_2D_3$ the rise in calcium absorption was not associated with a decrease in V_{o-}. Most of the additional amount of absorbed calcium was excreted in the urine (Vu), and VΔ was not modified

Recent investigations indicate very clearly that 1,25-$(OH)_2D_3$ deficiency cannot be the major cause of the EHDP-induced impairment of cartilage and bone mineralization. Indeed, in EHDP-treated rats a calcium kinetic and morphological study (BONJOUR et al., 1973c) showed that 1,25-$(OH)_2D_3$ given in doses (27 and 135 pmoles/day i.p.) normalized calcium absorption but did not prevent or correct the bone mineralization defect. Whether this defect results from a physicochemical inhibition of bone crystal formation remains to be established.

The study of the effect of 1,25-$(OH)_2D_3$ on control animals receiving a normal oral supply of vitamin D_3 (635 pmoles/day) with a calcium intake of about 50 mg revealed that the enhancement of intestinal calcium absorption was associated with an increase in urinary calcium excretion without alteration in the retention of calcium. In relation to this it is of interest to compare, in vitamin D_3-repleted control rats, the changes in some variables of calcium metabolism which accompany an augmentation in the net intestinal calcium absorption brought about either by raising the calcium intake, or by dosing the animals with 1,25-$(OH)_2D_3$. This comparison is illustrated in Fig. 1. It can be seen that bone resorption declined when the net calcium absorption was enhanced by increasing the dietary intake. This inverse relationship which has been described by several authors (AUBERT et al., 1961; SAMMON et al., 1970; MORGAN et al., submitted) can be interpreted as an homeostatic response. The administration of 1,25-$(OH)_2D_3$ given at 27 pmoles/day i.p. to vitamin D-repleted normal rats did not lead to a decrease in bone resorption. Thus the coupling between bone resorption and calcium absorption seems to be abolished by the administration of 1,25-$(OH)_2D_3$. This effect could explain both the trend to hypercalcemia and the hypercalciuria which was observed in this condition. These results also suggest that in normal growing rats, variation of calcium release from bone in response to change in calcium intake might be regulated by the 1,25-$(OH)_2D_3$ production.

References

AUBERT, J.-P., MOUKHTAR, M.S., MILHAUD, G.: Étude du métabolisme du calcium chez le rat à l'aide du calcium 45. III. Les relations entre les différents processus chez le rat normal. Rev. Franc. Etud. Clin. Biol. 6, 1034 (1961).

BAXTER, L.A., DELUCA, H.F., BONJOUR, J.-P., FLEISCH, H.A.: Inhibition of vitamin D metabolism by ethane-1-hydroxy-1,1-diphosphonate Arch. Biochem. Biophys. (in press).

BONJOUR, J.-P., RUSSELL, R.G.G., MORGAN, D.B., FLEISCH, H.A.: Intestinal calcium absorption, Ca-binding protein, and Ca-ATPase in diphosphonate-treated rats. Amer. J. Physiol. 224, 1011 (1973a).

BONJOUR, J.P., DELUCA, H.F., FLEISCH, H., TRECHSEL, U., MATEJOWEC, L.A., OMDAHL, J.L.: Reversal of EHDP inhibition of calcium absorption by 1,25-dihydroxycholecalciferol. Europ. J. Clin. Invest. 3, 44 (1973b).

BONJOUR, J.-P., DELUCA, H., FLEISCH, H., TRECHSEL, U.: Influence of 1,25-dihydroxycholecalciferol and diphosphonate on calcium metabolism. Experientia 29, 740 (1973c).

BOYLE, I.T., GRAY, R.W., DELUCA, H.F.: Regulation by calcium of in vivo synthesis of 1,25-dihydroxycholecalciferol and 21,25-dihydroxycholecalciferol. Proc. nat. Acad. Sci. (Wash.) 68, 2131 (1971a).

BOYLE, I.T., GRAY, R.W., OMDAHL, J.L., DELUCA, H.F.: The mechanism of adaptation of intestinal calcium absorption to low dietary calcium. J. Lab. Clin. Med. 78, 813 (1971b).

GASSER, A.B., MORGAN, D.B., FLEISCH, H.A., RICHELLE, L.J.: The influence of two diphosphonates on calcium metabolism in the rat. Clin. Sci. 43, 31 (1972).

HILL, L.F., LUMB, G.A., MAWER, G.A., STANBURY, S.W.: Indirect inhibition of the biosynthesis of 1,25-dihydroxycholecalciferol in rats treated with a diphosphonate. Clin. Sci. 44, 335 (1973).

MORGAN, D.B., BONJOUR, J.-P., GASSER, A.B., O'BRIEN, K., FLEISCH, H.: The influence of a diphosphonate on the intestinal absorption of calcium. Israel J. Med. Sci. 7, 384 (1971).

MORGAN, D.B., GASSER, A.B., LARGIADÈR, U., FLEISCH, H.: Effect of a diphosphonate on calcium metabolism in calcium deprived rats (submitted).

NICOLAYSEN, R.: The absorption of calcium as a function of the body saturation of calcium. Acta Physiol. Scand. 5, 200 (1943).
RECKER, R.R., HASSING, G.S., LAU, J.R., SAVILLE, P.D.: The hyperphosphatemic effect of disodium ethane-1-hydroxy-1,1-diphosphonate (EHDP™): Renal handling of phosphorus and the renal response to parathyroid hormone. J. Lab. Clin. Med. 81, 258 (1973).
SAMMON, P.J., STACEY, P.E., BRONNER, F.: Role of parathyroid hormone in calcium homeostasis and metabolism. Amer. J. Physiol. 218, 479 (1970).
SCHENK, R., MERZ, W.A., MÜHLBAUER, R.C., RUSSEL, R.G.G., FLEISCH, H.A.: Effect of ethane-1-hydroxy-1,1-diphosphonate (EHDP) and dichloromethylene diphosphonate (Cl_2MDP) on the calcification and resorption of cartilage and bone in the tibial epiphysis and metaphysis of rats. Calc. Tiss. Res. 11, 196 (1973).
TANAKA, Y., DELUCA, H.F.: The control of 25-hydroxy vitamin D metabolism by inorganic phosphorus. Arch. Biochem. Biophys. 154, 566 (1973).
TRÖHLER, U., TRELOAR, M., BONJOUR, J.-P., FLEISCH, H.: Net renal secretion of diphosphonate in rats: Inhibition by phosphate. Europ. J. Clin. Invest. 3, 272 (1973).

Effect of Various Doses of Disodium Ethane-1-Hydroxy-1,1-Diphosphonate (Sodium Etidronate, EHDP) on Suppression of the Biochemical and Histological Abnormalities in Paget's Disease of Bone*

R. G. G. Russell, C. Preston, R. Smith, R. J. Walton and C. G. Woods

Sodium etidronate (disodium ethane-1-hydroxy-1,1-diphosphonate, EHDP[TM]) is one of the diphosphonates which inhibit the growth and dissolution of calcium phosphate crystals in vitro (FLEISCH and RUSSELL, 1972; RUSSELL and SMITH, 1973; FLEISCH et al., 1972; FRANCIS et al., 1969; FLEISCH et al., 1969). In experimental animals EHDP inhibits pathological calcification and can reduce rates of bone resorption and bone turnover (FLEISCH and RUSSELL, 1972; RUSSELL and SMITH, 1973; FLEISCH et al., 1972; FRANCIS et al., 1969; FLEISCH et al., 1969; GASSER et al., 1972; JOWSEY et al., 1970; KING et al., 1971; RUSSELL et al., 1973). EHDP has been used successfully in preliminary studies in man to reduce ectopic calcification in myositis ossificans progressiva (RUSSELL and SMITH, 1973; GEHO and WHITESIDE, 1973; BASSETT et al., 1969; RUSSELL et al., 1972) and to reverse the biochemical changes associated with excess bone turnover in Paget's disease of bone (RUSSELL and SMITH, 1973; SMITH et al., 1972; BIJVOET et al., 1972). In this study we have investigated the effects of oral doses of EHDP of 0, 1, 5, 10 and 20 mg/kg/day in more than 40 patients with Paget's disease.

Some pharmacological information about EHDP is available. Thus absorption of EHDP from the gut varies in different species (MICHAEL et al., 1972); in man it ranges between 1 and 10% of an oral dose. EHDP is not metabolized and the only tissue in which it accumulates to an appreciable extent is bone (MICHAEL et al., 1972). The main route of elimination is the urine as unchanged diphosphonate. In rats the half-time of removal of the compound from the body is about 2 to 4 weeks (KING et al., 1971).

I. Plasma Phosphatase and Urine Hydroxyproline

Fig. 1 shows that the reduction in the initially elevated plasma alkaline phosphatase (AP) and urine total hydroxyproline (THP) is clearly related to the dose given. With no treatment at all, the values are remarkably constant over a 4 to 6 month period. At 1 mg/kg/day there was also very little change over a 4 to 6 month period, but with higher doses (FLEISCH et al., 1969; GEHO and WHITESIDE, 1973; SZYMENDERA et al., 1972) there were progressively larger effects. At a dose of 20 mg/kg/day a predictable suppression of these biochemical changes occured in nearly all patients treated, and normal values of AP and

+ This work has been supported by the Wellcome Trust, the National Fund for Research into Crippling Diseases and the Medical Research Council of Great Britain. Sodium etidronate (EHDP[TM]) was kindly supplied by Procter and Gamble, Cincinnati, Ohio, USA. We should like to thank all those who have helped us at various stages of this work.

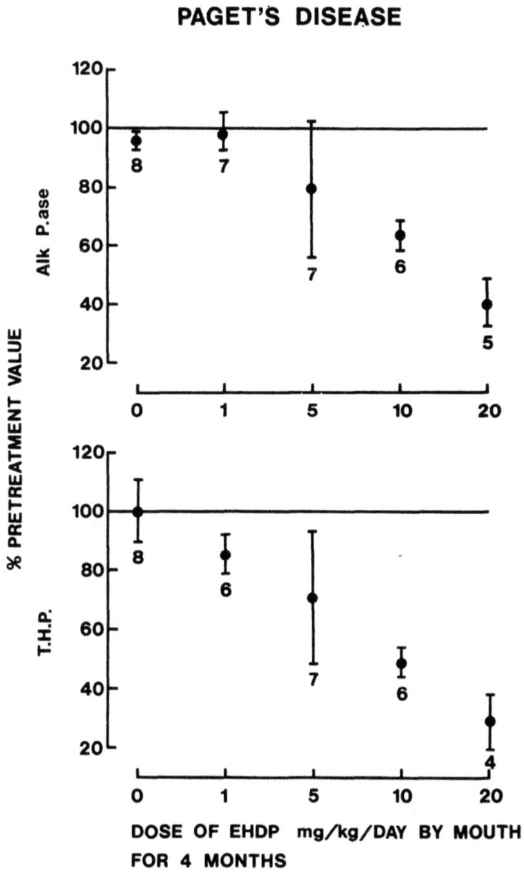

Fig. 1. Effect of EHDP at 0, 1, 5, 10 and 20 mg/kg/day on plasma alkaline phosphatase and urinary total hydroxyproline in Paget's disease. The values are shown (mean ± 1 S.E.M.) after four months of therapy expressed as percent of the mean value before treatment. The number of patients in each dosage group is shown

THP were attained in more than half of the patients after 3 to 6 months of treatment.

Long-term follow-up (Fig. 2) of the first patients treated reveals that normal biochemical values can be sustained in some patients for at least 1 year after stopping treatment, but that when biochemical values do begin to increase again the plasma alkaline phosphatase tends to rise rather more than hydroxyproline does. The reason for this is unclear.

II. Plasma Phosphate

During treatment with EHDP a rise in plasma phosphate occurs, the extent of which is dose-related. This effect begins to appear within

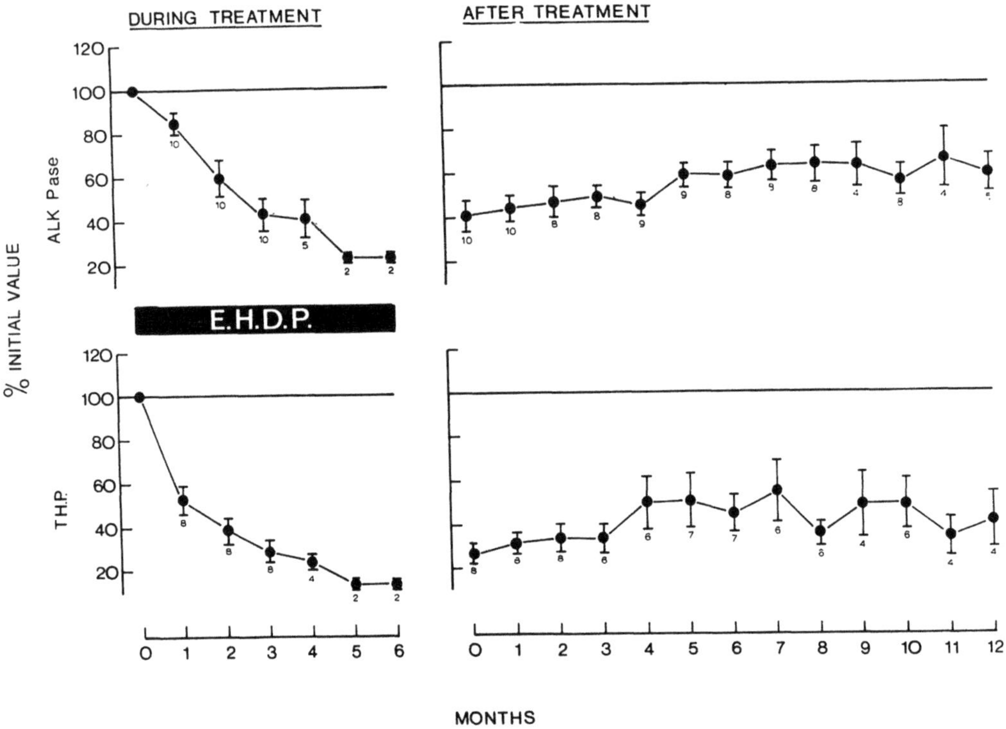

Fig. 2. Long term effects of sodium etidronate (EHDP) given orally at 20 mg/kg/day on the urinary excretion of total hydroxyproline and on plasma alkaline phosphatase in patients with Paget's disease of bone

2 to 3 days of starting treatment and takes about 2 weeks to reach a plateau and about 2 weeks to disappear when treatment stops. The effect does not depend directly on plasma concentrations of EHDP (kindly measured by Drs. Bisaz and Fleisch) and is probably renal in origin, since it is associated with a decreased clearance ratio of phosphate to creatinine (C_p/C_{Cr}) and an increase in the TmP/GFR value.

III. Histological Changes

Treatment with EHDP at all doses from 1 to 20 mg per kg body weight is associated with histological changes in the bone (Fig. 3). After treatment there is a reduction in active resorption and deposition as assessed by cellular morphology in the biopsies of both normal and diseased bone in patients with Paget's disease. There is an increase in "inactive" surfaces present and the fibrosis in the marrow space, which is typical of Paget's disease, also reverts to normal hematopoietic tissue. As reported previously (SMITH et al., 1972), some unmineralized osteoid accumulates with the highest dose of EHDP (20 mg/kg/day), but the extent of this change is usually small.

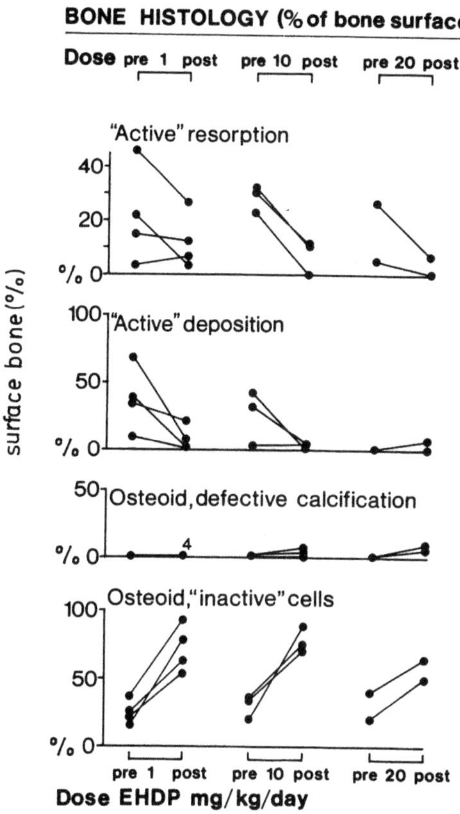

Fig. 3. Effect of EHDP on the histology of iliac crest bone biopsies. The values (%) refer to the proportion of the total bone surface showing the various appearances listed below: 1. 'Active resorption' refers to irregular surfaces in contact with multi-nucleate osteoclasts or mononuclear cells. 2. 'Active deposition' refers to regular surfaces in contact with plump osteoblasts. 3.'Osteoid with "inactive" cells' refers to osteoid in contact with flat cells. 4. 'Osteoid with defective calcification' includes both excessive numbers of osteoid lamellae (i.e. abnormal thickness) and separate foci of calcification within osteoid and is therefore recorded on surfaces already counted under (2.) and (3.) above. Where the sum of 'active resorption', 'active deposition' plus 'osteoid with inactive cells' add up to less than 100%, the deficit represents fully calcified surfaces in contact with flat cells

IV. Symptoms

It has been difficult to assess the effectiveness of EHDP in relieving symptoms, since without a double-blind controlled study it is impossible to distinguish true drug effects from placebo effects. However, about half of the patients with symptoms claimed benefit during treatment with the highest dose of EHDP (20 mg/kg/day), and since many of these patients had been tried with other treatments before, this may represent a real effect of the drug.

V. Side-Effects

Repeated examinations of the blood and urine in these patients have not revealed any toxic effects of EHDP on the liver, kidney or bone marrow, and the drug is well tolerated by patients.

VI. Discussion

EHDP in adequate dosage appears able to reduce the excessive rate of bone turnover in Paget's disease, as measured by changes in hydroxyproline in plasma and urine and in plasma alkaline phosphatase. Treatment with 20 mg/kg/day for 6 months or longer may be required before full biochemical response is reliably seen, but this response is often maintained for many months after treatment stops. One of its advantages over other effective agents, such as calcitonin, glucagon, mithramycin and actinomycin D, is that it can be given by mouth. Other possible advantages over calcitonin are its greater effect on the biochemical abnormalities, the apparent absence of a plateau effect, and the prolonged suppression after stopping treatment. However, it is not yet clear whether EHDP causes the marked relief in symptoms claimed for calcitonin and whether it can result in X-ray resolution of the disease as calcitonin sometimes does. Although doses lower than 20 mg/kg/day have not been so effective on the biochemical changes in our experience, they have been associated with apparent histological suppression of active Paget's disease without the accumulation of unmineralized osteoid. There therefore seems to be an interesting and unexplained discrepancy between the biochemical and histological effects, but more data are needed before this result can be verified.

There are other unresolved problems with the use of EHDP. Firstly, there is the rise in plasma phosphate which seems to be due to enhanced renal tubular reabsorption of phosphate. Since the effect is a large one, it is reasonable to suggest that the secretion and/or action of parathyroid hormone (PTH) might be interfered with. However, the patients do not exhibit other features of hypoparathyroidism (e.g. hypocalcemia) and reported PTH measurements are normal. Furthermore, PTH injected into people given EHDP produces a normal response (RECKER et al., 1973). The reason for this interesting effect therefore remains obscure. Strangely, it is not seen in other animal species, which unfortunately means there is no convenient animal model system to study.

Secondly, the histological information so far available and mentioned above points to a variable increase in osteoid content of bone biopsies after treatment with sodium etidronate. This is not unexpected, since similar effects have been observed in animals (JOWSEY et al., 1970; KING et al., 1971; RUSSELL et al., 1973). Whether this effect is large enough to be a severe deterrent to the use of EHDP remains to be established.

The changes in bone cells seen after EHDP are consistent with a reduction in rates of both bone formation and bone resorption as seen in animals (GASSER et al., 1972). On the basis of animal experiments (BONJOUR et al., 1973; HILL et al., 1973; BONJOUR et al., 1973) it is also possible that EHDP will interfere with the production of 1,25 dihydroxycholecalciferol (1,25 $(OH)_2$-D_3) in man. Although there are no direct data on this point, it does seem unlikely because balance and ^{47}Ca absorption studies in Paget's disease do not show the impaired intestinal absorption of calcium that might be expected if 1,25 $(OH)_2$-D_3 synthesis were inhibited (SMITH et al., 1972); indeed, in osteoporotics, calcium absorption may even be increased above normal by EHDP (SZYMENDERA et al., 1972).

Thirdly, there is the problem of toxicity, particularly in the long term. In this respect it is encouraging that EHDP is concentrated almost solely in the desired target tissue, bone, and that it does not appear to be metabolized.

The desirability of treating Paget's disease with diphosphonates or, indeed, with any other agent may be questioned, since so many of the patients have no symptoms. Although the prospect of merely treating the abnormal biochemistry may appear unattractive, it does seem reasonable to encourage the use of any procedure that prevents the complications of pain, deformity, fracture, neurological damage and sarcoma formation (NAGANT de DEUXCHAINES and KRANE, 1964). It appears logical to expect that agents that reduce the biochemical abnormalities will also suppress the disease and prevent such complications, but whether current treatment can achieve this remains to be evaluated. If future trials are satisfactory, EHDP could find a place as an easily administered form of treatment for any disease of bone, including Paget's disease, where a reduction in bone turnover is likely to be beneficial.

References

BASSETT, C.A.L., DONATH, A., MACAGNO, F., PREISIG, R., FLEISCH, H., FRANCIS, M.D.: Lancet. II, 845 (1969).

BIJVOET, O.L.M., FRÖHLING, P.G.A.M., VAN DER SLUYS VEER, J.: J.Bone Jt. Surg. 54B, 554 (1972).

BONJOUR, J.P., DELUCA, H.F., FLEISCH, H., TRECHSEL, U., MATEJOWEC, L.A., OMDAHL, J.L.: Europ. J. Clin. Invest. 3, 44 (1973).

BONJOUR, J.P., RUSSELL, R.G.G., MORGAN, D.B., FLEISCH, H.: Amer. J. Physiol. 224, 1011 (1973).

FLEISCH, H., RUSSELL, R.G.G.: J. Dent. Res. Supplement to 51, 324 (1972).

FLEISCH, H., RUSSELL, R.G.G., BISAZ, S., BONJOUR, J.P.: Ciba Foundation Symposium, No. 11. Hard Tissue Growth Repair and Remineralisation, 331 (1972).

FLEISCH, H., RUSSELL, R.G.G., FRANCIS, M.D.: Science. 165, 1262 (1969).

FRANCIS, M.D., RUSSELL, R.G.G., FLEISCH, H.: Science, 165, 1264 (1969).

GASSER, A.B., MORGAN, D.B., FLEISCH, H., RICHELLE, L.J.: Clin, Sci. 43, 31 (1972).

GEHO, W.B., WHITESIDE, J.A.: Clinical Aspects of metabolic bone disease in International Symposium on Clinical Aspects of Metabolic Disease. Detroit, 1972. Amsterdam: Excerpta Medica 1973.

HILL, L.F., LUMB, G.A., MAWER, E.B., STANBURY, S.W.: Clin. Sci. 44, 335 (1973).

JOWSEY, J., HOLLEY, K.E., LINMAN, J.W.: J. Lab. Clin. Med. 76, 126 (1970).

KING, W.R., FRANCIS, M.D., MICHAEL, W.R.: Clinical Orthopaedics. 78, 251 (1971).

MICHAEL, W.R., KING, W.R., WAKIM, J.M.: Toxicology and Applied Pharmacology. 21, 303 (1972).

NAGANT de DEUXCHAINES, C., KRANE, S.M.: Medicine, Baltimore. 43, 235 (1964).

RECKER, R.R., HASSING, G., Lau, J.R., SAVILLE, P.D.: J. Lab. Clin. Med. 81, 258 (1973).

RUSSELL, R.G.G., SMITH, R.: J. Bone Jt. Surg. 55B, 66 (1973).

RUSSELL, R.G.G., SMITH, R., BISHOP, M.C., PRICE, D.A., SQUIRE, C.M.: Lancet. I, 10 (1972).

RUSSELL, R.G.G., THORNTON, J., CASEY, P.A., MUHLBAUER, R.C., KISLIG, A., FLEISCH, H., WILLIAMS, D.A., SCHENK, R.: Calc. Tiss. Res. 11, 179 (1973).
SMITH, R., RUSSELL, R.G.G., BISHOP, M.C., WOODS, C.G., BISHOP, M.: Quart. J. Med. 42, 235 (1972).
SZYMENDERA, J., HEANEY, R., SAVILLE, P.: J. Clin. Lab. Med. 79, 570 (1972).

Chairmen: H. Czitober and G. J. Joplin

Different Behavior of ^{45}Ca and ^{89}Sr in Chronic Uremia in the Rat

W. REMAGEN, D. WEIDMANN, P. HEITZ, J. SCHUPPLER, H. LÜTHY AND E. DEL POZO

In earlier experiments presented at the 8th Symposium (MEYER et al., 1971), we have investigated the different handling of ^{45}Ca and ^{89}Sr as tracers in a kinetic model of Calcium metabolism in the rat. We have shown highly significant higher excretion rates for Sr than for Ca in the urine (v_u) and the endogenous fecal secretion (v_f), which is in agreement with the literature (BAUER et al., 1955; COHN and GUSMANO, 1967; GREENBERG and TROESBER, 1942; PANY, 1967). Yet in contradiction to the findings of other workers, we have found a highly significant discrimination in favor of Ca for the bone accretion rate (v_{O+}). These results have raised the question whether the lower accretion of Sr was only the consequence of the faster urinary excretion, or if it was really an independent process.

In order to investigate this problem we set up the following experiment: in rats of about 80 gr body weight we resected five sixths of their kidney parenchyma, and we irradiated the remaining kidney tissue with 700 r. Thus we reduced the excretory surface of the organ, and induced a chronic uremic syndrome. Five weeks later we controlled the degree of uremia by estimating the BUN and creatinine in the blood. Then ^{45}Ca and ^{89}Sr were injected simultaneously, and the specific radioactivity curves in the blood serum, the radioactivities excreted in urine and feces, and the overall Ca balance were determined. By applying the two-compartment model of AUBERT and MILHAUD (1960), we obtained the values for the different parameters listed in Table 1. The implications and criticisms of the model have been discussed in detail elsewhere (AUBERT and MILHAUD, 1960; AUBERT et al., 1963; REMAGEN, 1970). The radioactivities in the serum, and in the ashed feces, bones, and urine collected on ashfree filtration paper, were measured by liquid scintillation counting. As the tracers were injected into an identical group of 18 rats, we get the same mean values with Ca and with Sr for the first four parameters. The endogenous fecal excretion calculated with Sr was about 50% higher, as it was in the former experiments on normal rats. The urinary excretion rate calculated with Sr was significantly higher, but was reduced to about one half (6.2 times) of that in the normal rats (11.4 times). In contrast to this reduced urinary excretion, the discrimination of bone against Sr remained the same as in the normal rats. We controlled these calculated figures by directly measuring the deposition of ^{45}Ca and ^{89}Sr in the ashed bone tissue (Table 2). The ratio of the injected radioactivities divided by the one found in bone is 1.129, compared with 1.226 found in the normal rats. Setting the accretion rate (v_{O+}) calculated with Ca at 100%, the corresponding values for Sr are 82.8% in the uremic, and 82.9% in the normal rats. If we likewise set the deposition of Ca in bone at 100%, then that of Sr is 85.8% in the uremic, and 81.6% in the normal animals. The close correlation between these figures strongly supports the validity of the calculated bone accretion rates. Furthermore, it presents good evidence for the view that bone accretion and urinary excretion are independent processes.

There is a significant decrease in all parameters of the uremic animals, as compared with the normal ones, with the exception of the bone

Table 1. Parameters of the model; mean values ± SD, calculated with ^{89}Sr and ^{45}Ca; 18 animals within the group.

Parameter		^{89}Sr		^{45}Ca
Body weight	G	147.44 ± 3.46	gr	147.44 ± 3.46
Ca intake with food	v_i	64.21 ± 7.58	mg/d	64.21 ± 7.58
total fecal loss	v_f	56.00 ± 5.59	mg/d	56.00 ± 5.59
Urin. excret. chem. estim.	v_{uexp}	1.84 ± 0.20	mg/d	1.84 ± 0.20
Compartment fast exchange	P	17.03 ± 0.56	mg	18.02 ± 0.39
Compartment slow exchange	E	35.06 ± 1.77	mg	38.84 ± 1.68
Exchange P-E	V_e	62.31 ± 2.84	mg/d	69.74 ± 2.20
total loss from P	v_T	56.88 ± 1.28	mg/d	57.42 ± 1.21
Urinary excretion	v_u	5.63 ± 0.71	mg/d	0.90[a] ± 0.20
Endogenous fecal Ca	v_f	9.54 ± 0.66	mg/d	6.15[a] ± 0.62
Bone accretion	v_{o+}	41.71 ± 1.33	mg/d	50.37[a] ± 1.43
intestinal absorpt. percent	α	31.74 ± 2.49	%	25.68 ± 2.83
overall balance	Δ	+ 2.58 ± 3.61	mg/d	+ 7.32 ± 3.56
bone resorption	v_{o-}	39.12 ± 3.57	mg/d	43.05 ± 3.69

[a] Difference between mean values of ^{89}Sr and ^{45}Ca statistically significant at $p < 0.001$

resorption rate (v_{o-}). The reduced Ca intake (v_i) of the uremic animals (approximately 50% of that of the normal ones) is not compensated by an increased, but aggravated by a diminished absorption from the intestine (26% against 44%). This may indicate a disturbed vitamin D metabolism due to the reduced kidney parenchyma. It is compensated by a relative increase of the bone resorption rate, which is as high as in the normal rats. This is further evidence for the view that bone is the superordinate regulator organ which is able to adjust the oscillations of Ca metabolism provoked by any disturbance.

Supported by grant Nr. 3.310.70 of the Schweiz. Nationalfonds zur Förderung der wissenschaftlichen Forschung.

Table 2. Comparison of the bone accretion and urinary excretion rates in the uremic and in the normal rats of MEYER et al. (1971). Values calculated with ^{45}Ca set at 100% for the bone values.

	uremia	normal
$\frac{^{89}Sr/^{45}Ca \text{ injected}}{^{89}Sr/^{45}Ca \text{ in bone}}$	1.129	1.226
$v_{o+}{}^{45}Ca$	100 %	100 %
$v_{o+}{}^{89}Sr$	82.8%	82.9%
^{45}Ca deposited	100 %	100 %
^{89}Sr deposited	85.8%	81.6%
$v_u{}^{45}Ca$	0.90 mg/d	1.80 mg/d
$v_u{}^{89}Sr$	5.63 mg/d	20.52 mg/d
$v_u{}^{89}Sr/v_u{}^{45}Ca$	6.2	11.4

References

AUBERT, J.P., MILHAUD, G.: Biochim. Biophys. Acta 39, 122 (1960).
AUBERT, J.P., BRONNER, F., RICHELLE, L.J.: J. Clin. Invest. 42, 6 (1963).
BAUER, G.C.H., CARLSSON, A., LINDQUIST, B.: Acta Physiol. Scand. 35, 56 (1955).
COHN, S.H., GUSMANO, E.A.: Proc. Soc. Exp. Biol. Med. 126, 79 (1967).
COMAR, C.L., WASSERMAN, R.H., NOLD, M.M.: Proc. Soc. Exp. Biol. Med. 92, 859 (1956).
GREENBERG, D.M., TROESBER, F.M.: Proc. Soc. Exp. Biol. Med. 49, 488 (1942).
MEYER, G., REMAGEN, W., GROSSE, P.: Israel J. Med. Scie. 7, 393 (1971).
PANY, J.E.: Wien. Klin. Wschr. 79, 489 (1967).
REMAGEN, W.: "Calciumkinetik und Knochenmorphologie". Stuttgart: Thieme 1970.

The Effect of 1,25-Dihydroxycholecalciferol in Patients with Vitamin D Resistant States (Chronic Renal Failure, Familial Hypophosphatemia, Gluten Sensitive Enteropathy and Hypoparathyroidism)*

R. G. HENDERSON, R. G. G. RUSSELL, J. G. G. LEDINGHAM, D. O. OLIVER, C. PRESTON, R. SMITH, G. T. WARNER, R. J. WALTON AND A. W. NORMAN

Recent studies indicate that vitamin D undergoes a series of enzymatic hydroxylations before it exerts its biological effects. The metabolite 1,25-dihydroxycholecalciferol (1,25$(OH)_2$-D_3) produced by the kidney is probably the major physiologically active form that increases intestinal calcium absorption and bone resorption (DELUCA, 1973; OMDAHL and DELUCA, 1973; HOLICK et al., 1971; LAWSON et al., 1971; MYRTLE and NORMAN, 1971; NORMAN et al., 1971). In the light of this knowledge, it is conceivable that some disorders of calcium and phosphate metabolism that require large amounts of vitamin D for treatment could be due to failure to produce the natural metabolite in adequate amounts. We have therefore examined the effects of small (i.e. physiological) amounts of 1,25$(OH)_2$-D_3 in patients with a variety of vitamin D-resistant states.

The 1,25$(OH)_2$-D_3 was prepared biosynthetically (NORMAN et al., 1971). With the exception of patients on long-term hemodialysis for renal failure, all studies were done on a Metabolic Ward. The results are described and discussed below.

I. Chronic Renal Failure

A total of 11 patients, 8 on long-term hemodialysis, were studied. Intestinal calcium absorption, measured by 7-day retention of oral ^{47}Ca on a whole-body counter, was unchanged 6 hours after single oral doses of up to 2.7 µg (100 units) of 1,25$(OH)_2$-D_3 but increased in all those patients given 0.7 µg (25 units) or more daily by mouth for 4 to 8 days (Fig. 1). Intravenous 1,25$(OH)_2$-D_3 was more potent than oral. There was no significant change in plasma calcium, phosphate or alkaline phosphatase in these patients. Non-protein-bound plasma and total urinary hydroxyproline did not change during the time that 1,25$(OH)_2$-D_3 was given suggesting that bone resorption was not markedly increased in the short-term by doses which altered calcium absorption. In 2 patients with myopathy, there was no improvement after 1,25$(OH)_2$-D_3 as assessed by electromyography and muscle power measurements. Bone biopsies taken from 2 patients before and after 1,25$(OH)_2$-D_3 gave inconclusive results but no massive increase in osteoclastic resorption was seen.

Three normal persons given 0.7 µg 1,25$(OH)_2$-D_3 for 4 days by mouth also showed increased ^{47}Ca retention.

+ We are grateful for grants from the Wellcome Trust, The National Fund for Research into Crippling Diseases and The National Institute of Health, and to all the people who have helped at various stages of this work.

The ability of patients with chronic renal failure to respond to small doses of 1,25$(OH)_2$-D_3, compared with the large doses required of parent vitamin D or of dihydrotachysterol (DHT), supports the idea that endogenous production of 1,25$(OH)_2$-D_3 is impaired in this condition, perhaps due to the elevated plasma phosphate (TANAKA and DELUCA, 1973). BRICKMAN et al. (1972) have reported results similar to ours in renal failure.

II. Gluten-Sensitive Enteropathy

Two men with untreated gluten-sensitive enteropathy were given 2.1 µg (75 units) of 1,25$(OH)_2$-D_3 daily for 4 days. Even though given intravenously, the 1,25 $(OH)_2$-D_3 did not increase calcium absorption measured by ^{47}Ca retention or calcium balance. This suggests that villous atrophy in the intestine may impair the response of 1,25$(OH)_2$-D_3,

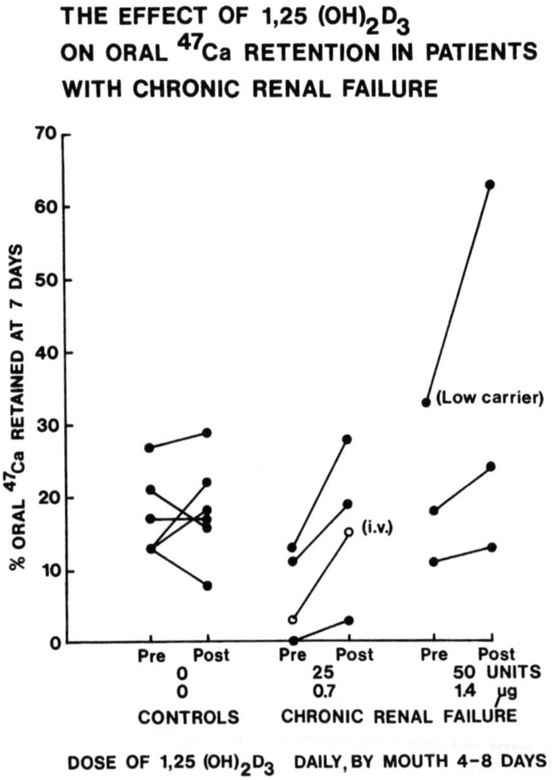

Fig. 1. The effect of daily oral or intravenous 1,25$(OH)_2$-D_3 for 4 to 8 days on the whole body retention of oral ^{47}Ca in patients with chronic renal failure. Control data are shown for patients with normal renal function given 2 consecutive oral doses of ^{47}Ca without 1,25$(OH)_2$-D_3. The difference in these responses between the two groups (i.e. with or without 1,25$(OH)_2$-D_3) is statistically significant at the 5% level (Wilcoxon's rank test). One patient shown was given 20 mg instead of 200 mg calcium carrier

probably because the cells which are the target for the hormone are much decreased in number. Again urinary total hydroxyproline was unaltered by 1,25$(OH)_2$-D_3 suggesting no effect on bone resorption at this fairly large dose.

III. Type I Inherited Hypophosphatemic Rickets

Three women with familial hypophosphatemia (DENT, 1952), none of whom had been treated with vitamin D for at least 6 years, were given 2.1 µg (75 units) of 1,25$(OH)_2$-D_3 daily by mouth for 4 to 12 days. Intestinal calcium absorption increased and all 3 showed a prompt rise in urinary calcium and magnesium. The hypophosphatemia was not corrected and there was no marked change in renal phosphate handling (TmP/GFR) (Fig. 2) or in hydroxyproline measurements. These results suggest that the defective tubular reabsorption of phosphate in this condition is not due to

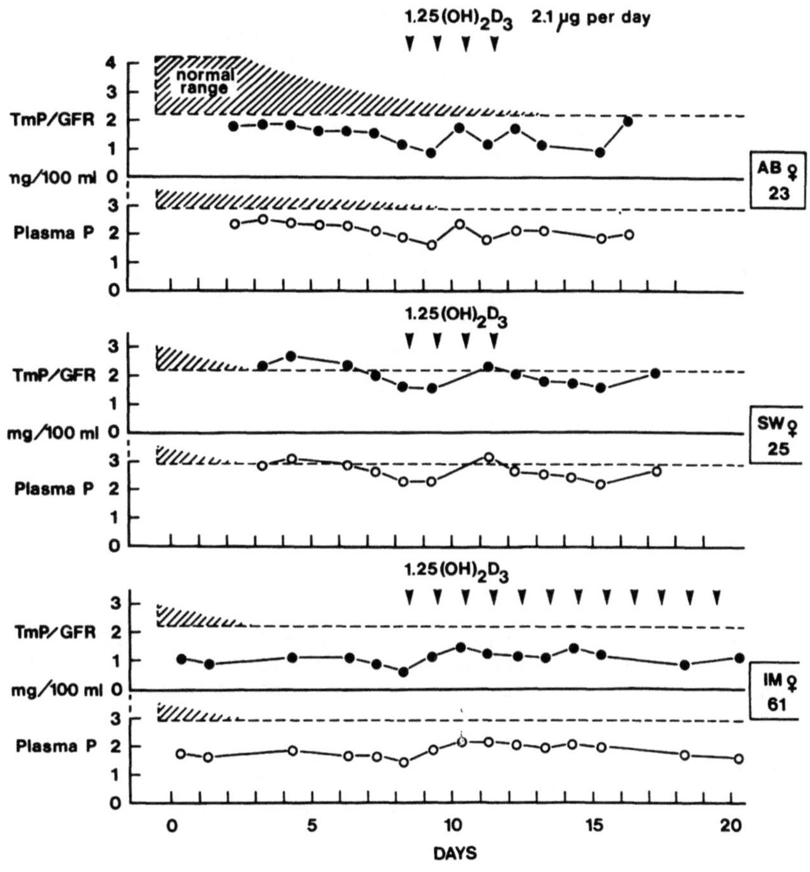

Fig. 2. The effect of 1,25$(OH)_2$-D_3 on plasma phosphate and TmP/GFR (BIJVOET et al., 1969) in 3 patients with Type I inherited hypophosphatemia

lack of 1,25(OH)$_2$-D$_3$. GLORIEUX et al. (1973) have come to the same conclusion independently. There may however still be some abnormality in the metabolism of vitamin D in patients with inherited hypophosphatemia since one might expect the persistently low plasma phosphate to stimulate 1,25(OH)$_2$-D$_3$ production (TANAKA and DELUCA, 1973) and thereby increase intestinal calcium absorption, whereas it is known that calcium absorption is usually below normal in this condition.

IV. Hypoparathyroidism

One woman with previously untreated post-thyroidectomy hypoparathyroidism showed a dramatic response to 1.4 - 2.7 µg of 1,25(OH)$_2$-D$_3$ daily. Her plasma calcium rose from 6.8 to 10.2 mg/100ml within a week of starting treatment, and was maintained in the normal range after the dose was reduced to 0.7 µg (25 units) daily. Her urine calcium rose from below 50 mg per day to more than 300 mg per day. The plasma and urine calcium both fell when the 1,25(OH)$_2$-D$_3$ was withdrawn. The mechanisms by which small amounts of 1,25(OH)$_2$-D$_3$ corrected the hypocalcemia are still being studied. Measurements available so far suggest the effect is mainly due to increased absorption of calcium together with a possible increase in bone resorption. Increased renal reabsorption of calcium has not been completely excluded but may not have to be invoked to explain the changes produced. In this particular patient, interpretation of the results is rather difficult since she may have sarcoidosis in addition to hypoparathyroidism. However, the ability of a hypoparathyroid patient to respond to minute doses of 1,25(OH)$_2$-D$_3$ compared with the large doses of DHT or vitamin D usually required in this condition may suggest that there is a block to the production of active metabolites. In view of current theories (DELUCA, 1973) about regulation of 1,25(OH)$_2$-D$_3$ production this might be due either to absence of parathyroid hormone or to the high plasma phosphate. This study also suggests that 1,25(OH)$_2$-D$_3$ maintains its biological activity in humans during long-term oral administration. This contrasts with the situation in rats where the potency of 1,25(OH)$_2$-D$_3$ becomes less during the continuous dosage by mouth (FROLICK and DELUCA, 1973).

V. Conclusions

Minute doses of 1,25(OH)$_2$-D$_3$ increase calcium absorption in normal persons and in those with uremia, hypoparathyroidism or familial hypophosphatemia, but do not correct the renal phosphate leak in the latter condition.

These short-term studies suggest a rapid turning on and off of calcium absorption in response to 1,25(OH)$_2$-D$_3$ without any marked short-term increase in bone resorption. Plasma calcium only rose significantly in hypoparathyroidism.

Defective endogenous production of 1,25(OH)$_2$-D$_3$ may contribute towards the abnormalities in calcium metabolism in chronic renal failure and hypoparathyroidism, but the value of 1,25(OH)$_2$-D$_3$ in the therapy of these conditions will become clear only after much further work.

References

BIJVOET, O.L.M., MORGAN, D.B., FOURMAN, P.: Clin. Chem. Acta 26, 15 (1969).
BRICKMAN, A.S., COBURN, J.W., NORMAN, A.W.: New Eng. J. Med. 287, 891 (1972).
DELUCA, H.F.: New Eng. J. Med. 289, 359 (1973).
DENT, C.E.: J. Bone Jt. Surgery 34B, 266 (1952).
FROLICK, C.A., DELUCA, H.F.: J. Clin. Invest. 51, 2900 (1972).
GLORIEUX, F.H., SCRIVER, C.R., HOLICK M.F., DELUCA, H.F.: Lancet II, 287 (1973).
HOLICK, M.F., SCHNOES, H.K., DELUCA, H.F.: Proc. Nat. Acad. Sci. USA 68, 803 (1971).
LAWSON, D.E.M., FRASER, D.R., KODICEK, E., MORRIS, H.R., WILLIAMS, D.H.: Nature (Lond.) 230, 228 (1971).
MYRTLE, J.F., NORMAN, A.W.: Science 171, 79 (1971).
NORMAN, A.W., MIDGETT, R.F., MYRTLE, J.F.: Biochem. Biophys. Res. Comm. 42, 1082 (1971).
OMDAHL, J.L., DELUCA, H.F.: Physiological Reviews 53, 327 (1973).
TANAKA, Y., DELUCA, H.F.: Arch. Biochem. Biophys. 154, 566 (1973).

Treatment of Renal Osteodystrophy Using Vitamin D or High Dialysate Calcium Concentration*

R. BOUILLON, R. VERBERCKMOES AND B. KREMPIEN

Most patients with chronic renal disease, treated or not with regular dialysis, develop some form of renal osteodystrophy. This metabolic bone disease, however, is usually mild and only detectable by its biochemical or endocrine features or by histological examinations of the bone. In some cases, however, the more severe stages of radiologic and clinical osteodystrophy occur (STANBURY, 1972). This presentation deals with the evolution and the treatment of the two forms of this disease. First the long-term influence of the change in the dialysate calcium concentration from 6 to 7.5 mg in patients without manifest clinical osteodystrophy is described. Secondly our experience with pharmacological doses of vitamin D_2 and D_3 in patients with severe radiological renal osteodystrophy is summarized.

I. Influence of a Higher Dialysate Calcium Concentration on the Secondary Hyperparathyroidism of Renal Insufficiency

Ten patients (6 males and 4 females), with an average age of 31 (range 20 to 43 yrs) had already been treated biweekly with regular dialysis for 22,6 months (range 2 to 75 m) on 2/11/72. On this date the dialysate calcium concentration was changed from 6 to 7.5 mg %, all other constituents remaining constant. Other treatment consisted of oral aluminiumhydroxide gels. Serum for radioimmunoassay of parathyroid hormone (BOUILLON and DE MOOR, 1973) was taken before the start of the dialysis 4, 3 and 1 months before the change in the dialysate calcium concentration and again 2, 3, 5, 6 and 7 months later. These measurements were compared with the mean of the predialysis serum concentrations of calcium, phosphorus and alkaline phosphatase, obtained during the month before the parathyroid hormone measurements. Statistical analysis included a paired Student's t-test for all parameters, and for the parathyroid hormone level additional calculations were done using a paired Wilcoxon rank sum test and a paired Student's t-test after logarithmic transformation of the data because there could be some doubt as to the normal distribution of the serum parathyroid hormone levels.

During the control period, when the dialysate calcium concentration was 6 mg %, all serum parameters (calcium, phosphorus, alkaline phosphatase and parathyroid hormone) remained relatively constant and no significant differences were detected during these three observation periods. (Fig. 1).

+ Miss M.P. QUINTENS and Mrs. P. DELVAUX-NOPPE are gratefully acknowledged for their technical assistance in the parathyroid hormone assays. The financial help of N.V. Sandoz-Belgium for the preparation and presentation of this manuscript on the Symposium is kindly appreciated.

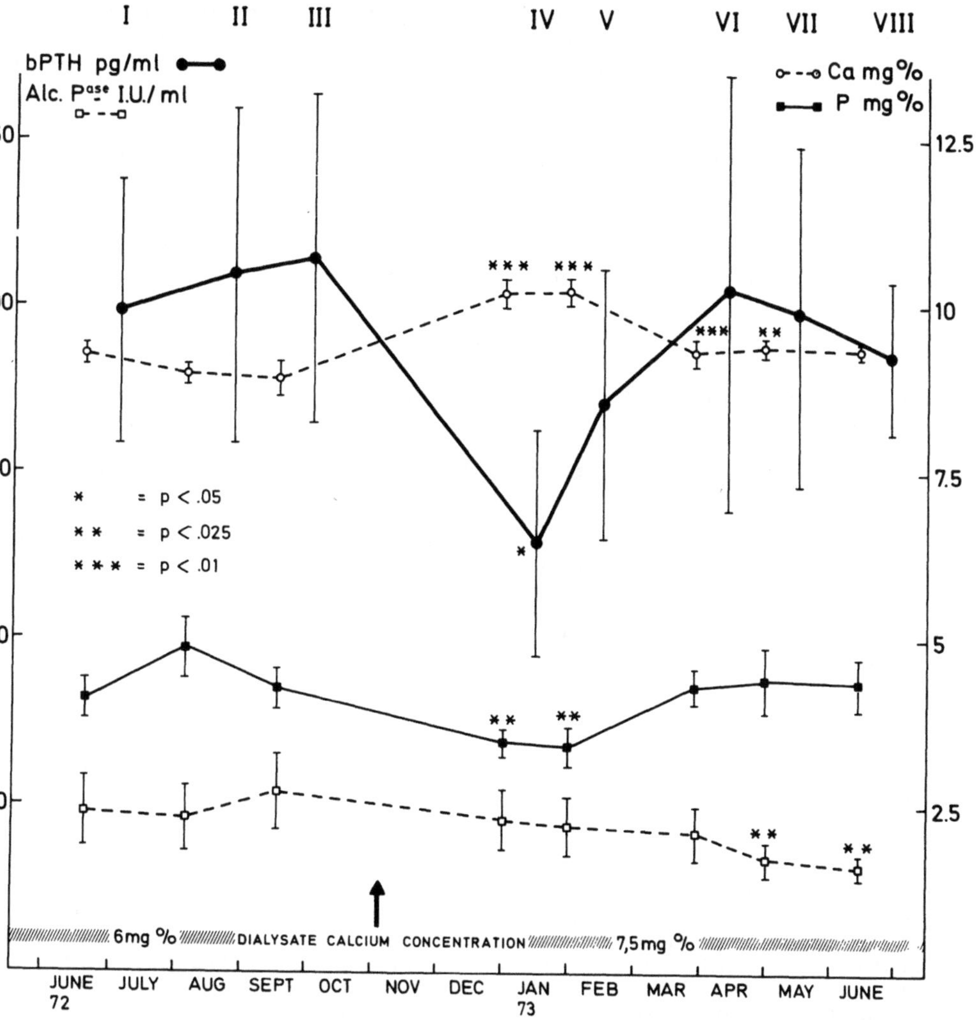

Fig. 1. Mean ± SEM values of predialysis serum calcium, phosphorus, alkaline phosphatase and parathyroid hormone, measured in ten patients on regular dialysis treatment before and after a rise in dialysate calcium concentration from 6 mg % to 7.5 mg %. Statistical analysis between period III and the others are indicated

Two months after the rise in bath calcium concentration to 7.5 mg % the serum calcium level was significantly increased (+ 1.3 mg % $p < 0.001$) and the serum phosphorus decreased (- 0.78 mg % $< p$ 0.025), but the serum alkaline phosphatase remained unchanged. Serum parathyroid hormone significantly declined (- 421 pg/ml $p < 0.05$) and this was most marked in patients with previously very high parathyroid hormone levels. During the following months of dialysis with the same dialysate calcium, however, the serum parathyroid hormone level progressively returned to the control values. Serum calcium decreased again although it still remained higher than in the control period. Serum phosphorus levels also returned to control levels after five months. Serum alkaline phosphatase levels, however, progressively declined and finally significantly lower levels were observed 6 and 7 months after the higher dialysate calcium concentration was used (Fig. 1).

A higher dialysate calcium concentration was advocated by FOURNIER et al., (1971a) because a retrospective study indicated that the major factor contributing to the production of bone disease was the use of a dialysate with a calcium concentration of less than 5.7 mg per 100 ml. These authors subsequently demonstrated (FOURNIER et al., 1971b) that predialysis serum parathyroid hormone levels were lowest when dialysate calcium was raised to 8 mg % and serum phosphorus lowered by oral treatment with phosphate binding gels. However, JOHNSON et al. (1972) could not find any difference in predialysis parathyroid hormone levels two months after an increase of dialysate concentration from 5.2 to 7 mg %.

Our own study confirms the short term suppression of serum parathyroid hormone, probably related to the simultaneous rise in predialysis serum calcium. After more than three months, however, serum parathyroid hormone levels increased again progressively, probably again related to a small decrease in predialysis calcium concentration. On a long-term basis (observation period of 7 months) no definite cure of the secondary hyperparathyroidism could be obtained. Especially the patients with initially very high hormone levels returned to high levels after the suppression period. This seems to imply that the use of a dialysate calcium concentration of 7.5 mg % is insufficient to cause a continuous involution of the preexisting hyperparathyroidism.

II. Influence of Pharmacological Doses of Vitamin D on Renal Osteodystrophy

Eight other patients (4 males, 4 females) with an average age of 29 at the start of treatment with regular dialysis, who presented severe radiologic manifestations of renal osteodystrophy were treated with vitamin D_2 or D_3 in a dose of usually 15 mg twice weekly, with an average duration of treatment of 8 months (range 3 to 15 1/2 m). In two patients, a relapse of bone disease became manifest one year after the end of the first period of treatment. They were treated again a second time. Dialysate calcium concentration was 5 mg % until August 1969, 6 mg % until November 1972 and 7.5 mg % thereafter. Aluminiumhydroxyde was administered to all patients. Repeated bone biopsies were obtained in 6 patients and analyzed on undecalcified sections. Predialysis serum samples for radioimmunoassay of parathyroid hormone were obtained repeatedly in six cases during eight treatment periods.

The radiologic defects before treatment consisted in subperiostal resorption in fingers, clavicula and pubis in 7 patients, ground glass appearances of the skull in 5 patients, important metastatic calcifications in two and Milkman-Looser pseudofractures in one. All these defects improved markedly or healed completely in all patients. Even, and this was rather unexpected, the vascular calcifications in the extremities of two patients disappeared during treatment.

On histological examination of the bone, marked improvement of the signs of osteitis fibrosa (decrease of osteoclastic activity, periosteocyte osteolysis and endosteal and marrow fibrosis) as well as a decrease of the osteoid matrix, an index of osteomalacia, was observed.

Predialysis serum parathyroid hormone levels were high in all patients before treatment with vitamin D. During treatment serum parathyroid hormone remained elevated but at the end of treatment, usually after a short period of hypercalcemia, lower but usually not entirely normal levels were found (Fig. 2). Radiologic and histologic improvement could occur even without substantial change in serum parathyroid hormone. A few months after the end of the vitamin D treatment, some patients again had very high serum levels of parathyroid hormone. This indicates that the present discontinuous treatment with vitamin D can indeed improve or cure the bone disease and suppress the parathyroid glands but that in the long run secondary hyperparathyroidism does not disappear. Careful observation for relapses of bone diseases during further regular dialysis and an active search for other treatment combinations to suppress parathyroid overactivity continuously are therefore indicated.

Treatment with high doses of oral vitamin D have previously been reported as being effective in both preventing and improving renal osteodystrophy (DENT et al., 1961; VERBERCKMOES, 1971; STANBURY and LUMB 1962; FLETCHER et al., 1963). When vitamin D_3 was used, similar doses as used by us were found necessary but lower doses of dihydrotachysterol are also effective (KAYE et al., 1970). More recently the kidney-produced 1,25 dihydroxy vitamin D has been found to increase the intestinal radiocalcium absorption in uremic patients, even when used in physiological doses (BRICKMAN et al., 1972; COBURN et al., 1973).

The nucleic receptor for vitamin D therefore seems to fit best with 1,25 dihydroxy vitamin D but its specificity can be overcome by massive doses of other components. Further development of analogs could result in dissociation of its avidity for different receptors in different organs (HOLICK et al., 1972).

Recently RITZ et al. reported their experience with vitamin D treatment in various forms of renal insufficiency. Intestinal calcium absorption and bone histology markedly improved even without changes in serum parathyroid hormone concentrations (RITZ and KREMPIEN, 1973). This is in concordance with our results, but we may add that further treatment or follow-up results in definite but only temporary suppression of parathyroid secretion.

III. Conclusions

Our therapeutic trial confirms the favorable effect of high doses of vitamin D on the renal osteodystrophy of patients with advanced renal failure. This beneficial effect on the radiologic and histologic appearance of the bone disease was obtained notwithstanding persistent high circulating PTH levels.

Although temporary suppression of hyperparathyroidism could be obtained by raising the dialysate calcium concentration or by treatment with vitamin D, definite suppression seems to be much more difficult to obtain. It is therefore our impression that for long-term cure not only the serum phosphorus must be lowered to normal (STANBURY and LUMB, 1962) but also normal to high normal serum levels of calcium must be obtained and maintained. The best method for a continuous normalizing of serum calcium is still not found but could be the use of a polar metabolite or analog of vitamin D in association with oral or dialysate calcium apport.

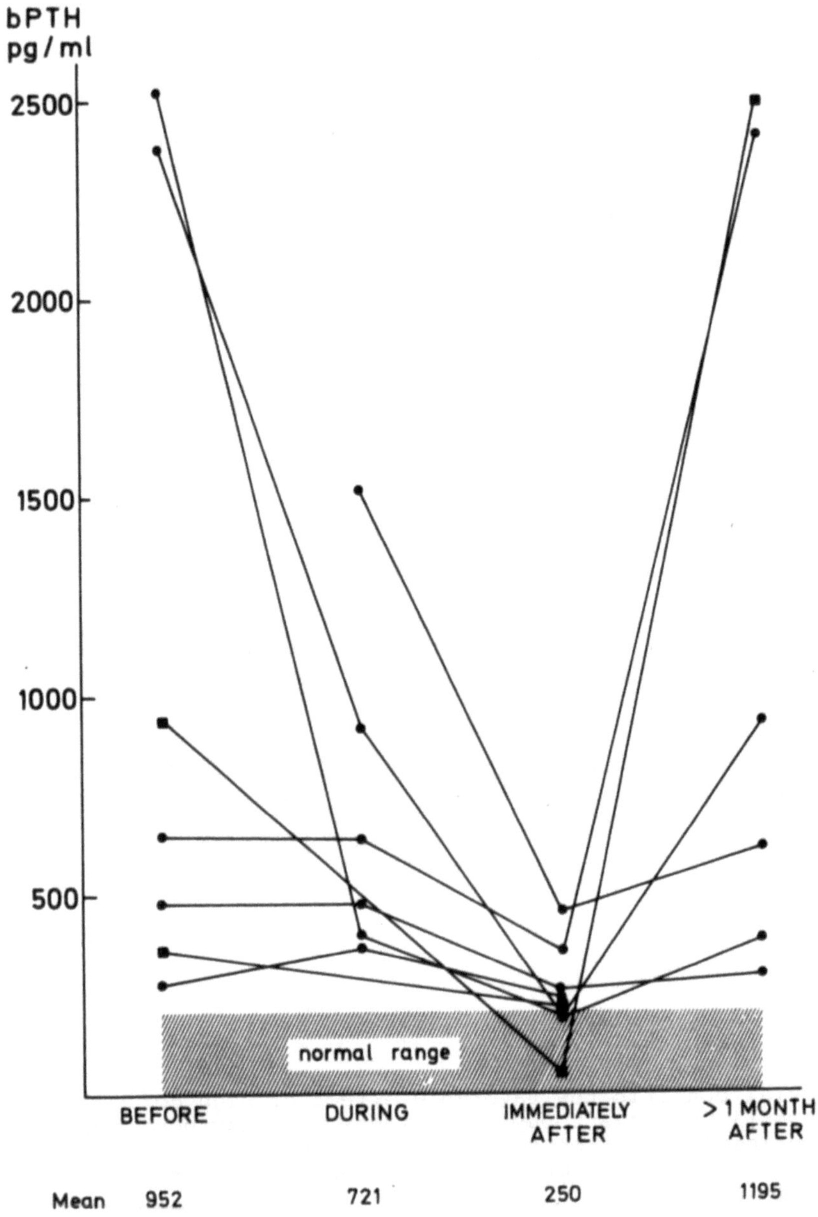

Fig. 2. Serum parathyroid hormone level in six individual patients on regular dialysis treatment, before, during and after eight treatment periods with vitamin D

References

BOUILLON, R. DE MOOR, P.: Pathophysiological data obtained with a radioimmunoassay·for human parathyroid hormone. Ann. d'Endocr. (Paris) 34, 657-667 (1973).

BRICKMAN, A.S., COBURN, J.W., NORMAN, A.W.: Action of 1,25 dihydroxycholecalciferol, a potent, kidney-produced metabolite of vitamin D_3, in uremic man. New Engl. J. Med. 287, 891-895 (1972).

COBURN, J.W., BRICKMAN, A.S., MASSRY, S.G., NORMAN, A.W.: Action of 1,25 dihydroxycholecalciferol in normal and uremic man. Abstract. Kidney International 3, 3A (1973).

DENT, C.E., HARPER, C.M., PHILPOT, G.R.: The treatment of renalglomerular osteodystrophy. Quart. J. Medic. 30,1-31 (1961).

FLETCHER, R.F., JONES, J.H., MORGAN, D.B.: Bone disease in chronic renal failure. Quart. J. Med. 32, 321 (1963).

FOURNIER, A.E., JOHNSON, W.J., TAUES, D.R., BEABOUT, J.W., ARNAUD, C.D., GOLDSMITH, R.S.: Etiology of hyperparathyroidism and bone disease during chronic hemodialysis I. Association of bone disease with potentially etiologic factors. J. Clin. Invest. 50, 592-598 (1971a).

FOURNIER, A.E., ARNAUD, C.D., JOHNSON, W.J., TAYLOR, W.F., GOLDSMITH, R.S.: Etiology of hyperparathyroidism and bone disease during chronic hemodialysis II. Factors affecting serum parathyroid hormone. J. Clin. Invest. 50, 599-605 (1971b).

HOLICK, M.F., GARABEDIAN, M., DELUCA, H.F.: The 5,6-trans isomers of cholecalciferol and 25-hydroxycholecalciferol: substitutes for 1,25 dihydroxycholecalciferol in anefric animals. Biochemistry 11, 2715-2719 (1972).

JOHNSON, J.W., HATTNER, R.S., HAMPERS, C.L., BERNSTEIN, D.S., MERRILL, J.P., SHERWOOD, L.M.: Effects of hemodialysis on secondary hyperparathyroidism in patients with chronic renal failure. Metabolism 21, 18-29 (1972).

KAYE, M., CHATTERJEE, G., COHEN, G.F.: Arrest of hyperparathyroid bone disease with dihydrotachysterol in patients undergoing chronic hemodialysis. Ann. Int. Med. 73, 225-233 (1970).

RITZ, E., KREMPIEN, B.: Skeletal changes after vitamin D treatment in various stages of renal insufficiency. Abstract 228 Europ. J. Clin. Invest. 3, 265 (1973).

STANBURY, S.W.: Azotaemic renal osteodystrophy. Clinics in Endorin. and Metabolism 1, 267-304 (1972).

STANBURY, S.W., LUMB, G.A.: Metabolic studies of renal osteodystrophy. Medicine 41, 1 (1962).

VERBERCKMOES, R.: Erfahrungen mit hochdosiertem Vitamin D bei Behandlung der renalen Osteodystrophie chronisch dialysierter Patienten. In: "Actuelle Probleme der Dialyseverfahren und der Niereninsuffizienz" (Ed. P.V. Dittrich und F.Skrabal) 147-163. Friedberg: Verlag C. Bindernagel 1971.

The Effects of Calcium Supplementation of the Diet on Bone Mass in Women*

D.A. SMITH, J.J.B. ANDERSON, J.M. AITKEN AND J. SHIMMINS

Low calcium diets, in the presence of adequate vitamin D, have been reported to produce osteoporosis in animals (POMMER, 1925; JAFFE et al., 1932; BELL et al., 1941). JOWSEY and GERSHON-COHEN (1964) repeated this work on cats and were able to demonstrate that the osteoporosis could be reversed by adding calcium to the diet. Positive calcium balances have been reported in patients when their diets were supplemented with calcium (OWEN et al., 1940; ANDERSON, 1950; WHEDON, 1959; HARRISON et al., 1961; NORDIN, 1962).

SMITH and FRAME (1965) investigated 2,000 ambulatory women, in whom they showed a progressive loss of bone density and cortical bone with age. However, they were not able to demonstrate any relation between osteoporosis and calcium intake, the calcium intake being calculated from the dietary history. DENT and WATSON (1966) were sceptical of the significance of calcium supplements being an effective therapy. Certainly, no convincing evidence that long-term calcium supplementation of the diet leads to increased bone density has yet been reported.

The object of the present study is to report the changes in bone density measurements in 94 women who were given calcium supplements for periods ranging from 18 to 54 months.

I. Procedure

All the patients had x-rays of their right hand taken before starting therapy, and at six-monthly intervals thereafter. The technique used was that devised by ANDERSON et al. (1966). From the investigations of SMITH (1971) it is evident that there can be a fall in whole bone density without a corresponding fall in the total amount of bone present because of the relative rates of increase in the internal and external diameter of bone with aging. Therefore, the results of both the whole bone densitiy (SAE) and the total cortical mineral per unit length of bone are reported.

In all 94 patients were studied. They were patients referred by physicians or surgeons who felt that they were osteoporotic after examination of the x-rays. All patients who had abnormal biochemistry or who had a history of renal, liver or endocrine disease were excluded from the trial, but the patients were not selected in any other way. The patients were treated with either calcium glycerophosphate or calcium lactate gluconate which supplemented the diet by 1.14 and 1.2 gm calcium per day, respectively.

+ We gratefully acknowledge the generous financial support of the National Fund For Research Into Crippling Diseases in connection with this project.

The control subjects consisted of a group of 317 normal female volunteers for whom the results have been reported elsewhere (SMITH et al., 1968; SMITH et al., 1969). Since marked changes in the rates of bone loss occur with aging, this must be allowed for in the assessment of any change during therapy. In order to do this, third order polynomials were fitted to the data to give a series of lines of best fit to derive the cumulative 5 to 95 percentile lines.

The absolute and relative changes in SAE and TCM in relation to the normal population have been derived and are reported (SMITH, 1971).

II. Results

The average age of the onset of the menopause was 46.6 years in the control group and 47 years in the patient group. The patients showed a similar distribution in social class. There was no significant difference in parity in the two groups. The figure shows the initial whole bone density in the patients in relation to the normal subjects. There was no significant difference in whole bone density (SAE) between the two groups when the patients were compared with the normal subjects of their own age. Eighty-seven per cent of the patients lay between 50 and 80 years of age.

Bone mass measurements were regressed against time in each of the patients. Of the 94 patients, 26 showed a significant fall in whole bone density, and 24 a significant fall in the total cortical mineral per unit length of bone. Three patients showed a significant rise in SAE and the same three patients showed a significant rise in TCM. The number of patients in these two groups was small, but no obvious difference in terms of the initial bone mass, age of onset of the menopause, social status, distinguished them from the other patients in the group. This group as a whole was examined further by estimating the percentile values and relating these to the time that the patients had been given therapy. There was a small fall in the percentile values of both the whole bone density and total cortical mineral per unit length of bone, but these were not statistically significant. However, there was a significant fall in both the SAE and the TCM when the change in percentile value from the initial reading was regressed against the time on therapy (Table a). This means that as a group, the patients treated with calcium supplements, showed a greater fall in whole bone density and total cortical mineral per unit length, than the fall which occurs in the normal population.

III. Discussion

Bone mass falls significantly with age in normal women. However, the rate of loss is a function of the measurement used (SMITH, 1971) and the bone measured (CONRAD JOHNSTON et al., 1968). Moreover, we have shown that the measurements on the metacarpal correlate with measurements in the spine (SMITH, 1971) and with other bone sites (SMITH et al., 1973), and though rates of change may differ, the pattern of fall is similar.

The 94 patients investigated were referred by physicians or surgeons who considered them to be osteoporotic. However, they were not found to differ from the subjects in the normal population of their own age

Table 1. The relation between the percentile value and the change in the percentile value of the SAE and the TCM, and the time on therapy of 94 patients treated for 18 to 54 months with calcium supplements

	Gradient	Intercept	r	p
SAE percentile v time on therapy	-0.075	35.73	0.039	n.s.
TCM percentile v time on therapy	-0.039	32.00	0.020	n.s.
Change in SAE percentile v time on therapy	-0.181	0.81	0.158	<0.01
Change in TCM percentile v time on therapy	-0.122	0.84	0.150	<0.01

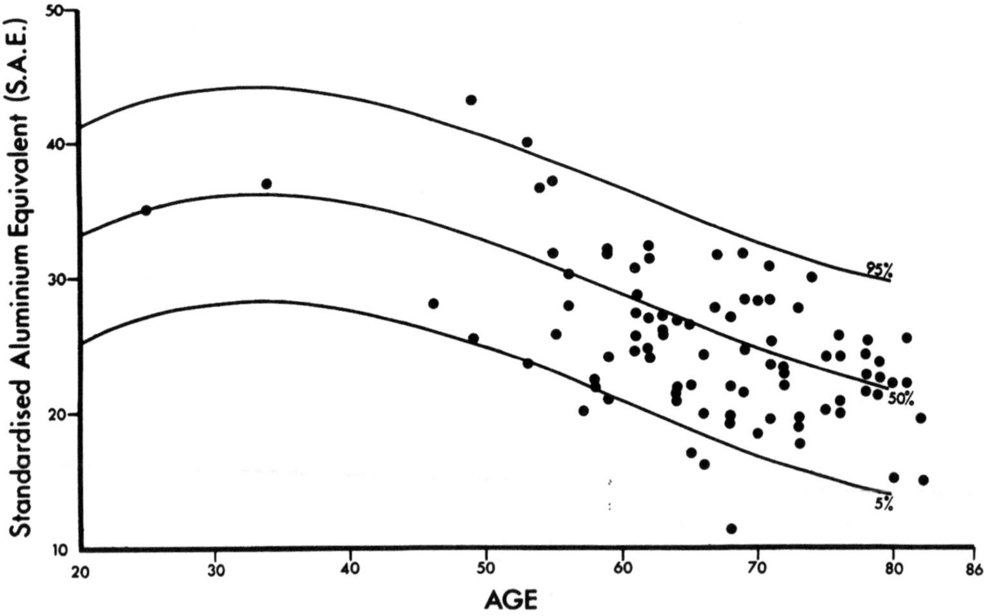

Fig. 1. The relation between the initial whole bone density (SAE) and age in female patients thought to be osteoporotic, compared with the normal range. The 5 and 95 percentile values for the normal subjects are shown

in their bone density measurements, social class, age of onset of the menopause, or parity. None of the patients has had any metabolic disorder that might have affected the bone density measurements.

The normal female population (SMITH et al., 1969) shows a highly significant fall in bone density with increasing age, and the rate of fall varies with age. The rate of fall in bone density is relatively slow between 40 and 45 years, accelerates between 45 and 65 years of age, and declines steadily thereafter (SMITH, 1971). Any change producing a more or less rapid fall in the patient group, when compared with normal subjects, will be evident as a changing percentile value. The percentile value shows no significant change and evidently the patients are showing a fall in whole bone density which is very close to that seen in a normal population. However, when the change in percentile value is measured for each patient as a change from the starting value, then there is a small but significant fall ($P<0.01$) in both the percentile values for whole bone density (SAE) and total cortical mineral (TCM) per unit length of bone with the time on therapy.

This must mean that either the patient population is different from the normal population, or that the calcium supplements are causing an accelerated rate of bone loss. The comparison between the two groups has not brought to light any significant difference between the two groups. The evidence therefore is in keeping with the calcium therapy being the cause of the acceleration in bone loss in the patient group. This is at variance with the reported positive balances obtained on calcium supplementation of the diet. The implication is that either reported balance studies must regularly and persistently overestimate balance results when calcium supplements are added, or positive balances must be relatively shortlived. Investigations (SMITH, 1971) of fecal fat excretion in patients on calcium supplements show a highly significant rise in fecal fat excretion, and this may be the result of long-term malabsorption of vitamin D or other nutrients which could lead to the development of a negative calcium balance.

References

ANDERSON, I.A.: Postmenopausal osteoporosis, clinical manifestations and treatment with estrogens. Quart. J. Med. 19, 67-96 (1950).

ANDERSON, J.B., SHIMMINS, J., SMITH, D.A.: A new technique for the measurement of metacarpal density. Brit. J. Radiol. 39, 443-450 (1966).

BELL, G.H., CUTHBERTSON, D.P., ORR, J.B.: The strength and size of bone in relation to calcium intake. J. Physiol. 100, 299-317 (1941).

CONRAD JOHNSTON, C., SMITH, D.M., YU, P.L., DEISS, W.S.: In vivo measurement of bone mass in the radius. Metabolism, 17, 1140-1153 (1968).

DENT, C.E., WATSON, L.: Osteoporosis. Postgrad.med. J. 583-608 (1966).

HARRISON, M., FRASER, R., MULLAN, B.: Calcium metabolism in osteoporosis. Acute and long-term responses to increased calcium intake. Lancet I, 1015-1019 (1961).

JAFFE, H.L., BODANSKY, A., CHANDLER, J.P.: Ammonium chloride decalcification, as modified by calcium intake. The relation between generalised osteoporosis and osteitis fibrosa. J. exp. Med. 56, 823-834 (1932).

JOWSEY, J., GERSHON-COHEN, J.: Clinical and experimental osteoporosis. In: Bone and Tooth, pp. 35-48, (Ed. H.J.J. Blackwood). Oxford: Pergamon Press 1964.

NORDIN, B.E.C.: Calcium balance and calcium requirement in spinal osteoporosis. Amer. J. clin. Nutr. 10, 384-390 (1962).

OWEN, E.C., IRVING, J.T., LYALL, A.: The calcium requirements of older male subjects with special reference to the genesis of senile osteoporosis. Acta med. Scand. 103, 235-250 (1940).
POMMER, G.A.: Über Osteoporose. Arch. klin. Chir. 136, 1-68 (1925).
SMITH, D.A., ANDERSON, J., SHIMMINS, J., SPEIRS, C., BARNETT, E.: Mineral and density changes in bone with age in normal and pathological states. In: Progress in Methods of Bone Mineral Measurements. (Ed. G.D. Whedon and J.R. Cameron), pp. 177-191. U.S. Department of Health. Washington: Education and Welfare, D.C. (1968).
SMITH, D.A., ANDERSON, J., SHIMMINS, J., SPEIRS, C., BARNETT, E.: Changes in metacarpal mineral content and density in normal male and female subjects with age. Clin. Radiol. 20, 23-31 (1969).
SMITH, D.A.S.: Osteoporosis. Thesis for M.D. Degree. University of Glasgow (1971).
SMITH, C.B., HORTON, P.W., AITKEN, J.M., SMITH, D.A.: The estimation of bone mineral content at selected skeletal sites by gamma ray absorption. Brit. J. Radiol. 47, 314-318 (1974).
SMITH, R.W., FRAME, B.: Concurrent axial and appendicular osteoporosis. New Engl. J. Med. 273, 73-78 (1965).
WHEDON, G.D.: Effects of high calcium intakes on bones, blood and soft tissues. Relationships of calcium intakes to balance in osteoporosis. Fed. Proc. 18, 1112-1118 (1959).

VII. Parathyroid Hormone and Calcitonin

Chairmen: D. H. Copp and I. MacIntyre

Chemical and Biologic Studies of Parathyroid Hormone, Proparathyroid Hormone, and Fragments of Parathyroid Hormone*

G.V. SEGRE AND J.T. POTTS, JR.

In the last few years there have been extremely rapid progress in our knowledge of the chemistry of the parathyroid hormones of various species and striking advances in our understanding of their complex biosynthesis, secretion, and metabolism. This brief review will summarize the current state of knowledge and will discuss some of the clinical implications resulting from these advances.

Biosynthesis of a true precursor to both human and bovine parathyroid hormone has now been demonstrated. The existence of a larger form of parathyroid hormone was first appreciated by Cohn and his coworkers (1972). More recently, formal kinetic proof of the existence of proparathyroid hormone was demonstrated in normal parathyroid glands from cows (KEMPER et al., 1972), man (CHU et al., 1973b), chickens (MACGREGOR et al., 1973), and rats (CHU et al., 1973a), and in human parathyroid adenomas (HABENER et al., 1973a). Studies of radioactive peptides of bovine prohormone generated by limited enzymatic cleavage indicated that the prohormone-specific sequence was added to the amino terminus of the parathyroid hormone sequence (HABENER et al., 1973b). Subsequently, this prohormone peptide was shown to be a hexapeptide (Fig. 1) having the same amino-acid sequence, lys-ser-val-lys-lys-arg, in both bovine and human hormone (COHN et al., 1974; HAMILTON et al., 1974; JACOBS et al., 1974). Because of the nature of these residues, the prohormone-specific hexapeptide is strikingly basic. The hormone, which is initially synthesized as a prohormone, is then converted into the stored-form of the hormone by a trypsin-like enzyme that cleaves the prohormone between an arginine and alanine bond in the bovine prohormone (or arginine and serine in the human prohormone). In the original report by COHN et al., analysis of the amino-acid composition suggested that the prohormone consisted of 25 additional residues (COHN et al., 1972). The finding of only six residues at the amino terminus has left open the possibility that there is a second addition, perhaps at the carboxyl terminus.

During the course of their studies of proparathyroid hormone, KEMPER and his coworkers (1974) found that bovine parathyroid tissue incubated _in vitro_ secretes a protein, comprising about 50% of the total secreted protein, that is distinct from both parathyroid hormone and proparathyroid hormone. This protein, which has been called "parathyroid secretory protein", appears to be an aggregate consisting of two or more subunits with a molecular weight of approximately 70,000. The

---

+ This work was supported by a grant from the John A. Hartford Foundation, Inc., grants AM-11794 and AM-04501 from the National Institutes of Health, and contract NAS 9-11011 from the National Aeronautics and Space Administration.
Dr. SEGRE is the George Morris Piersol Teaching and Research Scholar of the American College of Physicians and a Special Fellow of the National Institute of Arthritis, Metabolic, and Digestive Diseases of the National Institutes of Health.

secretion rate of this protein responds in parallel with that of parathyroid hormone to changes in the concentration of calcium in the medium.

At this time, the physiological roles of proparathyroid hormone and the parathyroid secretory protein are not known. One may speculate that it appears likely that proparathyroid hormone and the parathyroid secretory protein may participate in some manner in intracellular synthesis and transport of the hormone and, perhaps, in the storage process. It is still unknown whether either of these peptides or the prohormone-specific hexapeptide is secreted into the circulation in vivo. Radioimmunoassays that specifically measure the bovine proparathyroid hormone have been developed, but, thus far, proparathyroid hormone has not yet been detected in the circulation of cows (HABENER et al., 1974). With the development of more-sensitive radioimmunoassays, the nature of all the secretory products of the parathyroid glands will become better understood. Hopefully, with the availability of such assays, differences in the nature of these secretory products in the various forms of primary and ectopic hyperparathyroidism will become evident, and application of these assays will lead to improvements in our ability to diagnose and, consequently, treat these pathological disorders of parathyroid-gland function.

Within the last few years, sufficient quantities of highly purified

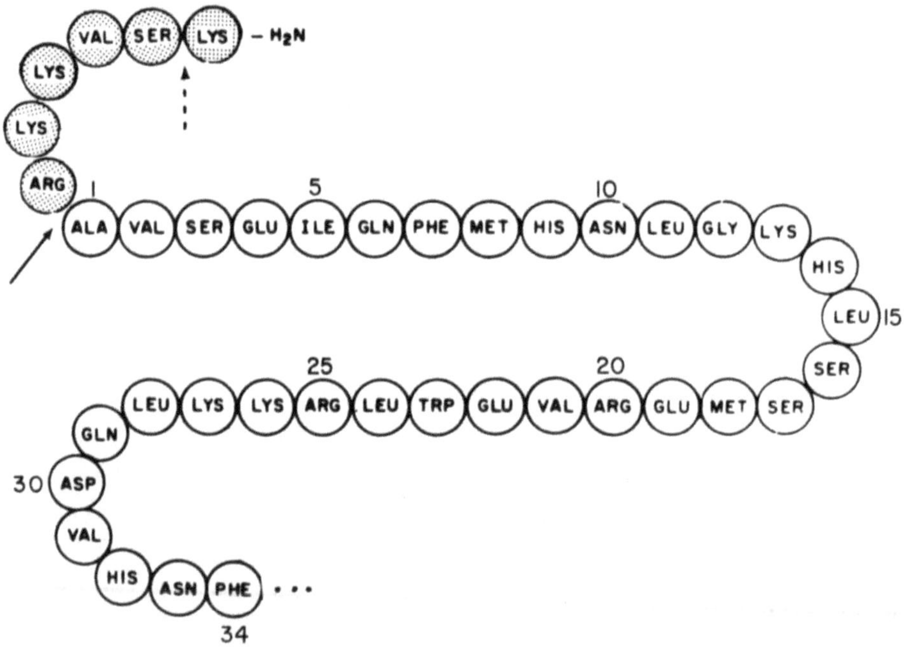

Fig. 1. Sequence of bovine proparathyroid hormone. The prohormone hexapeptide is represented by the shaded area. The arrows indicate the trypsin-sensitive cleavage site in the prohormone hexapeptide (from POTTS, J.T., Jr., NIALL, H.D., TREGEAR, G.W., VAN RIETSCHOTEN, J., HABENER, J.F., SEGRE, G.V., KEUTMANN, H.T.: Chemical and biologic studies of proparathyroid hormone and parathyroid hormone: analysis of hormone biosynthesis and metabolism. Mt. Sinai J. Med. 40, 448--461; 1973)

bovine, porcine, and human parathyroid hormone have been isolated to permit determination of the entire sequence of the non-human hormones (BREWER et al., 1970; NIALL et al., 1970; SAUER et al., 1974), and also the chemical composition (KEUTMANN et al., 1974) and the sequence of the amino-terminal region of the human hormone (BREWER et al., 1972; NIALL et al., 1974). All the hormones are 84-amino-acid, single-chain polypeptides with no disulfide linkages (BREWER et al., 1970; NIALL et al., 1970; SAUER et al., 1974; KEUTMANN et al., 1974). There are ten differences in the amino-acid composition between bovine and porcine hormones; the porcine differs from bovine at 7 (not 5) positions because of internal rearrangement of a serine and alanine residue (SAUER et al., 1974). Compositional analysis of human parathyroid hormone shows that the amino-acid composition differs in twelve residues from both bovine and porcine parathyroid hormones (KEUTMANN et al., 1974). We have found that the sequence of the amino-terminal 37 residues of human hormone differs at three amino-acid positions from that of the bovine hormone (substitution of serine for alanine at position 1, leucine for phenylalanine at position 7, and asparagine for serine at position 16) and at two positions from that of porcine parathyroid hormone (substitution of asparagine for serine at position 16 and of methionine for leucine at position 18) (Fig. 2). Therefore, the only unique residue thus far found in human parathyroid hormone is the asparagine at position 16 (NIALL et al., 1974). In several other respects, human parathyroid hormone is similar to the porcine hormone in that it, too, has serine at position 1 and it also lacks a tyrosyl residue. The compositional analysis, however, indicates that there will be several residues unique to human parathyroid hormone in the remaining sequence, including the presence of threonine and substi-

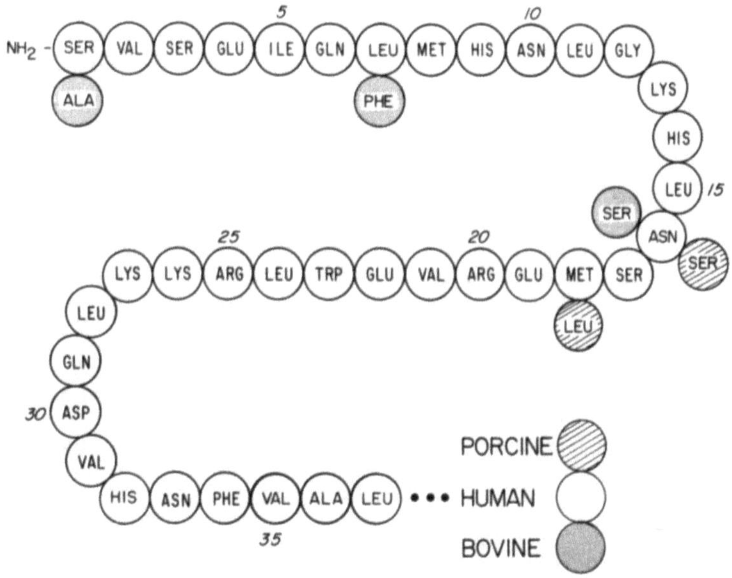

Fig. 2. Sequence of the amino-terminal portions of human, bovine, and porcine parathyroid hormones (from POTTS, J.T., Jr., and DEFTOS, L.J.: Parathyroid hormone, calcitonin, vitamin D, bone and bone mineral metabolism. In: Duncan's Diseases of Metabolism, 7th edition (Bondy, P.K., and Rosenberg, L.E., editors), pp. 1225--1430. Philadelphia: W.B. Saunders, 1972)

tutions for two isoleucine residues in the remaining portion of the sequence of the non-human hormones.

The recent structure reported by BREWER et al. (1972) of the amino-terminal 34 residues of human parathyroid hormone differs from the structure we have reported. BREWER et al. (1972) report residue 22 to be glutamine, residue 28 to be lysine, and residue 30 to be leucine. As will be discussed in further detail, the resolution of these differences and the determination of the remaining portion of the sequence of human parathyroid hormone will be critical to the development of homologous assays for the measurement of the biologically relevant forms of parathyroid hormone in the circulation of man.

Considerable information is now available concerning the structure-and-function relationships of the parathyroid hormones, particularly bovine and human parathyroid hormones. Earlier work with bovine parathyroid hormone 1--29 fragment, generated by dilute-acid cleavage of intact hormone, had shown that some portion of the amino-terminal sequence could account, at least qualitatively, for the biologic effects of the hormone on bone and kidney (KEUTMANN et al., 1972). In addition, it was shown that bovine parathyroid hormone 53--84 fragment, which was produced by tryptic digestion of native hormone after reversible blockade of the ε-amino groups of lysine residues, was biologically inactive (POTTS et al., 1972). In the last few years, the emphasis has shifted to studies of amino-terminal peptides and their analogs prepared by solid-phase methods. Several fragments representing progressively shorter segments of the amino-terminal region of bovine parathyroid hormone have been prepared and bioassayed in the rat renal adenylyl-cyclase assay (MARCUS and AURBACH, 1969) *in vitro* and in the chick hypercalcemia assay (PARSONS et al., 1973) *in vivo*. These studies have defined the minimal chain length required for biologic activity. Bovine parathyroid hormone 1--34 is 80% as potent, on a molar basis, as the intact hormone *in vitro* and 130% as potent *in vivo*. Deletion of the amino-terminal residue, alanine, results in a marked decrease in activity, and removal of the second residue, valine, completely abolishes activity in both assays (TREGEAR et al., 1973). Greater shortening, however, is tolerated at the carboxyl terminus. Bovine peptide 1--27 is still active *in vitro*, but 1--26 is inactive in both assay systems (TREGEAR et al., 1973). Further alterations of the structure of bovine parathyroid hormone 1--34 have shown that the addition of a single amino acid (tyrosine) or the addition of a tripeptide to the amino terminus results in loss of activity. Such studies have also shown that although oxidation of the methionine residues at positions 8 and 18 of the bovine molecule is accompanied by striking loss of activity, methionine itself is not required for activity. When the sterically similar amino acid, norleucine, is used to replace the methionine residues, there is no significant loss of activity.

Studies of the human hormone are less complete. Synthetically prepared human parathyroid hormone 1--34 has been found to be as active, on a molar basis, as intact human hormone in the rat renal adenylyl-cyclase assay *in vitro*. It has also been found to be as active as bovine parathyroid hormone 1--34 in the chick hypercalcemia assay *in vivo*, but only 20% as active as bovine parathyroid hormone 1--34 in the *in vitro* system. Substitution of serine at position 1 for the naturally occurring residue of the bovine hormone, alanine ([Ser^{1}]-bovine parathyroid hormone 1--34), results in a peptide that is 23% as active as bovine parathyroid hormone 1--34 or 120% as active as human parathyroid hormone 1--34 in the *in vitro* assay (SEGRE et al., 1974b). Alternatively, bioassay *in vitro* of human parathyroid hormone 1--34 in which the serine at position 1 has been replaced by alanine ([Ala^{1}]-human parathyroid hormone 1--34) showed a 400% increase in activity when compared with human parathyroid hormone 1--34 and was nearly 80% as active as bovine parathyroid hormone 1--34 (TREGEAR et al., 1974).

Several conclusions can be drawn from these studies. First, it appears that an amino-terminal fragment of the bovine and human hormones can account, quantitatively, for the biologic activity of the hormone. Second, the minimal sequence required for biologic activity of the bovine hormone extends from the valine at position 2 to the lysine at position 27 (Fig. 3). Third, extensions at the amino terminus result in loss of biologic activity. Fourth, the striking differences in biologic activity between human and bovine parathyroid hormones in the rat renal adenylyl-cyclase bioassay appear to be due to the fact that, in this model system, substitution of the serine residue for alanine at position 1 of the sequence is of critical importance. Since bovine and human parathyroid hormone 1--34 are equipotent in the chick hypercalcemia assay in vivo, this suggests that there are significant differences in the receptors for parathyroid hormone that depend on the particular species used for testing or on the particular assay system used, or both. Fifth, at this time, the function of the middle and carboxyl portions of the hormone is unknown.

These structure-and-function relationships of parathyroid hormone are of obvious importance, but they become of even greater significance when considered with respect to the fragments of parathyroid hormone that result from metabolism in vivo of the intact hormone after its secretion from the glands.

Several years ago, BERSON and YALOW (1968) demonstrated that plasma parathyroid hormone in man is immunologically distinct from the hor-

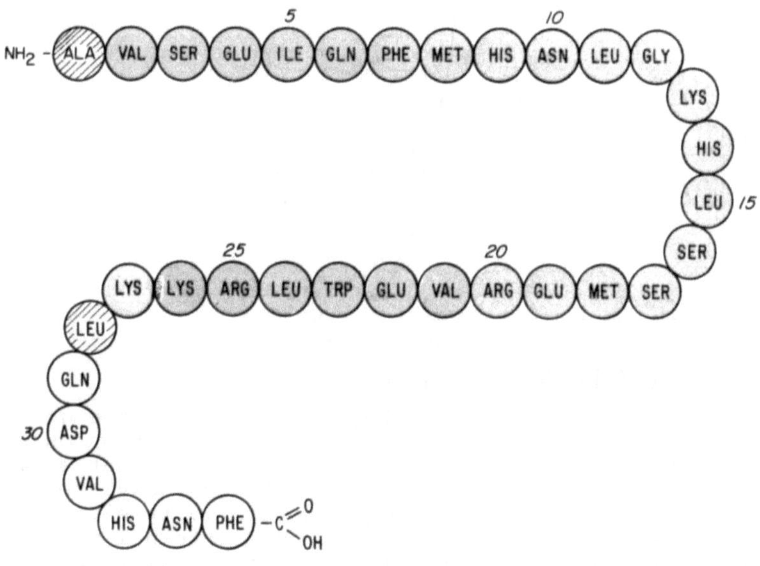

Fig. 3. Minimal sequence of bovine parathyroid hormone (2--27) required for biologic activity (from POTTS, J.T., Jr., and DEFTOS, L.J.: Parathyroid hormone, calcitonin, vitamin D, bone and bone mineral metabolism. In: Duncan's Diseases of Metabolism (P.K. Bondy and L.E. Rosenberg, editors), 7th edition, pp. 1225--1430. Philadelphia: W.B. Saunders, 1974)

mone extracted from human adenomas. This report, which was of central importance in establishing the immunochemical heterogeneity of plasma parathyroid hormone, led to intensive investigations of the properties, origin, and biologic significance of the multiple forms of parathyroid hormone in the circulation. These studies have dramatically altered our concepts of the physiology of parathyroid hormone and have led to several new questions concerning clinical use of immunoassays for parathyroid hormone.

Findings in several laboratories are in general agreement that the principal, circulating immunoreactive forms of the hormone appear to be hormonal fragments that can be distinguished from intact hormone by at least two criteria: greater retardation on gel-filtration than intact hormone, and altered immunological properties (HABENER et al., 1971, 1972; CANTERBURY and REISS, 1972; SEGRE et al., 1972; GOLDSMITH et al., 1973; SILVERMAN and YALOW, 1973; ARNAUD et al., 1974). From their elution positions, the major immunoreactive forms of the hormone appear to have a molecular size two-thirds as large as intact hormone extracted from glands. Immunochemical analysis, using radioimmunoassays with antisera that specifically recognize limited, defined portions of the hormonal sequence (sequence-specific radioimmunoassays)

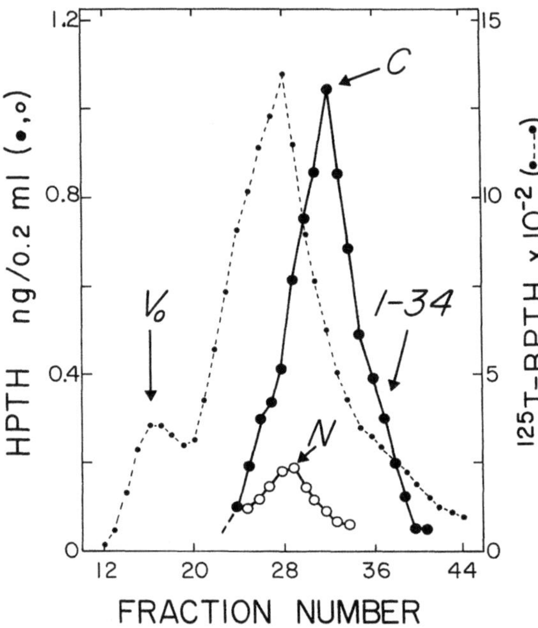

Fig. 4. Immunoreactive parathyroid hormone in the gel-filtration fractions of a sample from the peripheral circulation on Bio-Gel P-10. The sample was analyzed in assays using antisera that specifically measure the amino-terminal portions (N) and the carboxyl-terminal portions (C) of the molecule. Most of the immunoreactivity elutes later than the bovine ^{125}I-labelled parathyroid hormone internal marker (•----•). This late-eluting peak is devoid of amino-terminal reactivity. The elution position of bovine 1--34 and the void volume (V_o) are indicated (from POTTS, J.T., Jr., and DEFTOS, L.J.: Parathyroid hormone, calcitonin, vitamin D, bone and bone mineral metabolism. In: Duncan's Diseases of Metabolism (P.K. Bondy and L.E. Rosenberg, editors), 7th edition, pp. 1225--1430. Philadelphia: W.B. Saunders, 1974)

have shown that this late-eluting peak is completely lacking antigenic determinants within the amino-terminal portion of the sequence (Fig. 4). Because of the structure-and-function relationships just considered, this must mean that this smaller form of the hormone is biologically inactive (SEGRE et al., 1972). Therefore, it would appear that the dominant form of immunoreactive parathyroid hormone in the circulation consists of biologically inactive peptide (SEGRE et al., 1972).

More recently, the precise location of cleavage of the hormone has been determined by studying the metabolism of bovine ^{125}I-labelled parathyroid hormone in dogs. Evidence from studies using sequence-specific radioimmunoassays indicated that the large fragment seen in the general circulation included the middle and carboxyl portions of the hormone. Because the single tyrosyl residue of bovine parathyroid hormone is located at position 43, it should be contained within this fragment. Therefore, by subjecting the fragment of radioiodinated hormone, which has been principally labelled on the single tyrosyl residue at position 43, to sequence analysis using the automated Edman degradation, the position of the amino terminus of this large fragment and, thus, the site of cleavage of the hormone could be defined by counting the number of cycles of degradation before release of specific iodotyrosyl radioactivity. Four major radioiodinated fragments were seen, corresponding to cleavages of the injected hormone between residues 33 and 34, 36 and 37, 40 and 41, and 42 and 43. The changes in the concentration of these fragments with the time at which the samples were collected after intravenous injection into dogs suggested, but did not establish, that the initial cleavage occurs between residues 33 and 34 and that the other fragments result from secondary enzymatic events (SEGRE et al., 1974). In addition, hormone was not metabolized to a significant extent when incubated in plasma, indicating that hormonal cleavage is due to enzymes probably located in extravascular site(s) (SEGRE et al., 1974).

At present there is no consensus concerning the number and nature of other circulating forms of parathyroid hormone, some of which may include the amino-terminal portion of the sequence and, therefore, may be biologically active. Peaks of immunoreactive parathyroid hormone having molecular weights of approximately 2500 (CANTERBURY and REISS, 1972; GOLDSMITH et al., 1973) and 4500 daltons (CANTERBURY and REISS, 1972; GOLDSMITH et al., 1973; SILVERMAN and YALOW, 1973; ARNAUD et al., 1974) have been reported after gel filtration of human plasma. From their studies, SILVERMAN and YALOW (1973) deduced that the 4500-molecular-weight fragment consisted largely of amino-terminal peptide, but was biologically inactive, whereas CANTERBURY, LEVEY, and REISS (1973) recovered a fragment of similar size from concentrates of human plasma and found it to be active in the rat renal cortical adenylylcyclase assay. In our studies we have been unable to detect fractions of the hormone other than the intact and the large fragment comprising the middle and carboxyl portions of the sequence. Although we have failed to detect circulating amino-terminal fragments, our data, however, suggest that there are no cleavages nearer the amino terminus than is position 33, indicating that an amino-terminal fragment containing the necessary requirements for biologic activity may be produced within the peripheral tissues responsible for hormonal cleavage (SEGRE et al., 1974).

If there is a biologically active amino-terminal fragment, whether it is present in the circulation, as indicated by the studies of CANTERBURY and her associates (1973), or is not, as suggested by our studies (HABENER et al., 1972 a,b; SEGRE et al., 1972, 1974), its potential physiological significance is provocative. For example, if cleavage occurs adjacent to receptors in the target organs, this postulated active fragment, even without reentering the circulation, could constitute a form of the hormone whose spectrum of activity is different from that of the intact hormone, or, alternatively, the dominant, if not the sole, active molecular species of the hormone. Such an activation step might constitute a significant control point in the expression of parathyroid hormone action, and it would be important to determine whether disturbance of this reaction could account for certain disorders of parathyroid function.

On the other hand, it is important to emphasize that our findings provide no direct evidence against the concept that the principal, if not sole, active molecular species of parathyroid hormone is the intact, 84-amino-acid peptide secreted by the parathyroid glands. Further study is needed to clarify the essential features of hormonal metabolism and to resolve whether this metabolism is concerned primarily with destruction or activation of parathyroid hormone. Only through a more complete understanding of the complex nature of parathyroid hormone metabolism and an accurate definition of the number and nature of the circulating hormonal fragments will it be possible to design more-specific and more-useful assays to measure the biologically important components of plasma immunoreactive parathyroid hormone and to assess changes, if any, in hormonal metabolism in different physiological and pathological states.

References

ARNAUD, C.D., GOLDSMITH, R.S., BORDIER, P.J., SIZEMORE, G.W., LARSEN, J.A., GILKINSON, J.: Influence of immunoheterogeneity of circulating parathyroid hormone on results of radioimmunoassays of serum in man. Am. J. Med. 56, 785 (1974).

BERSON, S.A., YALOW, R.: Immunochemical heterogeneity of parathyroid hormone in plasma. J. Clin. Endocrinol. Metab. 28, 1037 (1968).

BREWER, H.B., Jr., RONAN, R.: Bovine parathyroid hormone: amino acid sequence. Proc. Nat. Acad. Sci. U.S.A. 67, 1862 (1970).

BREWER, H.B. Jr., FAIRWELL, T., RONAN, R., SIZEMORE, G.W., ARNAUD, C.D.: Human parathyroid hormone: amino-acid sequence of the amino-terminal residues 1--34. Nat. Acad. Sci., U.S.A. 69, 3585 (1972).

CANTERBURY, J.M., REISS, E.: Multiple immunoreactive molecular forms of parathyroid hormone in human serum. Proc. Soc. Exp. Biol. Med. 140, 1393 (1972).

CANTERBURY, J.M., LEVEY, G.S., REISS, E.: Activation of renal cortical adenylate cyclase by circulating immunoreactive parathyroid hormone fragments. J. Clin. Invest. 52, 524 (1973).

CHU, L.L.H., MACGREGOR, R.R., ANAST, C.S., HAMILTON, J.W., COHN, D.V.: Studies on the biosynthesis of rat parathyroid hormone and proparathyroid hormone: adaptation of the parathyroid gland to dietary restriction of calcium. Endocrinology 93, 915 (1973a).

CHU, L.L.H., MACGREGOR, R.R., LIU, P.I., HAMILTON, J.W., COHN,D.V.: Biosynthesis of proparathyroid hormone and parathyroid hormone by human parathyroid glands. J. Clin. Invest. 52, 3089 (1973).

COHN, D.V., MACGREGOR, R.R., CHU, L.L.H., HAMILTON, J.W.: In: Calcium, Parathyroid Hormone and the Calcitonins (Talmage, R.V., and Munson, P.L., editors) 1972, 173, Amsterdam: Excerpta Medica.

COHN, D.V., MACGREGOR, R.R., CHU, L.L.H., HUANG, D.W.Y., ANAST, C.S., HAMILTON, J.W.: Biosynthesis of proparathyroid hormone and parathyroid hormone: chemistry, physiology and role of calcium in regulation. Am., J. Med. 56, 767 (1974).

GOLDSMITH, R.S., FURSZYFER, J., JOHNSON, W.J., FOURNIER, A.E., SIZEMORE, G.W., ARNAUD, C.D.: Etiology of hyperparathyroidism and bone disease during chronic hemodialysis, J. Clin. Invest. 52, 173 (1973).

HABENER, J.F., POWELL, D., MURRAY, T.M., MAYER, G.P., POTTS, J.T., Jr.: Parathyroid hormone: secretion and metabolism in vivo. Proc. Nat. Acad. Sci. U.S.A. 68, 2986 (1971).

HABENER, J.F., MAYER, G.P., SEGRE, G.V., DEE, P., POTTS, J.T., Jr.: Metabolic fate of immunoreactive parathyroid hormone (IRPTH) in the bovine. Clin. Res. 20, 865 (1972a) (Abstr.).

HABENER, J.F., SEGRE, G.V., POWELL, D., MURRAY, T.M., POTTS, J.T., Jr.: Immunoreactive parathyroid hormone in circulation of man. Nature New Biol. 238, 152 (1972b).

HABENER, J.F., KEMPER, B., POTTS, J.T., Jr., RICH, A.: Proparathyroid hormone: biosynthesis by human parathyroid adenomas. Science 178, 630 (1973a).

HABENER, J.F., KEMPER, B., POTTS, J.T., Jr., RICH, A.: Bovine proparathyroid hormone: structural analysis of radioactive peptides formed by limited cleavage. Endocrinology 92, 219 (1973b).

HABENER, J.F., TREGEAR, G.W., STEVENS, T.D., DEE, P., POTTS, J.T., Jr.: Radioimmunoassay for proparathyroid hormone. Endocrine Res. Comm. 1, 1 (1974).

HAMILTON, J.W., NIALL, H.D., JACOBS, J.W., KEUTMANN, H.T., POTTS, J.T., Jr., COHN, D.V.: The N-terminal amino-acid sequence of bovine proparathyroid hormone. Proc. Nat. Acad. Sci. U.S.A. 71, 653 (1974).

JACOBS, J.W., KEMPER, B., NIALL, H.D., HABENER, J.F., POTTS, J.T., Jr.: Structural analysis of human proparathyroid hormone by a new microsequencing approach. Nature (London) 249, 155 (1974).

KEMPER, B., HABENER, J.F., POTTS, J.T., Jr., RICH, A.: Proparathyroid hormone: identification of a biosynthetic precursor to parathyroid hormone. Proc. Nat. Acad. Sci. U.S.A. 69, 643 (1972).

KEMPER, B., HABENER, J.F., RICH, A., POTTS, J.T., Jr.: Parathyroid secretion: discovery of a major, calcium-dependent protein. Science 184, 167 (1974).

KEUTMANN, H.T., DAWSON, B.F., AURBACH, G.D., POTTS, J.T., Jr.: A biologically active amino-terminal fragment of bovine parathyroid hormone prepared by dilute acid hydrolysis. Biochemistry 11, 1973 (1972).

KEUTMANN, H.T., BARLING, P.M., HENDY, G.N., SEGRE, G.V., NIALL, H.D., AURBACH, G.D., POTTS, J.T., Jr., O'RIORDAN, J.L.H.: Isolation of human parathyroid hormone. Biochemistry 13, 1646 (1974).

MACGREGOR, R.R., CHU, L.L.H., HAMILTON, J.W., COHN, D.V.: Partial purification of parathyroid hormone from chicken parathyroid glands. Endocrinology 92, 1312 (1973).

MARCUS, R., AURBACH, G.D.: Bioassay of parathyroid hormone in vitro with a stable preparation of adenyl cyclase from rat kidney. Endocrinology 85, 801 (1969).

NIALL, H.D., KEUTMANN, H.T., SAUER, R., HOGAN, M.L., DAWSON, B.F., AURBACH, G.D., POTTS, J.T., Jr.: The amino-acid sequence of bovine parathyroid hormone I. Hoppe-Seyler's Z. Physiol. Chem. 351, 1586 (1970).

NIALL, H.D., SAUER, R.T., JACOBS, J.W., KEUTMANN, H.T., SEGRE, G.V., O'RIORDAN, J.L.H., AURBACH, G.D., POTTS, J.T., Jr.: The amino-acid sequence of the amino-terminal 37 residues of human parathyroid hormone. Proc. Nat. Acad. Sci. U.S.A. 71, 384 (1974).

PARSONS, J.A., REIT, B., ROBINSON, C.J.: A bioassay for parathyroid hormone using chicks. Endocrinology 92, 454 (1973).

POTTS, J.T., Jr., DEFTOS, L.J.: Parathyroid hormone, calcitonin, vitamin D, bone and bone mineral metabolism. In: Duncan's Diseases of Metabolism (P.K. Bondy and L.E. Rosenberg, editors), 7th edition, pp. 1225-1430. Philadelphia: W.B. Saunders Co., 1974.

POTTS, J.T., Jr., KEUTMANN, H.T., NIALL, H.D., TREGEAR, G.W., HABENER, J.F., O'RIORDAN, J.L.H., MURRAY, T.M., POWELL, D., AURBACH, G.D.: Parathyroid hormone: chemical and immunochemical studies of the active molecular species. In: Endocrinology 1971: Proceedings of the Third International Symposium (S. Taylor, editor), pp. 333-349. London: Heinemann Medical Books, Ltd. 1972.

SAUER, R.T., NIALL, H.D., HOGAN, M.L., KEUTMANN, H.T., O'RIORDAN, J. L.H., POTTS, J.T., Jr.: The amino-acid sequence of porcine parathyroid hormone. Biochemistry 13, 1994 (1974).

SEGRE, G.V., HABENER, J.F., POWELL, D., TREGEAR, G.W., POTTS, J.T., Jr.: Parathyroid hormone in human plasma: immunochemical characterization and biological implications. J. Clin. Invest. 51, 3163 (1972).

SEGRE, G.V., NIALL, H.D., HABENER, J.F., POTTS, J.T., Jr.: Metabolism of parathyroid hormone: physiologic and clinical significance. Am. J. Med. 56, 774 (1974).

SEGRE, G.V., TREGEAR, G.W., KEUTMANN, H.T., POTTS, J.T., Jr.: Further immunologic and biologic characterization of the amino-terminal 1--34 peptide of human parathyroid hormone. Submitted to Endocrinology (1974b).

SILVERMAN, R., YALOW, R.S.: Heterogeneity of parathyroid hormone: clinical and physiologic implications. J. Clin. Invest. 52, 1958 (1973).

TREGEAR, G.W., VAN RIETSCHOTEN, J., GREENE, E., KEUTMANN, H.T., NIALL, H.D., REIT, B., PARSONS, J.A., POTTS, J.T., Jr.: Bovine parathyroid hormone: minimum chain length of synthetic peptide required for biological activity. Endocrinology 93, 1349 (1973).

TREGEAR, G.W., VAN RIETSCHOTEN, J., GREENE, E., NIALL, H.D., KEUTMANN, H.T., PARSONS, J.A., O'RIORDAN, J.L.H., POTTS, J.T., Jr.: Solid-phase synthesis of the biologically active N-terminal 1--34 peptide of human parathyroid hormone. Hoppe-Seyler's Z. Physiol. Chem. 355, 415 (1974).

Parathyroid Hormone: Structure and Immunoheterogeneity*

C. D. Arnaud and H. B. Brewer, Jr.

From the Mineral Research Laboratory and the Department of Endocrine Research, Mayo Clinic and Mayo Foundation, Rochester, Minnesota 55901, and the National Institutes of Health, Bethesda, Maryland 20014. This investigation was supported in part by Research Grants AM-12302 from the National Institutes of Health, Public Health Service, and by a grant from the Mayo Foundation. Reprint requests and correspondence should be addressed to Dr. Arnaud, Mayo Clinic, Rochester, Minnesota 55901. Read at the X European Symposium on Calcified Tissue, Hamburg, Germany, September 17 to 21, 1973.

Abstract

The amino acid sequence of the NH_2-terminal 34 residues of human parathyroid hormone (HPTH) has been determined and duplicated synthetically to produce a peptide that is biologically active. In the amino acid sequences of the bovine and porcine hormones, the glutamic acid function at position 22 has been revised to glutamine. Among these initial 34 residues, HPTH differs from bovine PTH by 5 residues and from porcine PTH by 4 residues.

A synthetic preparation of HPTH 1-34 was used to characterize two antisera to PTH. One antiserum (CH 14M) reacted almost equally well with the synthetic peptide (HPTH 1-34) as with HPTH 1-84 obtained from natural sources. The other antiserum (GP 1M) did not react with the synthetic peptide (HPTH 1-34) but reacted with HPTH 1-84 to an identical degree as antiserum CH 14M. On this basis, antiserum CH 14M was designated "anti N" to describe its NH_2-terminal specificity, and GP 1M was designated "anti C" to describe its COOH-terminal specificity.

+ The work on the amino acid sequence of human PTH could not have been done without the generous donations of human parathyroid tissue to us by more than 150 physicians from 12 different nations. We thank our colleagues Ms. Rosemary Ronan and Drs. Thomas Fairwell, Glen W. Sizemore, and Francis P. Di Bella who actively contributed to the work on the amino acid sequence of human PTH and Dr. Werner Rittel whose group at the Ciba-Geigy Co. synthesized human PTH 1-34. Dr. E. Travis Littledike provided the porcine parathyroid hormone for sequence analysis. Dr. Ralph S. Goldsmith was actively involved in the work on the application of region-specific radioimmunoassays of PTH to the study of parathyroid function in hyperparathyroid patients. We are especially grateful to Dr. Philippe J. Bordier who provided the osteoclast counts and serum for PTH analysis from his patients in Paris. Dr. Thomas Dousa generously carried out the renal cortical membrane adenylate cyclase assays of synthetic bovine and human PTH 1-34. The constantly superb technical assistance of Ms. Julianna Gilkinson, Ms. Judith A. Larsen, Ms. Kathleen A. Safford, and Ms. Judith M. Verheyden is greatly appreciated.

Radioimmunoassays using these antisera were systematically applied to the measurement of serum immunoreactive PTH (iPTH) in patients with primary or secondary (renal) hyperparathyroidism. The results showed that the "anti C" assay system was superior to the "anti N" assay system in assessing both chronic parathyroid hyperfunction and the biologic effects of excess circulating PTH (osteoclast number in bone biopsies). In contrast, reference is made to the superiority of the "anti N" assay system in assessing acute changes in parathyroid function and in localizing abnormal parathyroid tissue by the method of differential catheterization of the venous drainage of the neck and mediastinum.

An explanation for these important differences in the results obtained with the two assay systems is based on observations that the concentration of COOH-terminal fragments and the serum survival time of these fragments are much greater than those of the native, secreted whole molecule, PTH 1-84.

During the past 5 years there has been a great increase in our understanding of the chemistry, biosynthesis, secretion, and pathophysiology of parathyroid hormone (PTH). PTH has been shown by HAMILTON, COHN, KEMPER, and colleagues (HAMILTON et al., 1971; COHN et al., 1971; COHN et al., 1972; KEMPER et al., 1972) to be synthesized as a precursor with a molecular weight of approximately 12,000. Structural studies of the bovine proparathyroid hormone (pro-PTH) have shown six additional amino acid residues attached to the NH_2-terminal end of the molecule (HAMILTON et al., 1973). Because amino acid compositional data suggest that pro-PTH has amino acid residues not yet accounted for by this hexapeptide sequence, the possibility that there is an additional sequence of amino acids attached to the COOH-terminal region is now being considered (HABENER et al., 1973; COHN et al., 1974). It is likely that the pro-PTH is converted, in the parathyroid gland, to the storage form of the hormone, an 84 amino acid polypeptide with a molecular weight of 9,500. It is not known presently if pro-PTH is released into the circulation.

The 84 amino acid polypeptide is a major form of the hormone secreted by the gland into the circulation after appropriate physiologic stimuli (HABENER et al., 1971). However, recent studies by a number of groups (HABENER et al., 1973; BERSON and YALOW, 1968; ARNAUD et al., 1970; ARNAUD et al., 1971a; CANTERBURY and REISS, 1972; HABENER et al., 1972; SEGRE et al., 1972; ARNAUD et al., 1973; SILVERMAN and YALOW, 1973; ARNAUD, 1973a; ARNAUD et al., in press; SEGRE et al., in press) indicate that there are multiple forms of the hormone in the blood of patients with hyperparathyroidism. These multiple forms include the 84 amino acid polypeptide and fragments of it. The major fragment has a molecular weight of 6,000 to 7,000, is COOH-terminal, and has a very long half-life. Canterbury et al. (1973) have demonstrated that a fragment(s) in the plasma of hyperparathyroid patients was biologically active in the renal adenylate cyclase system, which suggests that some of the circulating plasma fragments retain biologic activity. Indisputable immunologic evidence of this latter observation is still lacking, although it is supported by work, presented in the present paper, using an antiserum of great specificity for NH_2-terminal PTH.

The present paper describes recent studies on the chemistry of human, bovine, and porcine PTH as well as some partical consequences of the immunoheterogeneity of PTH in serum when this radioimmunoassay is used in the routine evaluation of patients with hyperparathyroidism.

I. Chemistry of Parathyroid Hormone

1. Human Parathyroid Hormone

The human PTH (HPTH) used in our studies was isolated from parathyroid adenomas obtained from patients undergoing surgery for hyperparathyroidism. A trichloroacetic acid extract (RASMUSSEN et al., 1964) of the human adenomas was purified by gel filtration followed by ion exchange chromatography on CM-Sephadex. The isolated hormone was shown to be homogeneous by disc gel electrophoresis and Edman NH_2-terminal analysis (BREWER et al., 1972).

The NH_2-terminal sequence analysis of HPTH was performed on a Beckman Sequencer (model 890B) using 1 M Quadrol buffer. The phenylthiohydantoin derivatives of the amino acids obtained from the automated Edman degradations were identified by gas-liquid chromatography (PISANO et al., 1972) and mass spectroscopy (FALES et al., 1971; FAIRWELL and BREWER, 1973). Chemical ionization (CI) mass spectroscopy was performed on a Finnigan quadrapole mass spectrometer equipped with a PDP-8/e digital computer and a Complot plotter. Electron impact (EI) mass spectrometry was performed on an LKB mass spectrometer (model 9000).

The sequence of the initial 34 amino acids of HPTH is shown in Fig. 1. A detailed description of the sequence analysis of HPTH including the

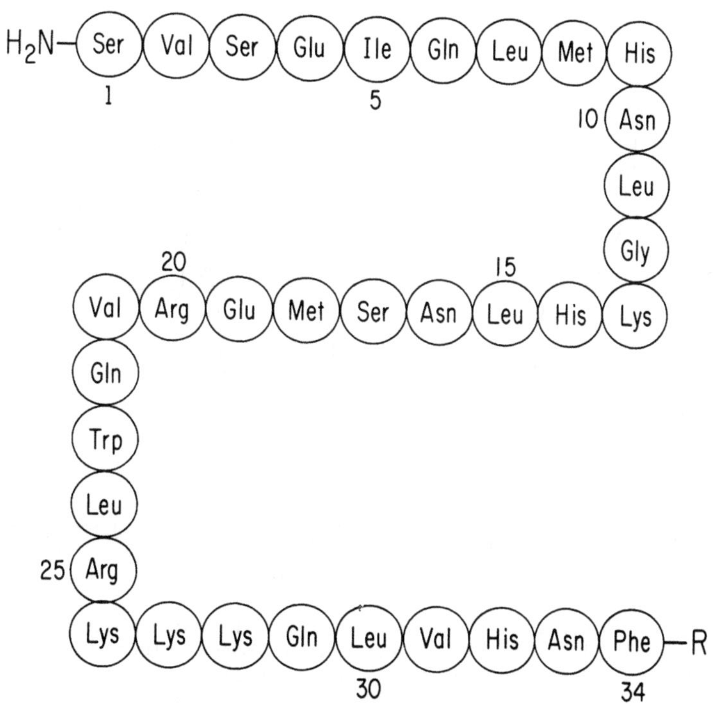

Fig. 1. Initial 34 amino acid residues of human parathyroid hormone. (From BREWER, H.B., Jr., et al.: Recent studies on the chemistry of human, bovine and porcine parathyroid hormones. Am. J. Med. 56, 759-766 (1974). By permission of Dun·Donnelley Publishing Corporation)

CI mass spectra of each of the individual steps in the sequence has been published (BREWER et al., 1972).

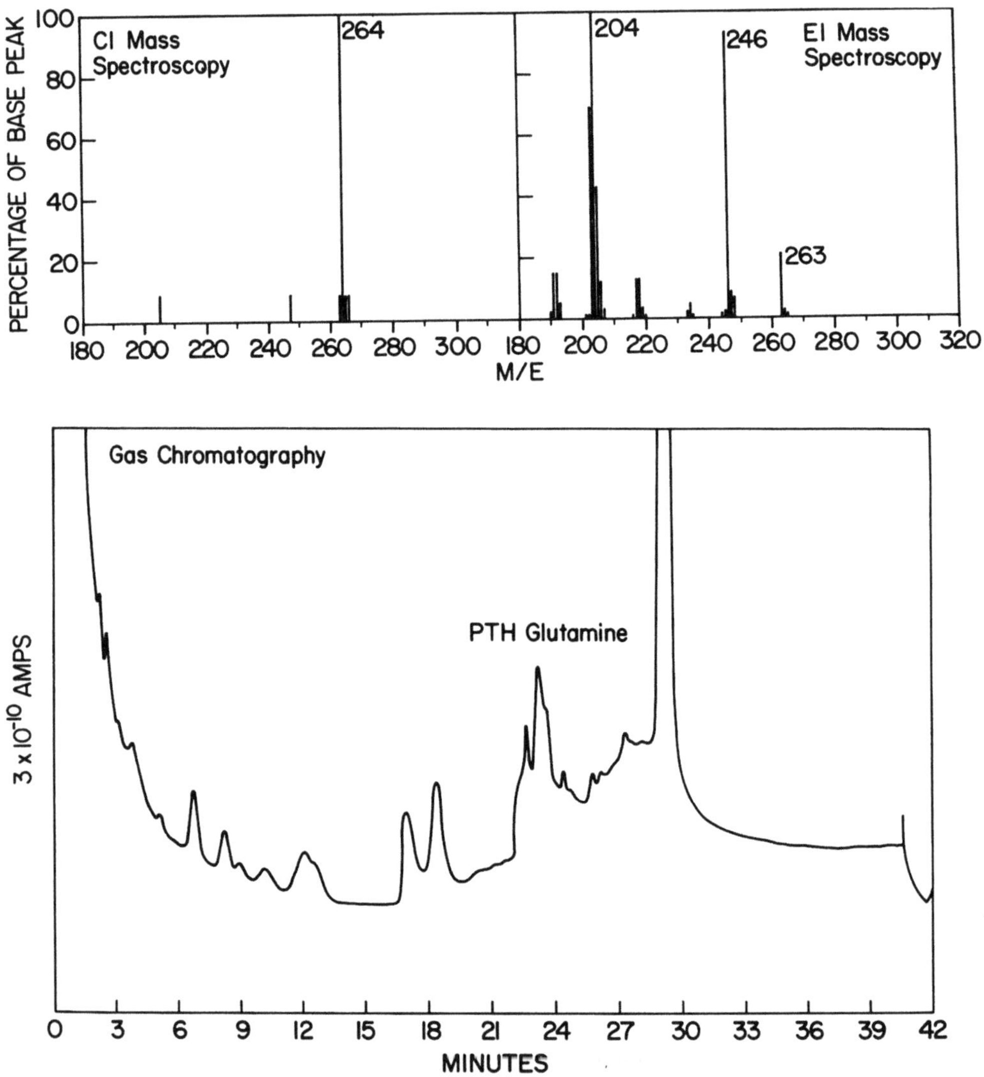

Fig. 2. Identification of phenylthiohydantoin amino acid derivative obtained at step 22 in automated Edman degradation of bovine parathyroid hormone. Glutamine was identified in the CI spectrum at mass 264 and in the EI spectrum at mass 263. The fragmentary ions at m/e 204 and 246 in the EI spectrum are derived from the thermal decomposition of glutamine. (From BREWER, H.B., Jr., et al.: Recent studies on the chemistry of human, bovine and porcine parathyroid hormones. Am. J. Med. 56, 759-766 (1974). By permission of Dun·Donnelley Publishing Corporation)

2. Bovine Parathyroid Hormone

The complete sequence of the 84 amino acid bovine hormone has been reported (BREWER and RONAN, 1970; NIALL et al., 1970). In those studies we identified residue 22 as glutamic acid. However, the human sequence contained glutamine at position 22, and so we reexamined position 22 in the bovine hormone (BREWER et al., 1974). The possibility existed that the glutamine was deamidated during the purification or the Edman degradation.

Bovine PTH was isolated from defatted parathyroid tissue with care to avoid acidic conditions for prolonged periods in order to minimize possible deamidation of glutamine. In addition, the temperature and cleavage time for conversion of the thiazolinone to the thiohydantoin were decreased during the Edman degradation at step 22 in the sequence. Three separate automated degradations were performed on the bovine hormone - one on the intact hormone and two on the isolated COOH-terminal cyanogen bromide peptide (residues 19 through 84). The cyanogen bromide peptide was utilized in these studies in order to decrease the number of Edman cycles required to reach step 22 in the sequence (22 cycles for the intact hormone; 4 cycles for the cyanogen bromide peptide) and thereby decrease the number of times that the polypeptide was exposed to heptafluorobutyric acid (HFBA). HFBA is the acid used during the cleavage step in the automated Edman procedure, and repeated exposure to it at elevated temperature (55°C) has been associated with deamidation of glutamine residues. In all three of these automated degradations, glutamine was identified at position 22. Three techniques - CI mass spectroscopy, EI mass spectroscopy, and gas-liquid chromatography - were used to confirm the identification of glutamine at this position (Fig. 2).

3. Porcine Parathyroid Hormone

O'RIORDAN et al. (1971) have previously reported the presence of a glutamic acid residue at position 22 in the sequence of porcine PTH. The presence of glutamine at position 22 in both human and bovine PTH suggested that the porcine hormone also might have glutamine at this position. We therefore reexamined the sequence of porcine PTH, using the same precautions to minimize deamidation that were used for the bovine hormone. However, porcine PTH was difficult to purify to homogeneity, and repetitive chromatography on CM-Sephadex was required to obtain hormone that was homogeneous by disc gel electrophoresis and NH_2-terminal analysis.

Automated Edman sequence analysis of purified porcine PTH revealed glutamine at position 22. CI mass spectroscopy and gas-liquid chromatography indicated that approximately two-thirds of the glutamine residues at position 22 had undergone deamidation to glutamic acid during purification or sequence analysis. EI mass spectroscopy (Fig. 3) revealed only glutamic acid (m/e 264), reflecting the extensive deamidation. These results indicated that the glutamic acid function reported in porcine PTH at position 22 was also glutamine. Furthermore, the glutamine at position 22 appears to be relatively labile and may undergo extensive deamidation during structural analysis.

The revised sequences of bovine and porcine PTH are shown in Fig. 4. The bovine and porcine hormones differ in only 7 of the 84 positions: residues 1, 7, 18, 42, 43, 47, and 74.

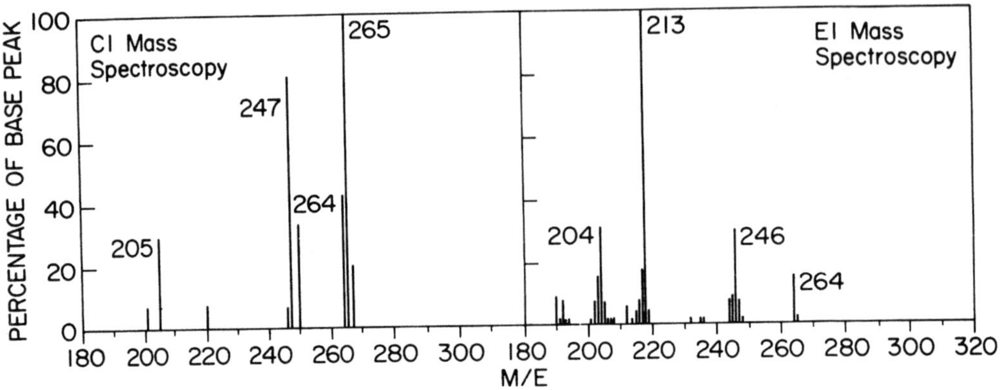

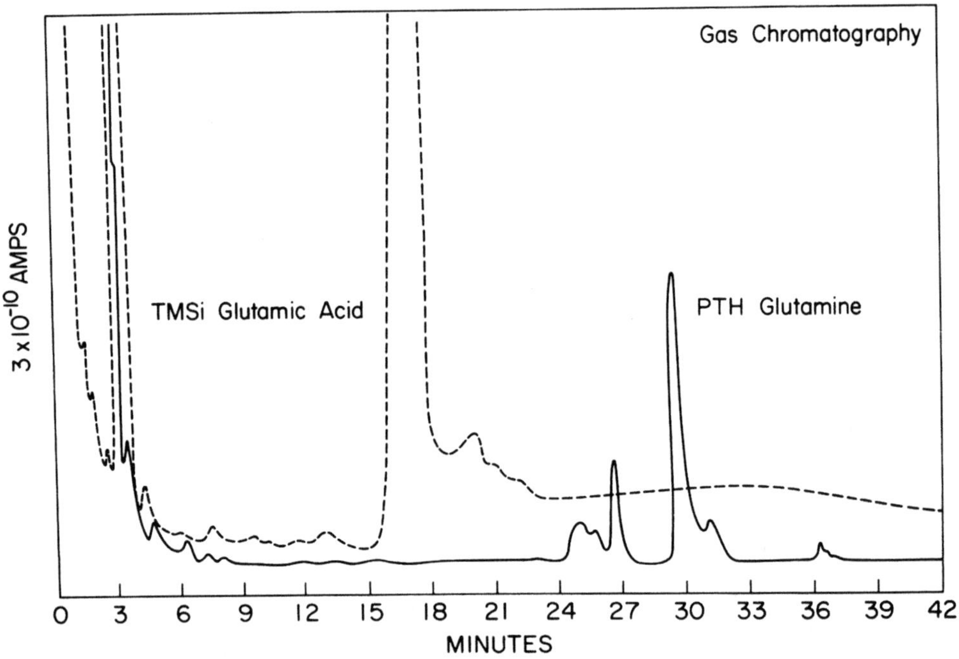

Fig. 3. Identification of phenylthiohydantoin amino acid derivatives obtained at step 22 in automated Edman degradation of porcine parathyroid hormone. Glutamine and glutamic acid were identified at m/e 264 and 265, respectively, in the CI spectrum. In the EI spectrum, only glutamic acid (m/3 264) was observed, due to the extensive deamidation of the glutamine residue. Fragmentary ions derived from the thermal decomposition of glutamine are seen at m/e 205 and 247 in the CI spectrum and at m/e 204, 213, and 246 in the EI spectrum. Glutamine and glutamic acid were observed in the gas chromatogram of step 22. In the gas chromatogram, the solid line represents the profile obtained after silylation of the sample (From BREWER, H.B., Jr., et al.: Recent studies on the chemistry of human, bovine and porcine parathyroid hormones. Am. J. Med. 56, 759-766 (1974). By permission of Dun·Donnelley Publishing Corporation)

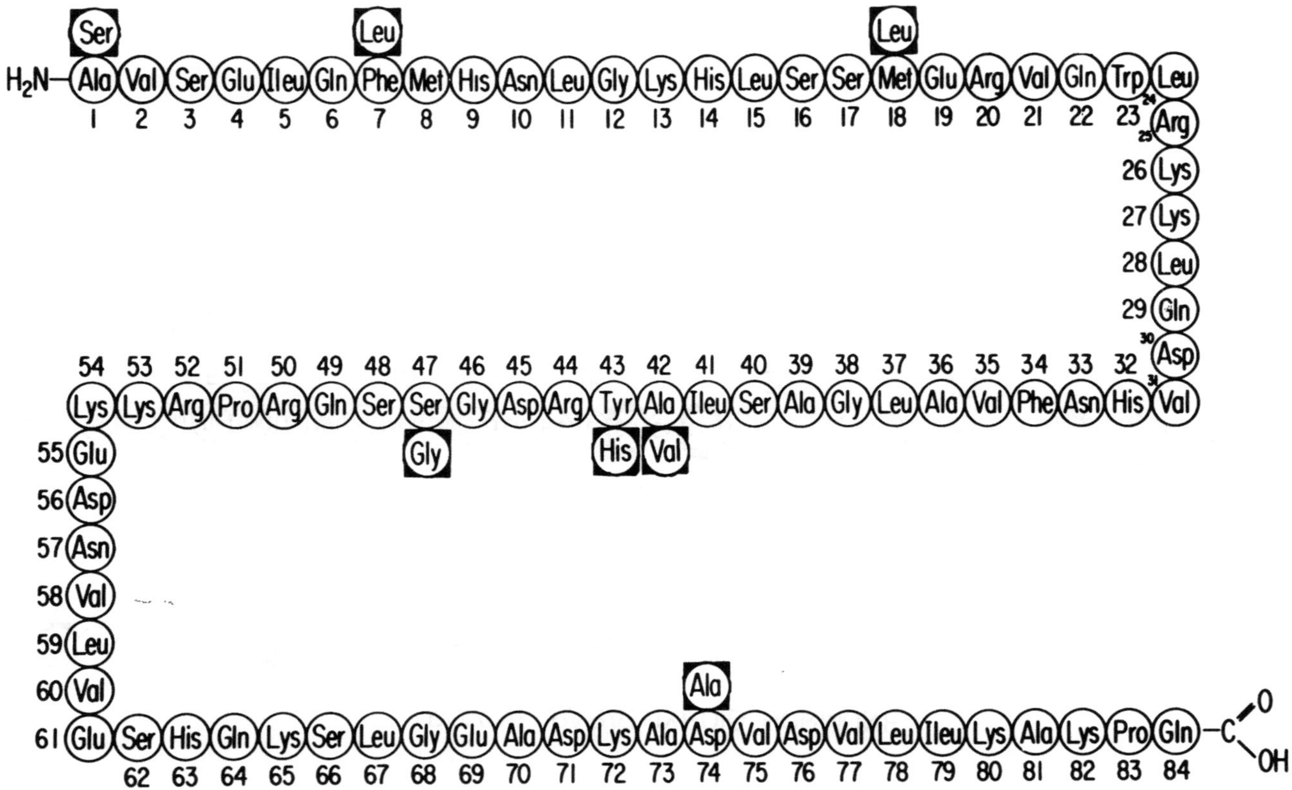

Fig. 4. Revised sequences of bovine (circles) and porcine (squares) parathyroid hormone, showing glutamine at position 22. The structures of the bovine and porcine hormones differ by only seven amino acid residues. (From BREWER, H.B., Jr., et al.: Recent studies on the chemistry of human, bovine and porcine parathyroid hormones. Am. J. Med. 56, 759-766 (1974). By permission of Dun·Donnelley Publishing Corporation)

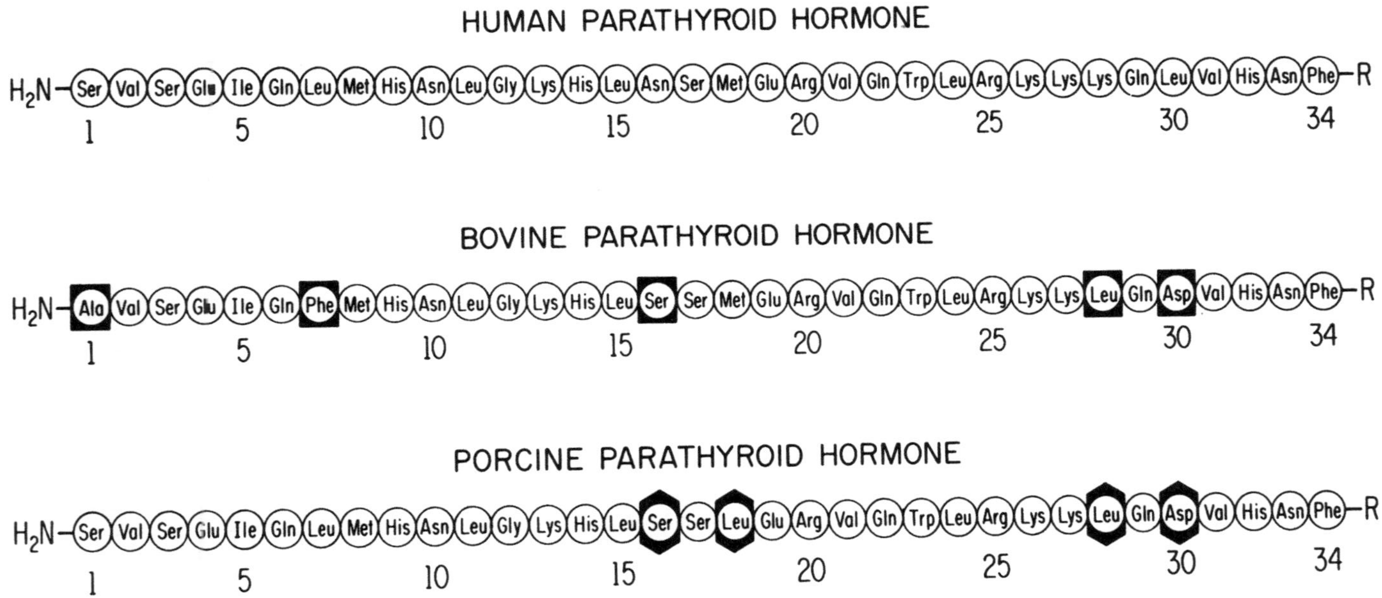

Fig. 5. Amino acid sequences of NH_2-terminal 34 residues of human, bovine, and porcine parathyroid hormones. (From BREWER, H.B., Jr., et al.: Recent studies on the chemistry of human, bovine and porcine parathyroid hormones. Am. J. Med. 56, 759-766 (1974). By permission of Dun·Donnelley Publishing Corporation)

4. Structural Differences in Human, Bovine, and Porcine PTH 1-34

Previous studies indicated that the intact 84 amino acid polypeptide is not required for biologic activity; the biologic activity appears to be within the first 27 residues of bovine PTH and the first 30 residues of porcine PTH (TREGEAR et al., 1973). In the first 34 residues, HPTH differs from bovine PTH by 5 residues and from porcine PTH by 4 residues (Fig. 5). The initial 15 residues of human and porcine PTH are identical; bovine PTH differs by an alanine residue at position 1 and a phenylalanine residue at position 7. The human and bovine hormones contain two methionine residues, whereas the porcine hormone contains a single methionine, at position 8. All three hormones have a glutamine at position 22. The amino acid residues of the NH_2-terminal region that are unique to the human sequence are an asparagine at position 16, a lysine at position 28, and a leucine at position 30.

POTTS et al. (1973) recently described a sequence for the initial 37 residues of HPTH that differs in three positions from our sequence. The differences include glutamic acid rather than glutamine at position 22, leucine rather than lysine at position 28, and aspartic acid rather than leucine at position 30. In addition, alternate residues, threonine at position 22 and serine at position 32, were identified. The differences between these sequences are as yet unexplained but may reflect isohormonal variation or differences in methods.

5. Biologic Activity of Synthetic Human PTH 1-34

ANDREATTA et al. (1973) synthesized, by classical techniques, a biologically active peptide containing the initial 34 residues of HPTH, based on the sequence shown in Fig. 1. When perfused into the conscious thyroparathyroidectomized rat, this synthetic peptide produced hypercalcemia (6 to 11 mg/dl), hyperphosphaturia (0.3 to 2.4 mg P/h), and hypocalciuria (0.3 to 0.1 mg Ca/h) (Fig. 6). However, on in vitro assay of the synthetic peptides in human renal cortical membranes, bovine PTH 1-34 stimulated adenylate cyclase 10 to 30 times greater than did HPTH 1-34 (Fig. 7). These studies indicate that use of different assay systems to measure the biologic activity of PTH may give discordant results. Therefore, one must be cautious in interpreting the biologic activity of different hormone preparations based on a single type of assay system.

II. Immunoheterogeneity of PTH in Human Circulation

1. History and Statement of Problem

In 1968, BERSON and YALOW (1968) published an extremely important paper in which they presented strong evidence in support of the immunoheterogeneity of PTH in human plasma. They showed that one of their anti-bovine-PTH sera distinguished between the hormone extracted from human parathyroid tumors and that present in the peripheral plasma of patients with hyperparathyroidism and that values for the concentration of immunoreactive PTH (iPTH) in plasma differed markedly depending on the antiserum used in the radioimmunoassays. Further support for their suggestion that these phenomena reflected immunoheterogeneity of plasma PTH was obtained by them with their observation that the measured half-life of disappearance of iPTH from plasma after parathyroidectomy in a patient with secondary hyperparathyroidism and chronic renal failure was different when different antisera were used in the assays of iPTH.

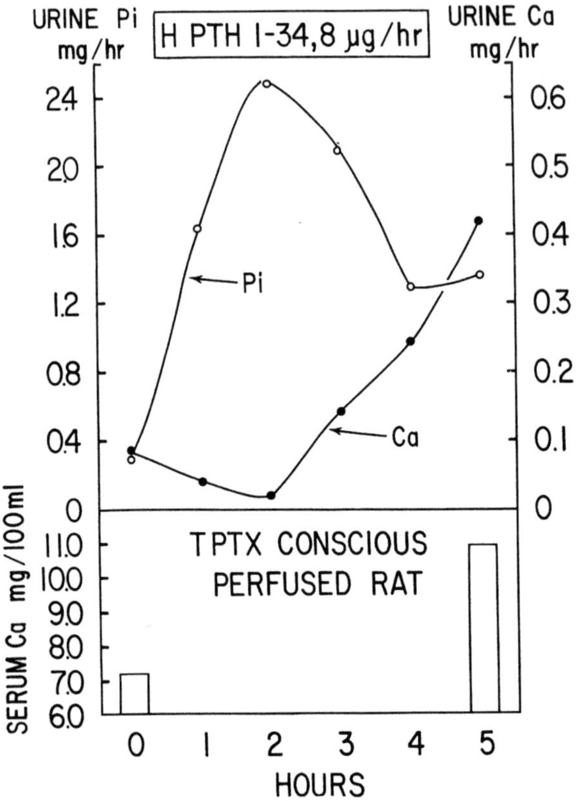

Fig. 6. Typical responses of urinary phosphorus (o-o), urinary calcium (●-●), and serum calcium (bar graph, lower panel) to intravenous infusion of synthetic human PTH 1-34 (Brewer et al. sequence) in a thyroparathyroidectomized rat. (From BREWER, H.B., Jr., et al.: Recent studies on the chemistry of human, bovine and porcine parathyroid hormones. Am. J. Med. 56, 759-766 (1974). By permission of Dun.Donnelley Publishing Corporation)

They interpreted their data as indicating the presence of metabolites of PTH rather than isohormones of PTH in the peripheral circulation of hyperparathyroid man.

At about the time that their paper appeared, REISS and CANTERBURY (1968) reported the development of a radioimmunoassay for PTH in human serum which was capable of measuring iPTH in all normal subjects and of distinguishing clearly between normals and patients with adenomatous hyperparathyroidism. This latter observation was in sharp contrast with that of Berson and Yalow (1966) who found a large overlap in plasma iPTH values between these two groups. It is likely that this apparent discrepancy was the first clue that the phenomenon of immunoheterogeneity of plasma PTH has great practical significance in the application of the radioimmunoassay of PTH to problems of clinical diagnosis and investigation.

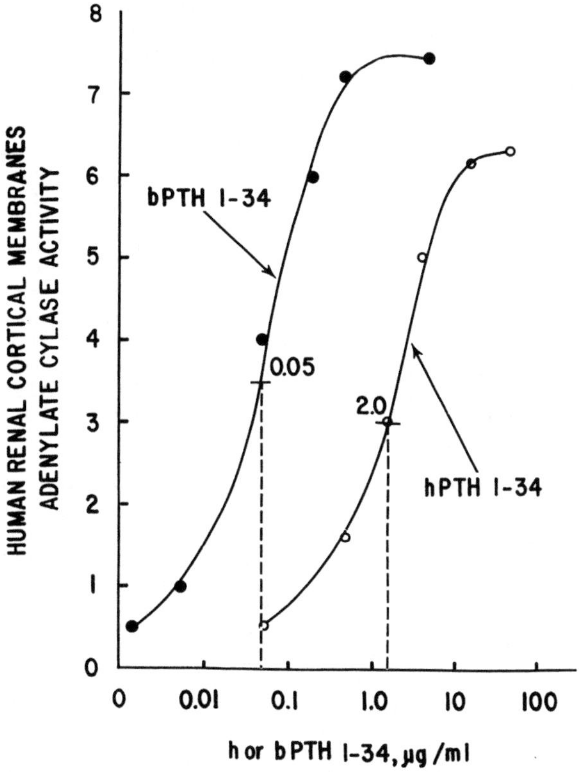

Fig. 7. Comparative dose-response curves of synthetic bovine PTH 1-34 (glutamic acid residue at position 22 (●-●) and human PTH 1-34 (BREWER et al. sequence) (o-o) in human renal cortical adenylate cyclase assay system

The second clue came from another discrepancy between the results obtained with different antisera in different laboratories. REISS and CANTERBURY (1969) reported that it was possible to distinguish between adenoma and primary hyperparathyroidism due to hyperplasia by measurement of serum iPTH during induced hypercalcemia. With this manipulation, hormone concentrations decreased in patients with hyperplasia but did not in patients with adenoma. In contrast, using a radioimminoassay with another antiserum, POTTS and co-workers (1971) found that serum iPTH increased with induced hypocalcemia and decreased with induced hypercalcemia in patients with adenomatous hyperparathyroidism.

We do not think that these discrepancies were due to differences in the experimental procedures used in the different laboratories or to the use of reagents derived from nonhuman species (bovine in each instance). As will be shown below, it is likely that the key variable in the explanation is antibody specificity and the differences in the rates of metabolism of the various molecular species of PTH in the circulation.

Initial attempts to characterize and identify the molecular species of PTH present in the circulation of hyperparathyroid man were made by ARNAUD and his co-workers (1970, 1971a) using cultures of parathyroid tumor explants in vitro. They showed that multiple immunoreactive species of PTH were present in the media of these cultures and ranged in size from molecular weights of 5,000 to >10,000; they suggested at that time that one source of the immunoheterogeneity of serum PTH might be parathyroid tissue itself.

Direct study of this problem was first reported by CANTERBURY and REISS (1972). They showed, by gel filtration of Amicon-filtered, pooled, peripheral hyperparathyroid serum, that there are at least three immunoreactive forms of PTH in this serum. One form was eluted from Bio-Rad P-10 columns with ^{131}I-labeled bovine PTH (84 amino acids; mol wt 9,500) and had a half-life of less than 30 minutes, whereas the other two forms (approximate mol wt 7,000 to 8,000 and 4,000 to 5,000) had half-lives in the range of hours. HABENER and associates (1971) gel-filtered whole serum from hyperparathyroid subjects and confirmed the presence of the 9,500 mol wt and the 7,000 to 8,000 mol wt forms of the hormone in peripheral serum (but not the 4,000 to 5,000 mol wt species of CANTERBURY and REISS, 1972) and also showed that, in serum obtained from small thyroid veins, the 9,500 mol wt form of the hormone predominated.

Recently, HABENER and co-workers (1972) and SEGRE and co-workers (1972) studied hyperparathyroid sera by using radioimmunoassays based on anti-bovine sera preadsorbed with fragments of bovine PTH and presumably specific for the NH_2-terminal (residues 14 through 19) and COOH-terminal (residues 53 through 84) regions of the bovine PTH molecule. They found that most of the immunoreactivity in these sera was of the COOH-terminal type. Because the synthetic NH_2-terminal fragment (residues 1 through 34) of bovine PTH is biologically active but the COOH-terminal fragment (residues 53 through 84) obtained from natural sources is biologically inert (POTTS et al., 1972), these workers concluded that the major portion of circulating PTH is biologically inactive. Although this work is interesting and important, it should be recognized that neither study (HABENER et al., 1972; SEGRE et al., 1972) detected the 4,000 to 5,000 mol wt circulating iPTH fragment previously reported by CANTERBURY and REISS (1972) and, in a recent paper, CANTERBURY and associates (1973) report that this fragment has biologic activity in the rat renal adenylate cyclase assay system. A model encompassing our interpretation of current knowledge in this field and incorporating information presented below in presented in Fig. 8.

2. Practical Importance in Assessment of Parathyroid Function in Man

We have assessed parathyroid function in patients with hyperparathyroidism by measuring serum iPTH with two radioimmunoassays for PTH - one with high relative sensitivity and specificity for the NH_2-terminal region and one with high relative sensitivity and specificity for the COOH-terminal region of the human PTH molecule - and have correlated these results with data on the number of osteoclasts in bone biopsy specimens from the same patients (ARNAUD et al., in press).

The methods we use in the radioimmunoassay of PTH in human serum have been described (ARNAUD et al., 1971b). Fig. 9 shows the specificity characteristics of the two assay systems with respect to their ability to react with synthetic human and bovine PTH 1-34. The assay system that uses CH 14M antiserum (chicken anti-bovine-PTH serum) reacts almost equally well with the two synthetic peptides. (In studies not shown here, it reacted almost as well with the two synthetic peptides as with the respective native hormones obtained from natural sources.) In contrast, the assay system that uses GP 1M antiserum (guinea pig anti-porcine-PTH serum) does not react with either synthetic human or

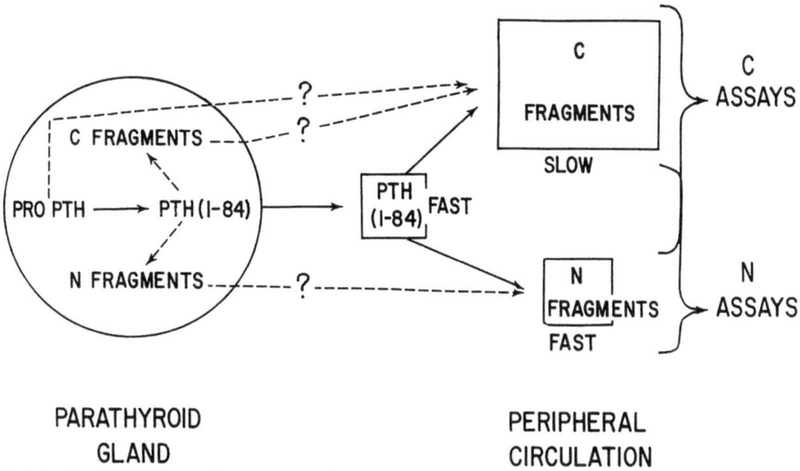

Fig. 8. Simplified schema of current concept of the metabolic alterations in PTH (1-84) that result in immunoheterogeneity. Proparathyroid hormone (proPTH) (mol wt ≧ 11,000) is converted to PTH (1-84) (mol wt = 9,500) by specific enzymatic cleavage. It is not presently known if proPTH is released from the gland into the blood. PTH (1-84) is the major secreted species of PTH, but there is evidence that the gland contains a calcium-regulated enzyme that can cleave PTH (1-84) to COOH-terminal (C) and NH_2-terminal (N) fragments (ARNAUD, 1973b). These fragments also might be released into the circulation, but there is no direct evidence for this. Once secreted, PTH (1-84) is converted peritherally into C and N fragments. The C fragments have a slow turnover rate. N-specific radioimmunoassays of serum iPTH give low values and C-specific radioimmunoassays give high values because of the differences in the pool sizes of these species of iPTH. This scheme is based on data obtained from the study of serum from hyperparathyroid patients. Data on normal serum are not available. (From ARNAUD, C.D.: Parathyroid hormone: coming of age in clinical medicine. Am. J. Med. 55, 577-581, 1973. By permission of Dun·Donnelley Publishing Corporation)

bovine PTH 1-34 but reacts with human PTH 1-84 with identical affinity as the assay system that uses CH 14M antiserum. From these and other studies, we have concluded that the assay system using CH 14M antiserum has relatively high specificity for the NH_2-terminal region of human PTH and is identified as "anti N" and that the assay system using GP 1M antiserum has relatively high specificity for the COOH-terminal region of human PTH and is identified as "anti C" in the figures and text of this paper.

The "anti N" and "anti C" assay systems were used to identify the various immunoreactive species of PTH in the effluent fractions from the gel filtration of whole peripheral serum from a patient with primary hyperparathyroidism (Fig. 10). As expected from the specificity characteristics of the two assay systems, both reacted equally well with an iPTH component eluting in a position similar to that of ^{131}I-labeled bovine PTH 1-84. However, only the "anti C" assay system reacted with an iPTH component that eluted between ^{131}I-labeled bovine PTH 1-84 and bovine PTH 1-34 and that was present in much greater quantities than the component eluting in the position of PTH 1-84. Both the "anti N" and the "anti C" system reacted with a component that eluted immediately after ^{131}I-labeled bovine PTH 1-34.

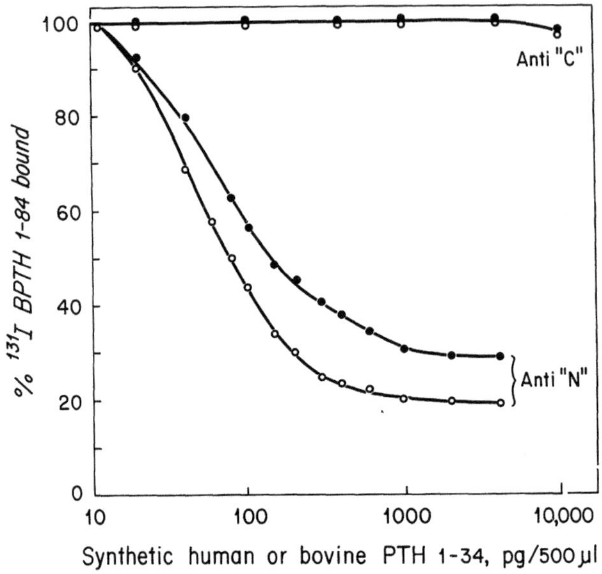

Fig. 9. Standard curves demonstrating the relative reactivity of GP 1M (anti-C) antiserum and CH 14M (anti-N) antiserum with synthetic human (●-●) and bovine (o-o) PTH (1-34). (From ARNAUD, C.D., et al.: Influence of immunoheterogeneity of circulating parathyroid hormone on results of radioimmunoassays of serum in man. Am. J. Med. 56, 785-793 (1974). By permission of Dun·Donnelley Publishing Corporation)

Although the resolution of these components by these methods is incomplete and the general configurations of their individual elution profiles suggest further heterogeneity within the components, we have concluded that the peripheral serum of patients with primary hyperparathyroidism contains at least three (and probably more) immunoreactive species of PTH. The first component probably represents endogenous PTH 1-84, although it is possible that it is heterogeneous and contains some secreted biosynthetic precursor of PTH as well. The second component is probably a large COOH-terminal fragment of PTH 1-84. This component has consistently comprised at least 5 to 20 times the quantity of iPTH eluting with PTH 1-84 in sera from patients with primary or secondary hyperparathyroidism; however, in the sera of patients with ectopic hyperparathyroidism due to nonparathyroid cancer, there is only one tenth to one seventh as much of this component as in sera from patients with primary hyperparathyroidism (BENSON et al., 1974). The third component appears to be recognized by both antisera. But, if these antisera have absolute "anti C" and "anti N" specificities for the human PTH molecule, it is necessary to conclude that this component is heterogeneous and consists of at least one COOH-terminal and one NH_2-terminal fragment that come off the column together. If, on the other hand, the antisera are not absolutely specific and have overlapping recognition sites, this third component might represent a single, small COOH-terminal fragment. We favor the former alternative.

The results of measurements of serum iPTH in hyperparathyroid and normal subjects with the "anti C" and "anti N" specific radioimmunoassays are shown in Fig. 11, 12, and 13. The standard curves in Fig. 11 compare the reactivities of the "anti C" and "anti N" assay systems

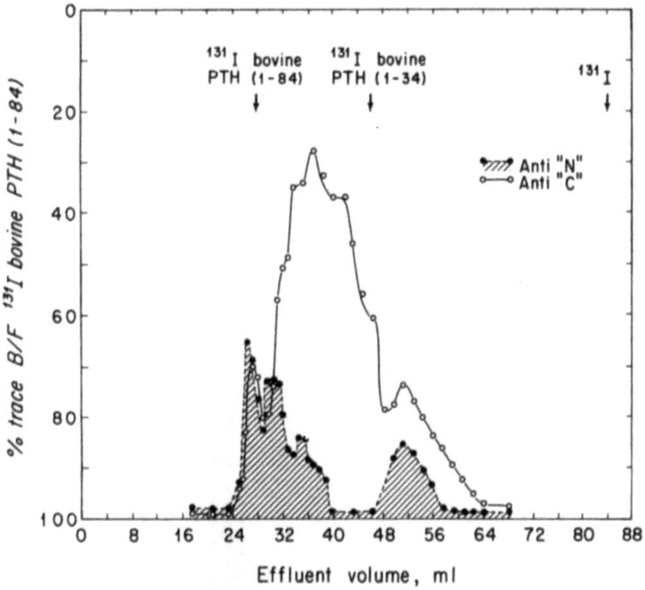

Fig. 10. Elution profile of iPTH assayed with GP 1M (anti-C) (o-o) and CH 14M (anti-N) (●-● and shaded area) after gel filtration of standard hyperparathyroid plasma (B1) on Bio-Gel P-30. Arrows indicate elution positions of ^{131}I-labeled bovine PTH (1-84), ^{131}I-labeled bovine PTH (1-34), and ^{131}I-. (From ARNAUD, C.D., et al.: Influence of immunoheterogeneity of circulating parathyroid hormone on results of radioimmunoassays of serum in man. Am. J. Med. 56, 785-793 (1974). By permission of Dun·Donnelley Publishing Corporation)

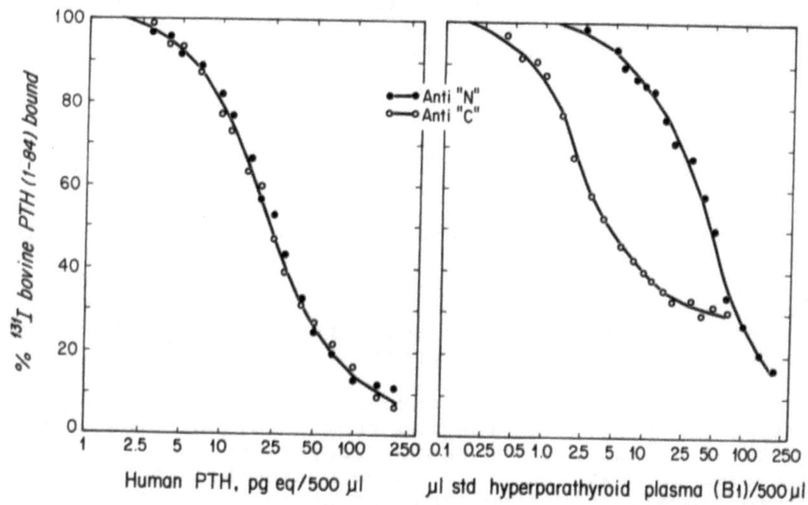

Fig. 11. Standard curves demonstrating the relative reactivities of GP 1M (anti-C) antiserum (o-o) and CH 14M (anti-N) antiserum (●-●) with human adenoma PTH (1-84) (Left) and with the PTH in hyperparathyroid plasma (Right). (From ARNAUD, C.D., et al.: Influence of immunoheterogeneity of circulating parathyroid hormone on results of radioimmunoassays of serum in man. Am. J. Med. 56, 785-793 (1974). By permission of Dun·Donnelley Publishing Corporation)

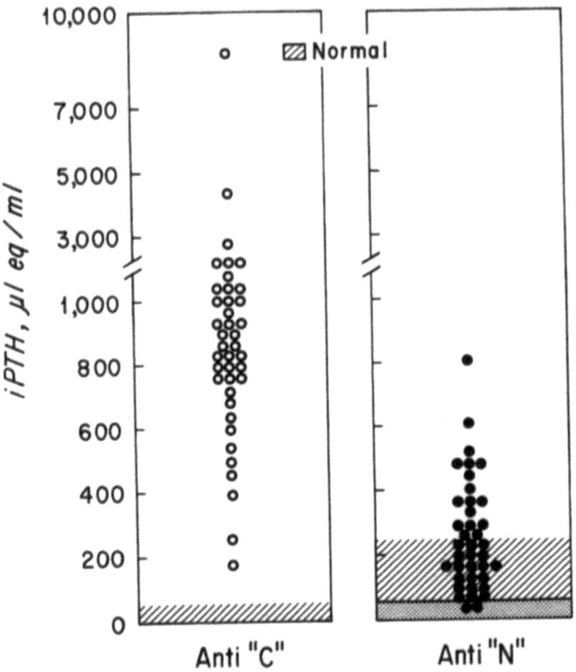

Fig. 12. Serum iPTH values in patients with chronic end-stage renal failure (creatinine clearance, <5 ml/min) with GP 1M (anti-C) and CH 14M (anti-N) assays. (From ARNAUD, C.D., et al.: Influence of immunoheterogeneity of circulating parathyroid hormone on results of radioimmunoassays of serum in man. Am. J. Med. 56, 785-793 (1974). By permission of Dun·Donnelley Publishing Corporation)

with highly purified human PTH 1-84 and the PTH in hyperparathyroid serum. It is clear that both assay systems have the same reactivities with human PTH 1-84 (Fig. 11 left). However, only one tenth to one fifth the amount of hyperparathyroid serum is required to inhibit the binding of ^{131}I-labeled bovine PTH 1-84 in the "anti C" assay system than in the "anti N" assay system (Fig. 11 right). This confirms, with whole serum, the observation made in the gel filtration studies (Fig. 9) that in hyperparathyroid serum there is much more COOH-terminal immunoreactivity than NH_2-terminal immunoreactivity.

In patients with end-stage chronic renal failure, the "anti C" assay system gave PTH values ranging from 3 to 200 times normal in all patients (Fig. 12). In contrast, assays with the "anti N" system showed increases in only 42% of the same patients, and these increases ranged between barely above the limit of normal and 3 to 4 times normal. Although the absolute values for the extremes of the normal ranges for both assays differ ("anti C", 10 to 50 µl eq/ml; "anti N", 50 to 280 µl eq/ml; based on a standard hyperparathyroid plasma assigned an arbitrary value of 1,000 µl eq/ml), their relative ranges do not. This phenomenon is related to the large quantity of COOH-terminal relative to NH_2-terminal immunoreactivity in the hyperparathyroid serum used as standard in both assay systems.

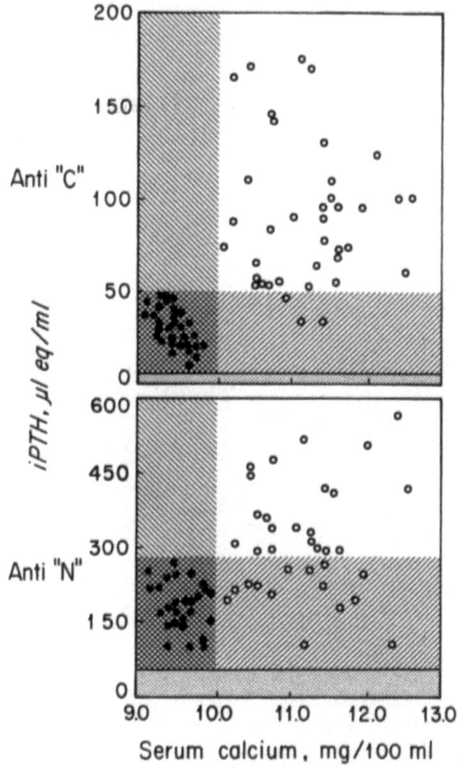

ig. 13. Relationships between serum calcium and serum iPTH values in ormals (●-●) and patients with primary hyperparathyroidism (o-o), sing GP 1M (anti-C) and CH 14M (anti-N) assays. (From ARNAUD, C.D., t al.: Influence of immunoheterogeneity of circulating parathyroid ormone on results of radioimmunoassays of serum in man. Am. J. Med. 56, 85-793 (1974). By permission of Dun·Donnelley Publishing Corporation)

hen serum iPTH values in normal subjects and in patients with primary yperparathyroidism were plotted as a function of the total calcium oncentration measured in the same serum sample (Fig. 13), a negative elationship (r = -0.568; P < 0.001) was observed over the normal serum alcium range with the "anti C" assay system. The relationship was not ignificant with the "anti N" assay system (r = -0.171; P < 0.1). Serum PTH values were in the normal range in 40% of patients with proved rimary hyperparathyroidism when the "anti N" system was used whereas his occured in only 10% of patients when the "anti C" system was used.

he plots in Fig. 14 show that, in patients with primary hyperparathy- oidism, osteoclast number increases uniformly with increases in serum PTH as assayed with the "anti C" assay system whereas this relation- hip is less consistent with serum iPTH values obtained with the "anti " assay system. Most important, however, is that there are significant ncreases in osteoclast number in 5 patients who had normal serum iPTH alues as measured with the "anti N" assay system. In sharp contrast, erum iPTH assayed with the "anti C" system was consistently increased n all patients, even in those who had normal numbers of osteoclasts.

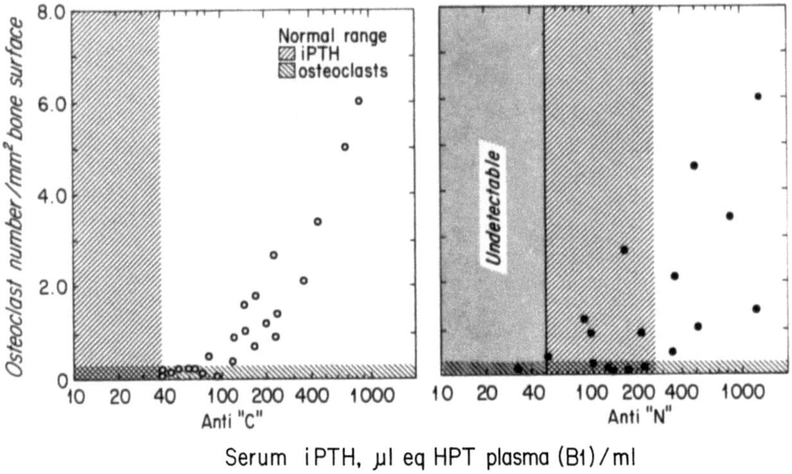

Fig. 14. Relationships between serum iPTH values and osteoclast counts in bone biopsy specimens obtained from patients with primary hyperparathyroidism, using GP 1M (anti-C) (o-o) and CH 14M (anti-N) (●-●) assays. For GP 1M, $r = 0.963$ ($P = 0.0001$); for CH 14M, $r = 0.669$ ($P < 0.01$). (From ARNAUD, C.D., et al.: Influence of immunoheterogeneity of circulating parathyroid hormone on results of radioimmunoassays of serum in man. Am. J. Med. 56, 758-793 (1974). By permission of Dun·Donnelley Publishing Corporation)

However incongruous these data appear, they clearly show that serum assays that are specific for the biologically inactive COOH-terminal region of PTH (compared with assays specific for the biologically active NH_2-terminal region) reflect better not only the state of chronic parathyroid hyperfunction but also the biologic effects of excess circulating PTH on one of its target cells (osteoclasts). The explanations for these phenomena are not clear but are probably related, at least in part, to the apparent long survival times of COOH-terminal fragments relative to PTH 1-84 in serum (CANTERBURY and REISS, 1972; SILVERMAN and YALOW, 1973; GOLDSMITH et al., 1973).

In the case of parathyroid function, measurement of serum concentrations of PTH 1-84 and COOH-terminal fragments in serum (as was done by the "anti C" assay system) probably is a better index of integrated PTH secretion over the previous hours than is measurement of PTH 1-84 alone (as was done by the "anti N" assay system). This would be true whether the COOH-terminal fragments were generated by the degradation of PTH 1-84 in peripheral organs of by parathyroid tissue itself.

In the case of biologic effects, COOH-terminal fragments of PTH 1-84 might be generated and released into the circulation after binding of the whole molecule to target cell receptors, and the measurement of these fragments in serum could be an indirect index of the size and activity of the pool of PTH receptors. Preliminary evidence in support of this latter mechanism has been obtained by DI BELLA and associates (1974) in their studies of receptors of PTH in isolated bovine renal cortical membranes.

The emphasis we have placed on the superiority of COOH-terminal specific assays of PTH in assessing chronic parathyroid hyperfunction should not be construed as a condemnation of NH_2-terminal specific assays. The COOH-terminal specific assays are much less useful than the NH_2-terminal specific assays in assessing acute changes in parathyroid function (as during induced hypercalcemia or hypocalcemia) and in demonstrating differences in iPTH in sera obtained during attempts to localize parathyroid lesions by differential catheterization of veins in the neck and mediastinum. This is because the major species of PTH secreted by parathyroid tissue is PTH 1-84 and it has a short survival in the circulation. The NH_2-terminal specific assays measure PTH 1-84 almost exclusively and essentially exclude the high "background" levels of COOH-terminal fragments (long survival time) that invariably are present in peripheral serum and may "contaminate" serum samples obtained from neck or mediastinal venous drainages. It therefore is logical to expect that NH_2-terminal specific assays would be particularly useful in the performance of these tests. Although not presented in this paper, systematic studies comparing the "anti C" and "anti N" assay systems in our laboratory strongly support these notions (BENSON et al., 1974) (unpublished data).

III. Comment

The elucidation of the amino acid sequence of the biologically active region of the human PTH will now permit detailed chemical and synthetic studies on the specific amino acid residues that are required for the biologic activity of the human hormone. Also, studies will now be required to evaluate the biologic and immunologic significance of the change from a glutamic acid to a glutamine residue at position 22 in the biologically active NH_2-terminal region of the bovine and porcine hormones.

We believe that the data we have presented concerning the practical consequences of the immunoheterogeneity of PTH in the circulation of hyperparathyroid man resolve, to some extent, the disagreements that have arisen about the separation of normal subjects from patients with hyperparathyroidism and about the regulation of PTH secretion in this disease. It is clear that the key variable responsible for the discordant results is antiserum specificity. Measurements of serum iPTH in both chronic and acute studies will differ, depending on this variable, not only because of the presence of multiple molecular species of PTH in the peripheral circulation but also because the ratios of the concentrations of these forms are different, depending on their metabolic turnover rates. These problems need not detract from the utility of the radioimmunoassay of PTH as an investigative and diagnostic tool. However, they dictate that knowledge of the specificity of an individual assay is essential.

References

ANDREATTA, R.H., HARTMANN, A., JÖHL, A., KAMBER, B., MAIER, R., RINIKER, B., RITTEL, W., SIEBER, P.: Synthese der Sequenz 1-34 von menschlichem Parat-hormon. Helv. Chim. Acta 56, 470-473 (1973).

ARNAUD, C.D.: Parathyroid hormone: coming of age in clinical medicine. Am. J. Med. 55, 577-581 (1973).

ARNAUD, C.D.: Immunochemical heterogeneity of circulating parathyroid hormone in man: sequel to an original observation by Berson and Yalow. Mt. Sinai J. Med. NY. 40, 422-432, 1973b.

ARNAUD, C.D., GOLDSMITH, R.S., BORDIER, P.J., SIZEMORE, G.W.: Influence of immunoheterogeneity of circulating parathyroid hormone on results of radioimmunoassays of serum in man. Am. J. Med. 56, 785-793 (1974).

ARNAUD, C.D., GOLDSMITH, R.S., SIZEMORE, G.W., OLDHAM, S.B., BISCHOFF, J., LARSEN, J.A., BORDIER, P.: Studies on characterization of human parathyroid hormone in hyperparathyroid serum: practical considerations. Excerpta Medica International Congress Series 270, 281-290 (1973).

ARNAUD, C.D., SIZEMORE, G.W., OLDHAM , S.B., FISCHER, J.A., TSAO, H.S., LITTLEDIKE, E.T.: Human parathyroid hormone: glandular and secreted molecular species. Am. J. Med. 50, 630-638 (1971a).
ARNAUD, C.D., TSAO, H.S., LITTLEDIKE, T.: Radioimmunoassay of human parathyroid hormone in serum. J. Clin. Invest. 50, 21-34 (1971b).
ARNAUD, C.D., TSAO, H.S., OLDHAM, S.B.: Native human parathyroid hormone: an immunochemical investigation. Proc. Natl. Acad. Sci. USA 67, 415-422 (1970).
BENSON, R.C. Jr., RIGGS, B.L., PICKARD, B.M., ARNAUD, C.D.: Immunoreactive forms of circulating parathyroid hormone in primary and ectopic hyperparathyroidism. J. Clin. Invest. 54, 175-181 (1974).
BERSON, S.A., YALOW, R.S.: Parathyroid hormone in plasma in adenomatous hyperparathyroidism, uremia, and bronchogenic carcinoma. Science 154, 907-909 (1966).
BERSON, S.A., YALOW, R.S.: Immunochemical heterogeneity of parathyroid hormone in plasma. J. Clin. Endocrinol. Metab. 28, 1037-1047 (1968).
BREWER, H.B., Jr., FAIRWELL, T., RITTEL, W., LITTLEDIKE, T., ARNAUD, C.D.: Recent studies on the chemistry of human, bovine and porcine parathyroid hormones. Am. J. Med. 56, 759-766 (1974).
BREWER, H.B., Jr., FAIRWELL, T., RONAN, R., SIZEMORE, G.W., ARNAUD, C.D.: Human parathyroid hormone: amino-acid sequence of the amino-terminal residues 1-34. Proc. Natl. Acad. Sci. USA 69, 3585-3588 (1972).
BREWER, H.B. Jr., RONAN, R.: Bovine parathyroid hormone: amino acid sequence. Proc. Natl. Acad. Sci. USA 67, 1862-1869 (1970).
CANTERBURY, J.M., LEVEY, G.S., REISS, E.: Activation of renal cortical adenylate cyclase by circulating immunoreactive parathyroid hormone fragments. J. Clin. Invest. 52, 524-527 (1973).
CANTERBURY, J.M., REISS, E.: Multiple immunoreactive molecular forms of parathyroid hormone in human serum. Proc. Soc. Exp. Biol. Med. 140, 1393-1398 (1972).
COHN, D.V., MACGREGOR, R.R., CHU, L.L.H., HAMILTON, J.W.: Studies on the biosynthesis in vitro of parathyroid hormone and other calcemic polypeptides of the parathyroid gland. Excerpta Medica International Congress Series 243, 173-182 (1971).
COHN, D.V., MACGREGOR, R.R., CHU, L.L.H., HUANG, D.W.Y., ANAST, C.S., HAMILTON, J.W.: Biosynthesis of proparathyroid hormone and parathyroid hormone: chemistry, physiology and role of calcium in regulation. Am. J. Med. 56, 767-773 (1974).
COHN, D.V., MACGREGOR, R.R., CHU, L.L.H., KIMMEL, J.R., HAMILTON, J.W.: Calcemic fraction-A: biosynthetic peptide precursor of parathyroid hormone. Proc. Natl. Acad. Sci. USA 69. 1521-1525 (1972).
DI BELLA, F.P., DOUSSA, T.P., MILLER, S.S., ARNAUD, C.D.: Parathyroid hormone receptors of renal cortex: specific binding of biologically active, ^{125}I-labeled hormone and relationship to adenylate cyclase activation. Proc. Natl. Acad. Sci. USA 71, 723-726 (1974).
FAIRWELL, T., BREWER, H.B., Jr.: Identification by chemical ionization mass spectrometry (CI-MS) of the phenyl (PTH) and methyl (MTH) thiohydantion and thiazolinone amino acids obtained during the automated Edman degradation of polypeptides and proteins (abstract). Fed. Proc. 32, 648 (1973).
FALES, H.M., NAGAI, Y., MILNE, G.W.A., BREWER, H.B. Jr., BRONZERT, T.J., PISANO, J.J.: Use of chemical ionization mass spectrometry in analysis of amino acid phenylthiohydantoin derivatives formed during Edman degradation of proteins. Anal. Biochem. 43, 288-299 (1971).
GOLDSMITH, R.S., FURSZYFER, J., JOHNSON, W.J., FOURNIER, A.E., SIZEMORE, G.W., ARNAUD, C.D.: Etiology of hyperparathyroidism and bone disease during chronic hemodialysis. III. Evaluation of parathyroid suppressibility. J. Clin. Invest. 52. 173-180 (1973).
HABENER, J.F., KEMPER, B., POTTS, J.R., Jr., RICH, A.: Bovine proparathyroid hormone: structural analysis of radioactive peptides formed by limited cleavage. Endocrinology 92, 219-226 (1973).

HABENER, J.F., POWELL, D., MURRAY, T.M., MAYER, G.P., POTTS, J.T., Jr.: Parathyroid hormone: secretion and metabolism in vivo. Proc. Natl. Acad. Sci. USA 68, 2986-2991 (1971).
HABENER, J.F., SEGRE, G.V., POWELL, D., MURRAY, T.M., POTTS, J.T., Jr.: Immunoreactive parathyroid hormone in circulation of man (letter to the editor). Nature (New Biol) 238, 152-154 (1972).
HAMILTON, J.W., MACGREGOR, R.R., CHU, L.L.H., COHN, D.V.: The isolation and partial purification of a non-parathyroid hormone calcemic fraction from bovine parathyroid glands. Endocrinology 89, 1440-1447 (1971).
HAMILTON, J.W., NIALL, H.D., KEUTMANN, H.T., POTTS, J.T., Jr., COHN, D.V.: Amino terminal sequence of bovine proparathyroid hormone (calcemic fraction A) (abstract). Fed. Proc. 32, 269 (1973).
KEMPER, B., HABENER, J.F., POTTS, J.T., Jr., RICH, A.: Proparathyroid hormone: identification of a biosynthetic precursor to parathyroid hormone. Proc. Natl. Acad. Sci. USA 69, 643-647 (1972).
NIALL, H.D., KEUTMANN, H., SAUER, R., HOGAN, M., DAWSON, B., AURBACH, G., POTTS, J., Jr.: The amino acid sequence of bovine parathyroid hormone. I. Hoppe Seylers Z. Physiol. Chem. 351, 1586-1588 (1970).
O'RIORDAN, J.H., WOODHEAD, J.S., ROBINSON, C.J., PARSONS, J.A., KEUTMANN, H., NIALL, H., POTTS, J.T.: Structure-function studies in parathyroid hormone. Proc. R. Soc. Med. 64, 1263-1265 (1971).
PISANO, J.J., BRONZERT, T.J., BREWER, H.B., Jr.: Advances in the gas chromatographic analysis of amino acid phenyl- and methylthiohydantoins. Anal. Biochem. 45, 43-59 (1972).
POTTS, J.T. Jr., KEUTMANN, H.T., NIALL, H.D., TREGEAR, G.W., HABENER, J.F., O'RIORDAN, J.L.H., MURRAY, T.M., POWELL, D., AURBACH, G.D.: Parathyroid hormone: chemical and immunochemical studies of the active molecular species. In: Endocrinology 1971 (Proceedings of the Third International Symposium) Edited by S. Taylor. pp 333-349. London: William Heinemann 1972.
POTTS, J.T., Jr., MURRAY, T.M., PEACOCK, M., NIALL, H.D., TREGEAR, G.W., KEUTMANN, H.T., POWELL, D., DEFTOS, L.J.: Parathyroid hormone: sequence, synthesis, immunoassay studies. Am. J. Med. 50, 639-649 (1971).
POTTS, J.T., Jr., NIALL, H.D., TREGEAR, G.W., VAN RIETSCHOTEN, J., HABENER, J.F., SEGRE, G.V., KEUTMANN, H.T.: Chemical and biologic studies of proparathyroid hormone and parathyroid hormone: analysis of hormone biosynthesis and metabolism. Mt. Sinai. J. Med. NY. 40, 448-461 (1973).
RASMUSSEN, H., SZE, Y.L., YOUNG, R.: Further studies on the isolation and characterization of parathyroid polypeptides. J. Biol. Chem. 239, 2852-2857 (1964).
REISS, E., CANTERBURY, J.M.: A radioimmunoassay for parathyroid hormone in man. Proc. Soc. Exp. Biol. Med. 128, 501-504 (1968).
REISS, E., CANTERBURY, J.M.: Primary hyperparathyroidism: application of radioimmunoassay to differentiation of adenoma and hyperplasia and to preoperative localization of hyperfunctioning parathyroid glands. N. Engl. J. Med. 280, 1381-1385 (1969).
SEGRE, G.V., HABENER, J.F., POWELL, D., TREGEAR, G.W., POTTS, J.T., Jr.: Parathyroid hormone in human plasma: immunochemical characterization and biological implications. J. Clin. Invest. 51, 3163-3172 (1972).
SEGRE, G.V., NIALL, H.D., HABENER, J.F., POTTS, J.T., Jr.: Metabolism of parathyroid hormone: physiologic and clinical significance. Am. J. Med. 56, 774-784 (1974).
SILVERMAN, R., YALOW, R.S.: Heterogeneity of parathyroid hormone: clinical and physiologic implications. J.Clin. Invest. 52, 1958-1971 (1973).
TREGEAR, G.W., VAN RIETSCHOTEN, J., GREENE, E., KEUTMANN, H.T., NIALL, H.D., REIT, B., PARSONS, J.A., POTTS, J.T. Jr.: Bovine parathyroid hormone: minimum chain length of synthetic peptide required for biological activity. Endocrinology 93, 1349-1353 (1973).

Immuno-Reactive Parathyroid Hormone during EDTA Infusion in Chronic Renal Failure*

P. BURCKHARDT, S. POLI AND J.-P. FELBER

I. Introduction

Increased plasma levels of immunoreactive parathyroid hormone (PTH) are found in most patients with chronic renal failure (CRF) which often exceed the values observed in primary hyperparathyroidism. Since the anti-sera used for radioimmunoassay usually also bind inactive fragments of the hormone which accumulate in CRF, the PTH values measured in CRF may not reflect secondary hyperparathyroidism.

In order to examine the parathyroid function in CRF by PTH measurements, we did an EDTA infusion in 10 patients with CRF and compared the increase in PTH to that observed in 8 controls given the same amount of EDTA. It has been shown that the rise in PTH during a transient hypocalcemia is greater in patients with hyperparathyroidism than in normals whose calcium was reduced by a similar amount (POTTS et al., 1969; MURRAY et al., 1972). This increase in PTH would therefore depend mainly on hormone secretion and should permit an assessment of parathyroid function in CRF.

II. Methods

Basal plasma calcium and PTH were measured in 42 normals, 13 hypocalcemic patients with normal renal function, and 22 patients with CRF in fasting (8 a.m.) venous blood. Plasma calcium was determined by flame photometry (normal range 9 to 11 mg/100ml). Plasma PTH was measured by a radioimmunoassay as previously described (SEGRE et al., 1972) using a guinea-pig anti-bovine PTH antiserum, 125 I-bovine PTH, and purified bovine PTH as standard. All samples were measured in quadruplicata. PTH values are expressed in ng bovine PTH/ml. The upper normal limit in this assay is 0.372 ng/ml. In 10 patients with CRF (7 on chronic hemodialysis) and in 8 normocalcemic patients with normal renal function and normal basal PTH levels, EDTA was infused over a two hour period (50 mg/kg). Plasma calcium and PTH were measured immediately prior to the infusion, at 30 minutes' intervals during the infusion, and two hours after the infusion. Plasma calcium was measured by spectrophotometry at 50090 A^{o} after coloration with murexide.

+ This work was supported by a grant from the Swiss National Foundation (No. 3. 917 0.72) and by the Foundation "Roche". The authors are grateful to Professor J.-T. POTTS Jr. for the anti-serum and the purified bovine PTH. The PTH measurements were performed by Miss E. FERGUSON.

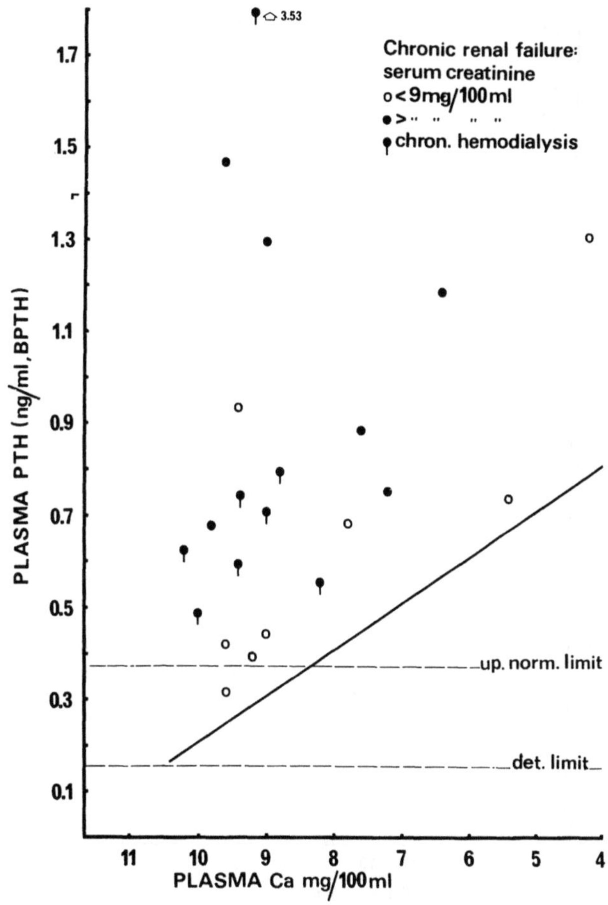

Fig. 1. Basal plasma PTH and plasma calcium in 22 patients with chronic renal failure. The regression line indicated is obtained from a group of 42 normals and 13 hypocalcemic patients with normal renal function (r = -0.722, P <0.001)

III. Results

In the group of the 42 normals and the 13 hypocalcemic patients with normal renal function a significant negative correlation was found between plasma calcium and plasma PTH. (r = -0.722, P < 0.001). Compared to this regression line, the PTH levels measured in 22 patients with CRF were higher than they would be predicted by the actual calcium level, except in patients with a moderately elevated serum creatinine (Fig. 1). The highest levels were found mainly in patients with high serum creatinine.

Plasma calcium and plasma PTH before and after the EDTA infusions of the eight controls and the 10 patients with CRF are indicated in Fig. 2. All 8 controls showed a similar response. The average increase was 0.143 ng/ml (63% of the basal values), whereas the average decrease

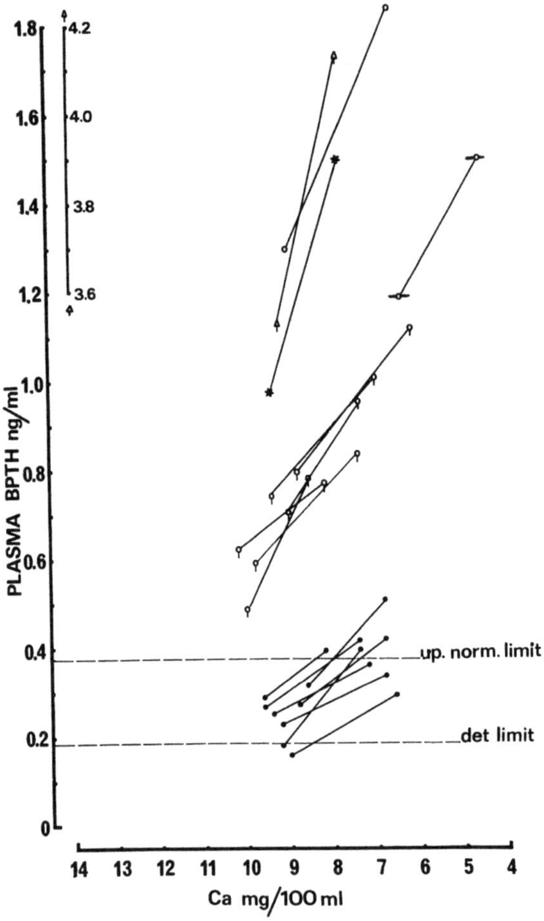

Fig. 2. Plasma PTH and plasma calcium in 8 controls (•) and in 10 patients with chronic renal failure, before and immediately after a 2-hr infusion of EDTA (50 mg/kg). ♀ ♀ patients on chronic hemodialysis, ✱ patient Er., -○- patient Fo., ♀ patient Za. (see text)

in calcium was -2.03 mg/100ml (-23%). Different responses were observed amongst the patients with CRF. In all patients the rise in PTH during the EDTA infusion was greater than the average response of the controls. The average increase was 0.405 ng/ml and the average decrease in plasma calcium was -1.96 mg/100ml (-22%), similar to that obtained in the controls. Expressed in per cent of the basal values however, the rise in PTH was less important in most patients with CRF than in the controls. The patients with the highest basal PTH also had the steepest increase in PTH during the EDTA infusions. An increased response was also found in patient Er. whose basal level was less elevated.

IV. Discussion

Compared to the normals and the hypocalcemic patients with normal renal function, most of the 22 patients with CRF have a higher plasma PTH level than would be predicted by their plasma calcium, especially patients with extremely elevated serum creatinine. Therefore the increase in immunoreactive plasma PTH is not only caused by hyperparathyroidism secondary to the hypocalcemic state, but also by the impaired renal function. This may be due to the increased concentration of immunoreactive, biologically inactive hormone fragments in CRF. The major part of immunoreactive hormones measured in the peripheral blood is composed of biologically inactive hormone fragments (SEGRE et al., 1972; HABENER et al., 1971). The disappearance rate of these fragments is slower than that of intact PTH, especially in CRF (SILVERMANN and YALOW, 1973). This explains the observations that the disappearance rate of immunoreactive PTH decreases in CRF (MASSRY et al., 1972; REISS et al., 1969). Therefore, the impaired renal function causes an accumulation of hormone fragments in the circulation.

The rise in PTH during the EDTA infusion was greater in the patients with CRF and a high basal PTH than in the 8 controls whose calcium was decreased by a similar amount. While the basal value of immunoreactive PTH might be falsely elevated by the accumulated inactive fragments, the increase during the EDTA infusion must rely on PTH secretion. This increase, related to the fall in plasma calcium, is exaggerated in primary hyperparathyroidism (MURRAY et al., 1972). The same observation was reported in parturient cows which have secondary hyperparathyroidism (POTTS et al., 1969). Since 9 out of 10 patients with CRF had a greater response in PTH than the 8 controls, one could conclude that secondary hyperparathyroidism is present in all of them.

However, the half life of the intact hormone is of some minutes only. (SILVERMANN and YALOW, 1973). Therefore the progressive increase of immunoreactive PTH during the two hours of EDTA infusion must be due mainly to the increased production of fragments. Since the half life of these fragments is prolonged in CRF, an increased assumulation may occur during the two hours of EDTA infusion. For that reason the increased rise in immunoreactive PTH observed in CRF does not necessarily reflect hyperfunction of the parathyroid glands. But differences within the group of patients with CRF may indicate parathyroid hyperfunction.

Secondary hyperparathyroidism was suspected in three patients with CRF. Two of them (Za. and Fo.) had an extremely elevated plasma alkaline phosphatase (137 and 167 I.U.) and radiological signs of bone resorption. In patient Er. hyperparathyroidism was suspected because of slightly elevated plasma calcium values despite the presence of hyperphosphatemia. These three patients had the highest basal PTH levels observed in the group but only two of them (Za. and Er.) had a greater response in PTH to the EDTA infusion than the other patients with CRF. Patient Fo. had the lowest basal plasma calcium of the group and was not on chronic hemodialysis. His high basal PTH and his PTH response are therefore more suggestive for secondary hyperparathyroidism than they would be in a normocalcemic patient on chronic hemodialysis. An increased response in PTH was also found in patients where secondary hyperparathyroidism was not suspected. All of them had a slightly elevated plasma alkaline phosphatase. Follow-up studies will show if hyperparathyroidism becomes evident at a later time. We conclude that both basal PTH levels and the parathyroid response to an EDTA infusion account for the diagnosis of secondary hyperparathyroidism in CRF. But both parameters must be brought into relation to the actual plasma calcium, the fall in plasma calcium induced by the EDTA infusion and the degree of renal failure.

References

HABENER, J.F., POWELL, D., MURRAY, T.M., MAYER, G.P., POTTS, J.T.: Parathyroid hormone: Secretion and metabolism in vivo. Proc. Nat. Acad. Sci. USA 68, 2986-2991 (1971).

MASSRY, S.G., COBURN, J.W., PEACOCK, M., KLEEMAN, C.R.: Studies of turnover of endogenous parathyroid hormone in patients with advanced renal failure. Clin. Res. 20, 218 (1972).

MURRAY, T.M., PEACOCK, M., POWELL, D., MONCHIK, J.M., POTTS, J.T.: Non-autonomy of hormone secretion in primary hyperparathyroidism. Clin. Endocrinology 1, 235-246 (1972).

POTTS, J. T., REITZ, R.E., DEFTOS, L.J., KAYE, M.B., RICHARDSON, J.A., BUCKLE, R.M., AURBACH, G.D.: Secondary hyperparathyroidism in chronic renal disease. Arch. Int. Med. 124, 308-412 (1969).

REISS, E., CANTERBURY, J.M., KANTER, A.: Circulating parathyroid hormone concentration in chronic renal insufficiency. Arch. Intern. Med. 124, 417-422 (1969).

SEGRE, G.V., HABENER, J.F., POWELL, D., TREGEAR, G.W., POTTS, J.T.: Parathyroid hormone in human plasma. Immunochemical characterization and biological implications. J. Clin. Invest. 51, 3163-3172 (1972).

SILVERMAN, R., YALOW, R.S.: Heterogeneity of parathyroid hormone: Clinical and physiological implications. J. Clin. Invest. 52, 1958-1971 (1973).

Preliminary Studies on the Effect of Calcium on the Enzymatic Degradation of Human Calcitonin on the Dog*

G. V. FOSTER, D. L. GILDERSLEEVE, CHI-LI KUO, D. T. KELLY, W. W. NICHOLS AND T. A. PEARSON

I. Introduction

In preliminary studies it has been observed that plasma from some patients with hypercalcemia degrades calcitonin *in vitro* more readily than plasma from normal subjects (BAYLIN et al., unpublished data). In this paper we report the results of investigations in animals which suggest that acute elevations in plasma calcium may stimulate destruction of the hormone by the kidney.

II. Effect of Calcium on Calcitonin Metabolism

Fasting mongrel dogs, weighing between 15 and 30 kg were anesthetized with sodium pentabarbitone (25 mg/kg) and, during fluoroscopy or surgery, catheters were placed in the aorta, renal vein, hepatic vein, portal vein and pulmonary artery. The animals were then infused at separate sites with synthetic human calcitonin (1.0 - 1.5 mg) and also calcium as calcium gluconate (10 mg elemental calcium/kg body weight), the total volume and rate of infusion at all times being maintained constant. Heparinized blood samples were collected from the catheters at intervals both in the presence and absence of calcium stimulation for determination of plasma calcitonin by radioimmunoassay (CLARK et al., 1969).

It was observed that hormone arterio-venous differences were significant across the kidney and liver but not across gut and lung, that of kidney and liver, the A-V difference across kidney was greater and at any given arterial level of calcitonin in any given animal, the arterio-venous difference across the kidney was greater during hypercalcemic stimulation. Findings in three experiments are shown in Fig. 1. Irrespective of whether the arterial level of calcitonin was relatively constant (Fig. 1a), rising (Fig. 1b) or falling (Fig. 1c), the renal calcitonin arterio-venous difference was greater during calcium infusion. In these particular studies the increase in the mean of post-calcium infusion values above the mean of pre-infusion values was 72%, 20% and 40% respectively. A good correlation between the increase in the mean of the A-V differences during calcium infusion however was not observed. This can be explained in one of three ways. First, the measurement of arterial and venous levels by radioimmunoassay may have been insufficiently precise to relate their differences to changes

---

+ We thank Drs. W. RITTEL and R. MAIER of Ciba-Geigy for the synthetic human calcitonin and Miss Mary Alice MALOY, Mr. Floyd BARNETT and Mr. Alexander WRIGHT for their expert assistance. This work was supported by grants from the United States Public Health Service (R01 AM 15462-01 GMB) and the American Heart Association (71-962).

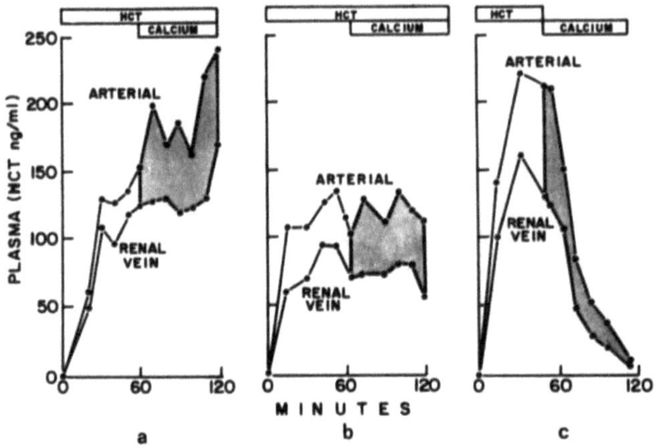

Fig. 1. Arterial and renal venous calcitonin levels in three dogs during intravenous infusions of calcitonin. The shaded areas indicate arterio-venous differences during hypercalcemia

in calcium. Second, the effect of calcium on the kidney was indirectly mediated. Third, the changes produced by calcium are inconstant. None of these possibilities needs to be mutually exclusive.

III. Mode of Action of Calcium on Kidney

Calcium may affect the renal handling of calcitonin in a number of ways. We have examined two possibilities: first, that calcium may increase blood flow to the kidney and, second, that it may increase the urinary excretion of the hormone.

CHARBON and his colleagues have reported that both parathyroid hormone and calcitonin affect renal blood flow (CHARBON and PIEPER, 1970). Since these changes could be explained by changes in plasma calcium, we have carried out experiments to investigate this possibility. In these studies an electromagnetic probe was placed around one of the renal arteries to observe alterations in arterial blood flow to the kidney before and during calcium infusion. At concentrations from 7-12 mg percent no change was observed. A rise was not produced until the concentration was raised to 20 mg percent by a single bolus injection of calcium at the conclusion of the study. This finding suggests that the increased removal of calcitonin by the kidney during hypercalcemia is not mediated by vascular changes. To determine whether or not hypercalcemia promoted increased urinary excretion of calcitonin, studies were carried out in which urine was collected directly from cannulated ureters and hormone estimated by radioimmunoassay. Since the amounts of calcitonin in urine were high, samples for assay could be diluted sufficiently to obviate assay artifacts produced by interfering substances. The results confirmed our previous conclusion

(FOSTER et al., 1972) that less than 1% of infused hormone is immunologically detectable in urine and support the view that increases in detectable hormone in urine observed during calcium infusion are of insufficient magnitude to account for more than 10% of the total hormone removed by the kidney.

By excluding these two possibilities, namely that hypercalcemia promotes urinary excretion of the hormone or increases the rate of blood flow to the kidneys, we are left with a number of alternative explanations. Among some of the more likely are the following. Calcium may (1) activate a peptide degrading enzyme in the kidney, (2) promote de novo synthesis of renal degrading enzymes, (3) affect plasma protein in blood thereby affording less protection to the hormone against degradation, and (4) increase permeability of cellular membranes in the kidney thereby hastening access of the hormone to intracellular degrading enzymes. Studies are in progress to assess these possibilities.

IV. Summary

In conclusion we would like to make three points. First, as measured by radioimmunoassay, calcitonin arterio-venous differences were significant across the kidney and liver but not across gut or lung. Second, that of kidney and liver, loss by the kidney was increased during experimental hypercalcemia. Third, the loss of calcitonin by the kidney during hypercalcemia appears to be unrelated to changes in renal blood flow or urinary excretion of the hormone.

Addendum

Since this preliminary report, we have more critically examined the effects that altering the plasma calcium level have on the destruction of calcitonin. In these studies we have carried out two separate standard infusions of hormone in each animal until equilibrium was reached: one in the presence and the other in the absence of plasma calcium perturbation. We have to date found in one intact dog made hypocalcemic by infusion with EDTA and in one thyroparathyroidectomized dog made hypercalcemic by infusion with calcium gluconate that hormone plateau levels, renal arterio-venous hormonal differences and half-life of disappearance of the hormone were not markedly affected by changing plasma calcium levels. We therefore tentatively conclude that either our initial deduction that hypercalcemia promotes increased renal destruction of the hormone may be incorrect or that the relationship of the thyroid due to destruction of calcitonin is far more complex than previously supposed.

References

BAYLIN, S.B., MALOY, M.A., FOSTER, G.V.: unpublished data.

CHARBON, G.A., PIEPER, E.M.: Influence of calcitonin on parathyroid hormone-induced augmentations of arterial hepatic and renal blood flow. In: Calcitonin 1969: Proceedings of the Second International Symposium (S. Taylor and G.V. Foster, Eds.), pp. 451-457. London: Heinemann Medical Books 1970.

CLARK, M.B., BOYD, G.W., BYFIELD, P.G.H., FOSTER, G.V.: A radioimmunoassay for human calcitonin. Lancet II, 74-76 (1969).

FOSTER, G.V., CLARK, M.B., WILLIAMS, C., NATHANSON, B.M., HORTON, R., BURANAPONG, P., GLASS, H.I.: Metabolic fate of human calcitonin in the dog. In: Endocrinology 1971 (S. Taylor Ed.), pp. 71-78. London: Heinemann Medical Books 1972.

Factors Determining the Calcitonin Response in Man*

M.A. DAMBACHER, J. GUNČAGA, TH. LAUFFENBURGER AND H.G. HAAS

We showed in a previous study (DAMBACHER et al., 1969) that human calcitonin (hCT) lowers serum calcium (SCa) in patients with severe hypercalcemia, an effect which was thought to be partly due to accelerated calcium deposition into the bone (OLAH et al., 1970). Calcitonin (CT) effectiveness seemed to depend primarily on the degree of hypercalcemia, since no response was seen in normo- and hypocalcemic subjects.

From these results we expected that CT would lower SCa in moderate hypercalcemia, too, for example, in the 12 mg/100 ml range. A group of 7 patients with a mean SCa of 11.9 ± 0.3 mg/100 ml was studied. All subjects suffered from primary hyperparathyroidism of mild degree, as shown by hydroxyproline excretion (47 ± 8 mg/day). hCT was infused (1 mg over 3 h), but failed to lower SCa as compared with control infusions in the same patients. Hypercalcemia therefore is probably not the only factor determining CT response in adult man.

In 1968 BIJVOET et al. studied the effect of porcine calcitonin in 4 patients with Paget's disease. They reported that SCa fell more in the polyostotic type than in the monostotic form. Since Paget's disease is known to be due to high bone turnover, we investigated the correlation between the rate of bone turnover and the degree of the calcium-lowering effect of the hormone.

Ten patients with Paget's disease of widely varying extent were studied. The initial hydroxyproline excretion ranged from 34 to 455 mg per day on a gelatin-free diet. The SCa of all subjects was normal. On experimental days patients received alternately 100 MRC U of hCT (1,000 µg) and of salmon CT (sCT) (33 µg) i.m. Control studies were done with a 1‰ albumin solution. Changes in SCa were recorded 7 h and 10 h after injection of the hormone. A fall in SCa (from -2 to -16%) was recorded at both times of observation, the degree depending on the baseline hydroxyproline excretion. The correlation between the △ SCa and the urine hydroxyproline was significant for both hCT and sCT. No differences were observed between the salmon and the human peptide in equipotent doses (100 MRC U), indicating a weight-dose equivalence of the two hormones of about 1 : 30. These data fit well with the results obtained in animal experiments (GALANTE et al., 1973; GUTTMAN et al., 1969).

Our data confirm the earlier hypothesis of BIJVOET et al. who suggested that bone turnover may be the main determinant for the hypocalcemic action of CT. Differences in bone turnover explain the less marked or absent response to CT in primary hypercalcemic hyperparathyroidism as compared to the effect of the peptide on SCa in subjects with normocalcemic Paget's disease.

+ A study comparing hypercalcemic hyperparathyroid patients with normocalcemic subjects suffering from Paget's disease of bone.

References

BIJVOET, O.L.M., VAN DER SLUYS VEER, J., JANSEN, A.P.: Lancet 1, 876 (1968).
DAMBACHER, M.A., GUNCAGA, J., LAUFFENBURGER, Th., HASS, H.G.: DMW 94, 798 (1969).
GALANTE, L., JOPLIN, J.F., MACINTYRE, J., WOODHOUSE, N.J.Y.: Clin. Sci. 33, 605 Short Communication (1973).
GUTTMAN, S.I., PLESS, J., HUGUENIN, R.L., SANDRIN, E., BOSSERT, H., ZEHNDER, K.: Helv. Chim. Acta 52, 1789 (1969).
OLAH, A.J., DAMBACHER, M.A., HAAS, H.G.: Calc. Tissue Res. 4, (Supple.), 154 (1970).

Separation of the Hypocalcemic and Hypophosphatemic Effects of Calcitonin

R.V. TALMAGE, J.H. ROYCROFT AND J.J.B. ANDERSON

That calcitonin produces a hypophosphatemia has been known since the existence of this hormone was first demonstrated. However, it has been generally assumed that the hypophosphatemia and the hypocalcemia were both concurrent effects resulting from decreases in the rate of bone resorption. We have recently demonstrated that calcitonin causes an immediate movement of phosphate out of plasma in contrast to its inhibition of the entry of calcium into plasma. These and other studies have led us to postulate that the hypocalcemic effect of calcitonin is not produced by decreases in the rate of bone resorption but rather by the interaction of phosphate and calcium at the surface of bone (KENNEDY and TALMAGE, 1971).

As an extension of this postulate, the studies to be summarized here were carried out to determine if it were possible to separate physiologically these two results of calcitonin administration by inhibiting the action of the hormone on one parameter without influencing its effect on the other; or to produce temporal differences in the two effects following hormone administration.

The first approach used was to compare changes in the hypocalcemic and hypophosphatemic effects of calcitonin in aging rats. It has been adequately demonstrated in several species that the ability of the hormone to reduce plasma calcium concentrations decreases with the age of the animal. No previous study had examined the hypophosphatemic effect in the aging animals. The results of these studies are as follows: The observation was confirmed that during the aging process the hypocalcemia produced by a given dose of calcitonin is gradually reduced both in depth and duration. However, a markedly different change relative to age was seen in the hypophosphatemia produced by this hormone. In this parameter, the major change was a delay in the time necessary to produce the maximum hypophosphatemia, and a corresponding delay in the recovery from a single injection of the hormone. These data add further evidence for the hypothesis that the effects of calcitonin on plasma calcium and phosphate concentrations are the reflections of distinct and separate physiological responses (ROYCROFT and TALMAGE, 1973).

The second approach to a separation of these two actions of the hormone was by preadministration of rats with EHDP (disodium ethane-1-hydroxy-1,1-diphosphonate). This pyrophosphate analog is thought to inhibit bone resorption by "coating" appatite crystals. For these experiments, a toxic dose of the drug (40 mg/kg) was injected subcutaneously. As early as 24 hours after the first dose of EHDP, a reduction in the hypocalcemic effect of calcitonin was observed. By the 3rd day and thereafter, a dose of calcitonin normally producing a 25-30% decrease in plasma calcium concentrations in two hours, produced only a 5% drop in EHDP-treated rats (TALMAGE et al., 1974).

This same dose of the hormone produced, under these standardized conditions, a 30% decrease in plasma phosphate concentrations in control rats. EHDP treatment had no effect on this hypophosphatemia for up to five days; after which time a gradual reduction in this effect of calcitonin occurred. However, the hypophosphatemic effect was maintained thereafter at a 15% drop in plasma phosphate levels or 50% of that in control rats. Concurrently with this change in the hypophosphatemic effect (5-6 days of EHDP treatment), injection of calcitonin was followed by a phosphaturia believed to be caused by an increased sensitivity of the kidneys to endogenous PTH. No phosphaturic effect was noted following calcitonin injection to control animals or during the first 3 days of EHDP treatment. Since EHDP affected the hypophosphatemic effect of calcitonin differently, both temporally and in degree, from its influence on the hypocalcemic effect, it is again concluded that these two effects of calcitonin are distinct physiological entities.

These studies support the conclusion that the hypophosphatemic effect of calcitonin does not depend upon its hypocalcemic action, but leaves open the possibility that the reverse may be true; i.e., the hypocalcemic effect of the hormone may be an indirect result of its action on phosphate.

Reference

KENNEDY, J.W., TALMAGE, R.V.: Endocrinology 88, 1203 (1971).
ROYCROFT, J., TALMAGE, R.V.: Proceedings of the Society for Experimental Biology and Medicine 144, 17 (1973).
TALMAGE, R.V., ANDERSON, J.J.B., KENNEDY, J.W., III: Endocrinology 94, 413 (1974).

VIII. Varia

Chairmen: H. G. Haas and S. P. Nielsen

Influences of Endogenous Parathyroid Hormone upon Development and Treatment (Calcium, Calcitonin, Sodium Fluoride) of Experimental Bone Atrophy*

R. ZIEGLER, S. BELLWINKEL, A. SCHÄFER AND H. MINNE

I. Introduction

The problems of experimental disuse atrophy of bone are still unsolved with respect to pathogenesis and treatment. One year ago we presented data on unsuccessful efforts to prevent castration osteoporosis and immobilization atrophy in the rat by administration of sodium fluoride and/or calcitonin (ZIEGLER et al., 1972). As the ineffectiveness of the treatment could have been caused by secondary hyperparathyroidism or insufficient calcium supply, additional experiments were performed in parathyroidectomized rats as well as in animals nourished with a calcium-rich diet.

II. Materials and Methods

In femal intact (sham operated) and parathyroidectomized (by electrocautery = PTX) albino rats (strain: FW 49 Kirchb., Lemgo, Bib.) of about 120 gm body weight, the sciatic and femoral nerves of the left hind leg were transsected during ether anesthesia. According to treatment, the following groups of 6 to 8 rats were studied: Groups II (Int.) and VII (PTX): vehicle alone (= 1 ml 10% gelatin s.c. daily).- Groups III (Int.) and VIII (PTX): NaF (50 ppm in drinking water).- IV (Int.) and IX (PTX): NaF plus calcitonin++ (200 MRCmU daily in 10% gelatin s.c.).- Groups V (Int.) and X (PTX): NaF plus calcitonin plus calcium-rich diet (containing 4.45% Ca and 3.25% P instead of 0.95% Ca and 0.87% P in the normal diet). One intact group (I) and one PTX (VI) without paresis and treatment served as controls.

The duration of the experiment was 20 weeks. Then the following parameters were determined: Body weight, serum calcium and phosphorus, femur fresh weight (immediately after removal of soft tissue), dry weight (in a drying oven at 105° C until constant weight is reached) and ash weight (muffle furnace at 580° C for 24 hrs); femur volume (principle of Archimedes); femur calcium and magnesium (atomic absorption spectrophotometry (EEL) after dissolution of ash in 5 N HCL); volumetric density of tibia metaphysis (fixation in Carnoy for 3 hrs and then in 70% ethanol, embedding in methylmetacrylate, integration of undecalcified sections of 5 µ by means of an integrating eye piece).

+ Supported by "SFB 87 Endokrinologie, Ulm", Projekt I.

++Extractive porcine calcitonin (Armour, Lot NN 2501); we thank Dr. J.W. Bastian for the generous supply.

III. Results

During the experiment, the rats uniformly gained weight by about 145 gm. There was no significant difference, especially between the intact and PTX animals. Serum calcium, being around 5 mEq/l in all intact groups, was decreased by 1.5 mEq/l in the PTX groups with the exception of group X (calcium-rich diet), the calcium of which was almost normal; serum phosphorus showed a reciprocal pattern.

Femur dry weight, ash weight, calcium content and volume showed a similar reduction in all paretic legs (Fig. 1 and 2), and the volumetric density of the tibia corresponded to these changes (Fig. 2, bottom). PTX rats had the same degree of paresis as the intact ones. All types of treatment neither prevented nor ameliorated the development of the paretic process. There were only some remarkable tendencies: Calcium-rich diet was accompanied by somewhat larger bone mass in the intact group V, but not in the PTX-group X; nevertheless the difference between the mobile and the immobile leg was not diminished. NaF-treatment induced a slight increase in bone mass (especially volume) in the PTX group VIII, but not in the intact group III.

IV. Discussion

The mechanism of immobilization bone atrophy is still unclear. The consideration that e.g. parathyroid hormone as an osteolytic factor could be part of this process was not supported by our results: PTX animals did not show less atrophy than intact ones. In contrast to this observation, CONAWAY et al. (1973) had seen less bone loss in PTX rats 10 to 14 days after denervation. Perhaps this difference passes out of existence during longer observation periods. Disuse osteopenia could not be prevented by administration of fluoride (MILICIC and JOWSEY, 1968) or calcitonin (CHIROFF and JOWSEY, 1970; DELLING et al., 1969, 1970; ORIMO et al., 1971; PENNOCK et al., 1972; SINGH and JOWSEY, 1970; ZIEGLER and DELLING, 1969). The combination of both substances was ineffective, too (ZIEGLER et al., 1972). Although the doses of NaF (50 ppm instead of 30 ppm (ZIEGLER et al., 1972)) and of calcitonin (200 MRCmU instead of 100 MRCmU (PENNOCK et al., 1972) or 150 MRCmU (ZIEGLER et al., 1972)) were increased during the present study and supplemented by a calcium-rich diet, the same disappointing results were obtained: the process of local bone atrophy did not respond to the combined treatment with different agents. Secondary hyperparathyroidism as a possible cause of the ineffectiveness of treatment was excluded since calcium supplementation in intact rats as well as parathyroidectomy did not lead to better results.

The increase in bone mass and volume after NaF in PTX rats could prove new bone formation which was not seen in the intact groups because of activation of the parathyroids. This finding needs confirmation in larger groups of rats, and this is also the case for the increase in bone mass during calcium-rich diet found in the intact rats, but not in the PTX rats. Therefore one could speculate on a need for endogenous parathyroid hormone <u>and</u> for calcitonin for effective new bone formation.

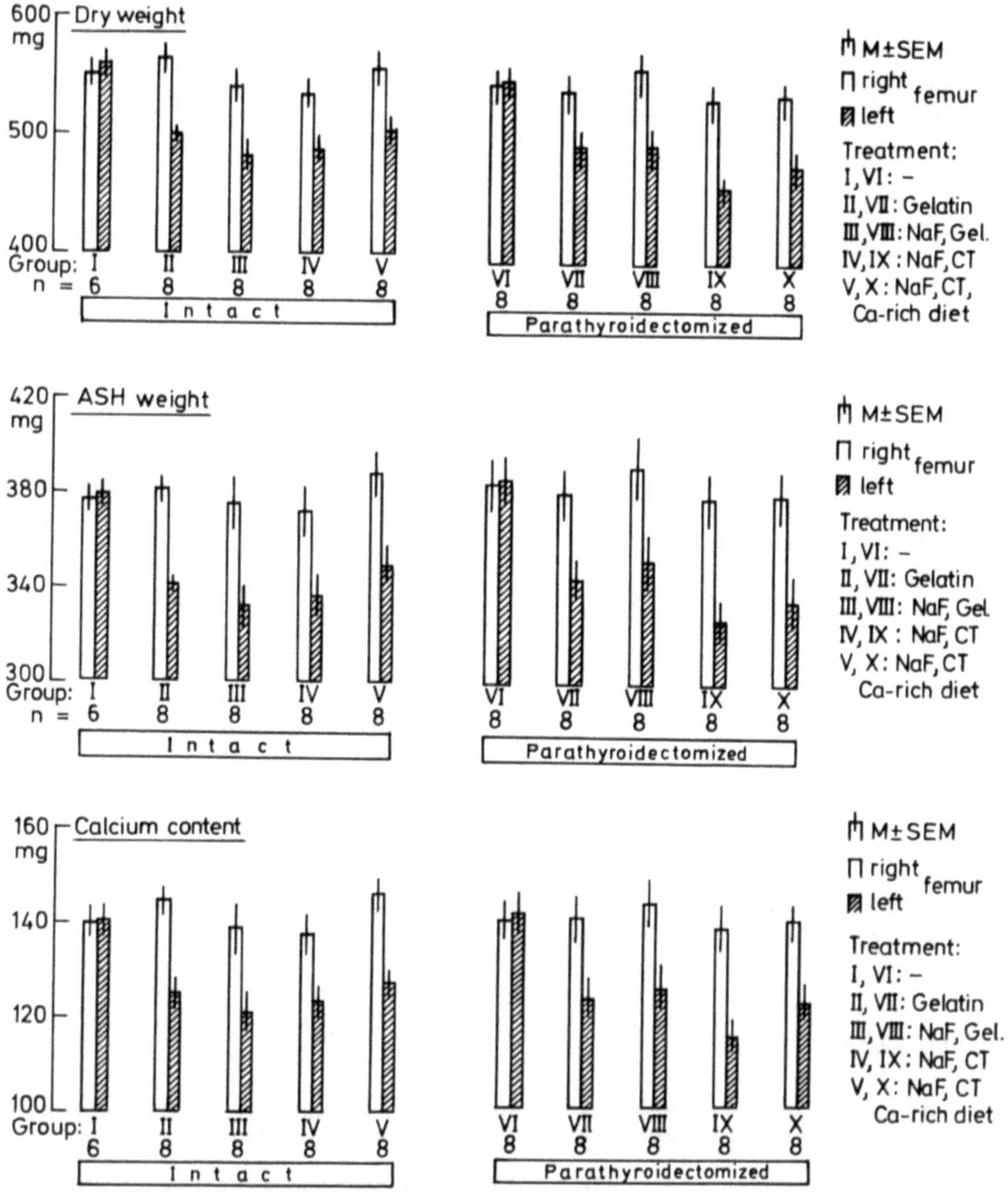

Fig. 1. Dry weight (top), ash weight (center) and calcium content (bottom) of the femora of intact and parathyroidectomized rats treated with the vehicle (1 ml 10% gelatin s.c. - groups II and VII), NaF (50 ppm in drinking water - groups III and VIII), NaF plus calcitonin (200 MRCmU/day in 10% gelatin s.c. - groups IV and IX) and NaF, calcitonin, and calcium-rich diet (groups V and X). In groups II to V and VII to X the left hind leg was immobilized by nerve transsection

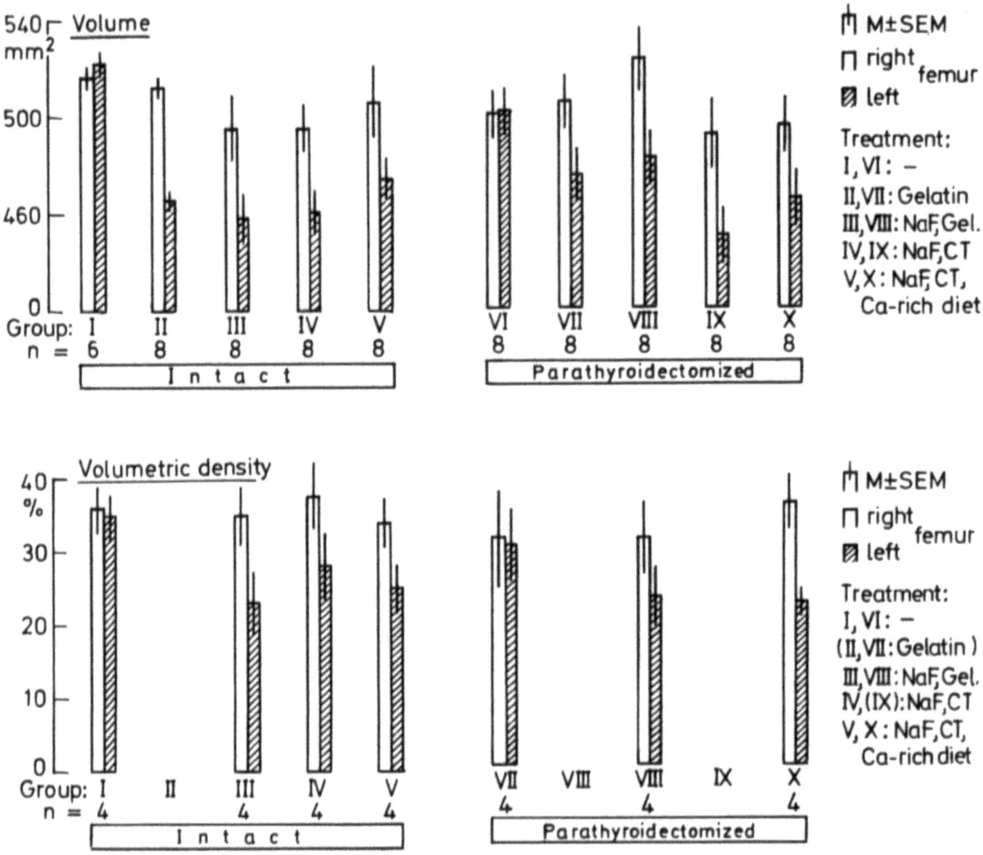

Fig. 2. Femur volume (top) and volumetric density of the proximal tibia metaphysis (bottom) in the same animals (see legend to Fig. 1)

V. Conclusions

Local disuse atrophy of bone seems to be a process which follows its own rules and does not depend upon functioning parathyroid glands.

As the treatment with NaF, calcitonin and a calcium-rich diet does not show better results in parathyroidectomized rats than in intact ones, the cause of the ineffectiveness of the treatment is not due to secondary hyperparathyroidism.

References

CHIROFF, R.T., JOWSEY, J.: The effect of calcitonin on immobilization osteopenia. J. Bone Jt. Surg. 52 A, 1138-1146 (1970).

CONAWAY, H.H., WAITE, L.C., KENNY, A.D.: Immobilization and bone mass in rats. Effects of parathyroidectomy and acetazolamide. Calc. Tiss. Res. 11, 323-330 (1973).

DELLING, G., SCHÄFER, A., SCHLEICHER, H.J., ZIEGLER, R.: Die Wirkung von Calcitonin auf die experimentelle lokalisierte Knochenatrophie. Symp. Dtsch. Ges. Endokrin. 15, 404-405 (1969).
DELLING, G., SCHÄFER, A., SCHLEICHER, H.J., ZIEGLER, R.: The effect of calcitonin on disuse atrophy of bone in the rat. Calc. Tiss. Res. 6, 143-150 (1970).
MILICIC, M., JOWSEY, J.: Effect of fluoride on disuse osteoporosis in the cat. J. Bone Jt. Surg. 50 A, 701-708 (1968).
ORIMO, H., FUJITA, T., YOSHIKAWA, M.: Effect of calcitonin on the development of immobilization osteoporosis in rat. Endocrinol. Japon. 18, 117-121 (1971).
PENNOCK, J.M., KALU, D.N., CLARK, M.B., FOSTER, G.V., DOYLE, F.H.: Hypoplasia of bone induced by immobilization. Brit. J. Radiol. 45, 641-646 (1972).
SINGH, M., JOWSEY, J.: Failure of calcitonin to prevent disuse osteopenia: an experimental study in rabbits. Endocrinology 87, 183-186 (1970).
ZIEGLER, R., BELLWINKEL, S., MINNE, H., DELLING, G.: Studies on the interrelations between castration osteoporosis and bone atrophy following nerve transsection in the rat - influences of treatment with fluoride and/or calcitonin. 9th Europ. Symp. Calcif. Tiss., Baden/Vienna (1972). Facta-Publication 1973, 163-167.
ZIEGLER, R., DELLING, G.: The effect of calcitonin on atrophy and new formation of bone. Acta endocr. (Kbh.), Supple. 138, 183 (1969).

Quantitative Analysis of Bone Changes in Hypercalcitonism (C-Cell Carcinoma)*

G. DELLING AND A. SCHULZ

I. Introduction

The skeleton is the target organ of calcitonin (CT). CT inhibits bone resorption and promotes the deposit of calcium and phosphorous into bone. The treatment of various diseases with CT was observed to be effective only in Paget's disease of bone. The effect of calcitonin in human bone physiology is still under investigation.

An increased CT production in man occurs only in patients with medullary carcinoma of the thyroid gland. This tumor is originated by the calcitonin secreting C-cells (WILLIAMS et al., 1966). Numerous cases of medullary carcinoma were examined with regard to their calcitonin blood levels measured by radioimmunological or bioassay methods (CUNLIFFE et al., 1968; TASHJIAN and MELVIN, 1968; TUBIANA et al., 1968). Very little is known about the morphological bone changes induced by CT in this disease. Therefore we studied the bone changes developing in a group of 4 patients with medullary carcinoma of the thyroid gland. Bone alterations were analyzed by histomorphometry of the iliac crest cancellous bone.

The bone material examined was obtained by iliac crest biopsies in three cases, and in one case by autopsy. The patients were aged 30, 41, 44, and 65 years, two of them had bone metastases of the carcinoma. At the same time when the biopsy was made we measured a pathologically increased activity of CT in blood serum by rat bioassay (Table 1).

Without previous decalcification the bone material was embedded in methylmethacrylate, cut (5 µ) and stained by different methods (Trichrom-Goldner, v. Kossa-Krutsay-modification). The histomorphometric method we used was proposed by MERZ (1967). By means of an integration eyepiece we measured 18 parameters of bone structure and bone remodeling. The values obtained were compared with those of a control collective of the same age group with healthy skeletons (DELLING, 1973).

II. Results

The histological feature of the cancellous bone of patients with bone metastases is somewhat different from that of patients without bone metastases: in patients with bone metastases we observed marrow spaces filled up by a fibrous tissue which is interspersed with tumor islands. Beside denser parts of cancellous bone there are also areas with advanced tumor infiltration destroying several bone trabeculae. The

+ Supported by "Deutsche Forschungsgemeinschaft, SFB 34 - Endokrinologie" metastases usually show the same structure as the original tumor. In patients without bone metastases bone structure is well preserved. The histological aspect does not reveal any conspicuous alterations. The more distinct results are given by histomorphometric analysis, which was performed in areas of the iliac crest cancellous bone without tumor infiltration. The histomorphometric results are summarized in Fig. 1 as follows: The volume density as a structural parameter of the total bone mass is slightly increased in one of the cases without bone metastases, in the other case it is between low normal ranges. One of the cases with bone metastases shows subnormal values. The same relations are shown by the parameter of surface density. Bone formation - represented by all osteoid parameters - is clearly enhanced in one case without bone metastases. But in the two cases with bone metastases the activity of the bone formation process seems to be reduced as demonstrated by the parameter of osteoblast-osteoid interface. In one of the latter cases bone resorption is between low normal ranges or slightly reduced, while the other cases are characterized by elevated resorption parameters. It is remarkable that the case with a reduced bone resorption is identical with the case with increased volume density.

III. Discussion

Our results show that the tumor mass - which is increased in the cases with bone metastases - is obviously without any influence on the bone formation process which is normal or diminished in these two cases. Only one of these two cases shows a reduced bone resorption, thus there is a certain agreement with the case reported by HAAS et al. (1972), which was characterized by a reduced bone remodeling activity or low bone turnover. In contrast to these authors, our results do not confirm a tendency to osteoporotic bone changes in these patients. More

Table 1. Cases studied in this investigation with their names, age, sex and serum values of calcitonin-like activity as measured by rat bioassay

Morphological bone changes in patients with medullary carcinoma of the thyroid

Case	Age (years)	Sex	Serum - Calcitonin (rat bioassay[a])
1 H.H.	30	M	40 mU CTLA
2 L.K.	41	F	?
3 G.E.	44	M	20 mU CTLA
4 E.K.	65	M	10 mU CTLA

[a] Normal serumvalues of Calcitonin not detectable CTLA = Calcitonin-like activity

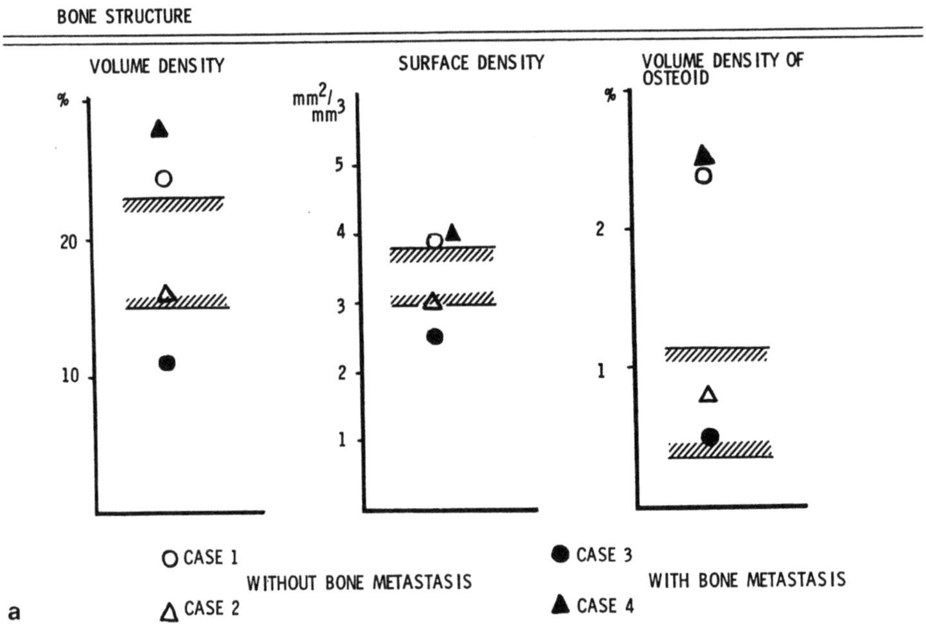

a

Fig. 1a-c. Demonstration of the most important histomorphometric parameters of bone structure and bone remodeling as measured in our cases. The normal ranges are represented by the space between the two marking lines

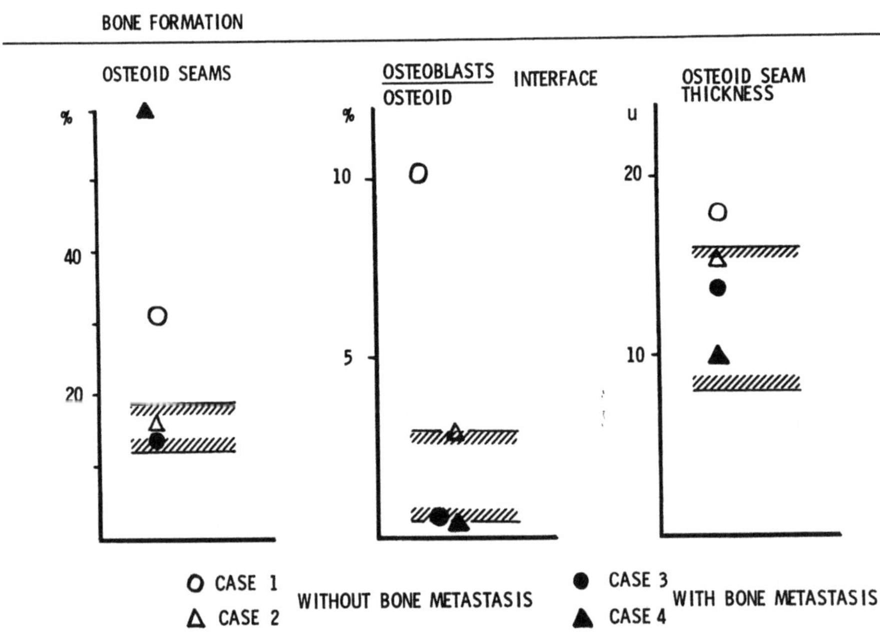

b

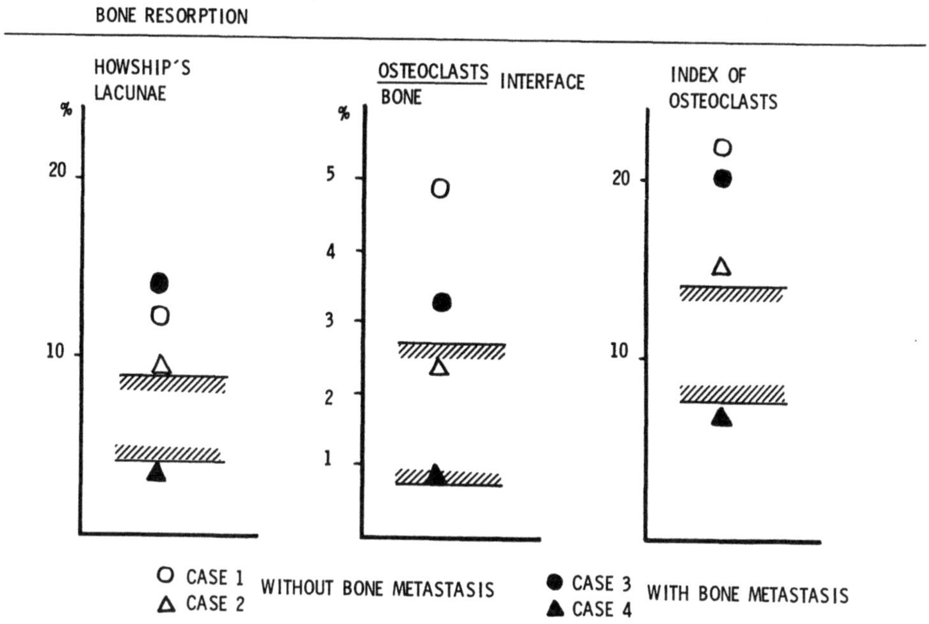

c

homogeneous are the alterations in bone resorption: Three of the four cases studied have a distinctly increased osteoclastic bone resorption activity. In our opinion this fact should be explained by the development of a secondary hyperparathyroidism in the cases without bone metastases. The enhancement of bone resorption in the case with bone metastases does not fit with the thesis that the parathyroid glands are inhibited by the removal of calcium from the skeleton in osteolytic tumor metastases. Because of the difficulty to quantify osteolytic tumor metastases as well as amounts of calcium deliberated, this point should remain under discussion until more experiences are available.

Finally, we conclude from our investigation that calcitonin produced by human medullary carcinoma of the thyroid gland has obviously no clear effect on bone formation. The influence on bone resorption might be indirect via secondary hyperparathyroidism, which is obviously not completely prevented by CT. Total bone mass is not characteristically altered, it seems to depend on the initial values of the patients before suffering from medullary carcinoma.

References

CUNLIFFE, W.T., HALL, R., HUDGSON, P., GUDMUNDSON, T.V., WILLIAMS, E.D., GALANTE, L., BLACK, M.M., JOHNSTON, I.D.A., SHUSTER, S., JOPLIN, G.F., WOODHOUSE, N.J.Y., MACINTYRE, I.: A Calcitonin - Secreting Thyroid Carcinoma. Lancet, 63 (1968).

DELLING,G.: Age Related Bone Changes. Curr. Top. Path. 58, 117 (1973).

HAAS, H.G., DAMBACHER, M.A., GUNCAGA, J., LAUFFENBURGER, T.H., LENTNER, C.H.: Fragen zur Calcitonin Forschung. Klin. Wschr. 50, 2 (1972).
MERZ, W.A.: Die Streckenmessung an gerichteten Strukturen im Mikroskop und ihre Anwendung zur Bestimmung von Oberflächen - Volumen - Relationen im Knochengewebe. Mikroskopie 22, 132 (1967).
TASHJIAN, A.H., MELVIN, K.E.W.: Medullary Carcinoma of the Thyroid Gland. New Engl. J. Med. 279, 279 (1968).
TUBIANA, M., MILHAUD, G., COUTRIS, G., LACOUR, J., PARMENTIER, C., BOK, B.: Medullary Carcinoma and Thyrocalcitonin. Brit. Med. J. 4, 87 (1968).
WILLIAMS, E.D., BROWN, C.L., DONIACH, I.: Pathological and Clinical Findings in a Series of 67 Cases of Medullary Carcinoma of the Thyroid. J. Clin. Path. 19, 103 (1966).

Intestinal Calcium Absorption 0–30 Years Post Gastrectomy; Study with a Double Isotope Radio Technique

D. G. Ikkos, A. Nikolaou and C. C. Alevizaki

Although a decrease in intestinal absorption of calcium has often been discussed as a consequence of gastric surgery, quantitative data on calcium absorption in gastrectomized patients published so far are conflicting, some workers even reporting increase of fractional intestinal

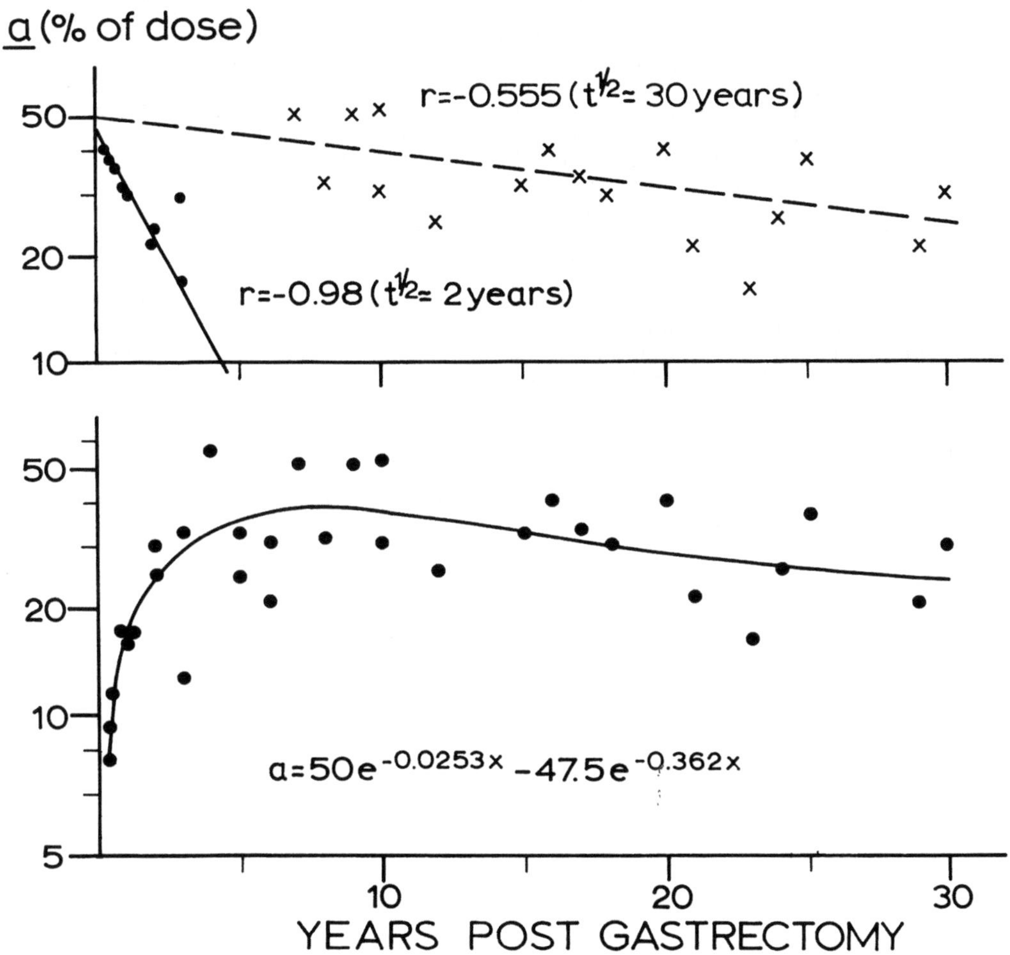

Fig. 1. Effect of time elapsed since the operation on intestinal absorption coefficient of calcium a in gastrectomized subjects. The two exponential components are plotted separately above

absorption of calcium in these subjects. The present study was undertaken in order to re-examine this question by measuring the true intestinal absorption coefficient of calcium on 31 consecutive patients aged 40 to 74 years, who had undergone partial gastrectomy (Billroth I or Polya) 3 months to 30 years previously. The coefficient of intestinal absorption of calcium, a-value, was determined by a double radioisotope ratio technique (DE GRAZIA et al., 1965). Results were compared to those obtained in a group of 52 normal controls.

It was found that:

1. Type of operation had no effect on a-value ($p = 0.9$).
2. Gastrectomized patients had as a group significantly ($p < 0.02$) lower values than normals, the mean $\log_{10} a$ being 1.433 ± 0.037 and 1.536 ± 0.024 (SEM) respectively.
3. The difference persisted even when age was taken into account with seven values lying below the -2σ limits for normals of corresponding age.
4. Time elapsed since operation had a very marked effect on a-value, as shown in Fig. 1. Thus, a-values were definitely low during the first year, improving as to reach a maximum between 5-10 years post operation and declining very slowly thereafter. As shown in Fig. 1, this relationship was fitted by hand by the difference of two negative exponential functions of time, one slow component with a half life of about 30 years minus one fast component with a half life of 1.9 years. Of these, the fast component represents the effect of the operation which, being very pronounced at the beginning, decreases rapidly leading to an improvement of a-value, whereas the slow component represents in all probability an advancing age effect, similar to that observed in normal subjects.

Although the phenomenon represents very probably a readjustment of the gastrointestinal tract compensating for the effects of operation on calcium absorption, no explanation as to the exact mechanism involved can be offered on the present evidence.

Reference

DE GRAZIA, J.A., IVANOVICH, P., FELLOWS, H., RICH, C.: J. Lab. Clin. Med. 66, 822 (1965).

The Morphology of the Intestinal Calcium Excretion by the Paneth Cell of Rat, Mouse and Man*

H.-J. SCHÄFER AND H. F. OTTO

I. Introduction

Salivary, gastric and pancreatic juice, bile, goblet cells and enterocytes contribute to intestinal calcium excretion. Additionally, the light microscopic GBHA technique (KASHIWA, 1966) drew attention to the Paneth cells. These excretory cells are found predominantly in the ileum in typical position at the bases of the crypts. Their physiological role is not yet clear. Probable functions are the production of lysozyme, proteolytic enzymes and nutritive factors for the other crypt cells, intraluminal acid-base-balance and elimination of heavy metals (OTTO, 1973). The high calcium content of these cells induced us to do further morphological studies on a probable calcium excretion by the Paneth cells in several species and under different conditions of calcium metabolism.

II. Materials and Methods

The investigations were carried out on rats, guinea-pigs, Syrian hamsters, mice and human biopsy material.++ The ileum of 5 untreated animals of each species and duodenal biopsies of 5 patients were stained by glyoxal bis (2-hydroxyanil) (= GBHA) for light microscopic demonstration of mobile calcium. According to KASHIWA (1966) the fresh tissue was immersed into the staining fluid rapidly after excision. Tissue samples of some rats were prefixed in 10% neutral formalin with or without addition of 1% sodium-oxalate, dehydrated and embedded with subsequent GBHA staining of the paraffin sections (for details see SCHÄFER and KLÖPPEL, 1974a). For electron microscopic calcium detection, tissue samples were fixed in glutaraldehyde and OsO_4, containing 2% potassium pyroantimonate (for details see SCHÄFER and KLÖPPEL, 1974b). 6 rats were given 0,3 mC ^{45}Ca i.v. 10, 60 and 120 minutes respectively before being killed. Samples of the ileum were excised, fixed in 0,1 m cacodylate-buffered 2,5% glutaraldehyde with addition of 1% sodium-oxalate, dehydrated in ethanol and embedded in Epon 812. 0,5 µ sections were covered by Kodak AR 10 stripping film or Ilford L4 emulsion.

In a further experiment 5 rats were made hypercalcemic by application of 3 x 100 USP-U. Para-thormone Lilly^R 8, 16 and 24 hours before being killed. 6 rats were given 2 x 400 mMRC-U. porcine calcitonin 6 and

+ Supported by DFG, Sonderforschungsbereich 34 "Endokrinologie", Hamburg.

++ Biopsy material was obtained by the courtesy of Prof. Dr. F. KUHLENCORDT, Department of Clinical Osteology, I. Clinic of Internal Medicine, University of Hamburg.

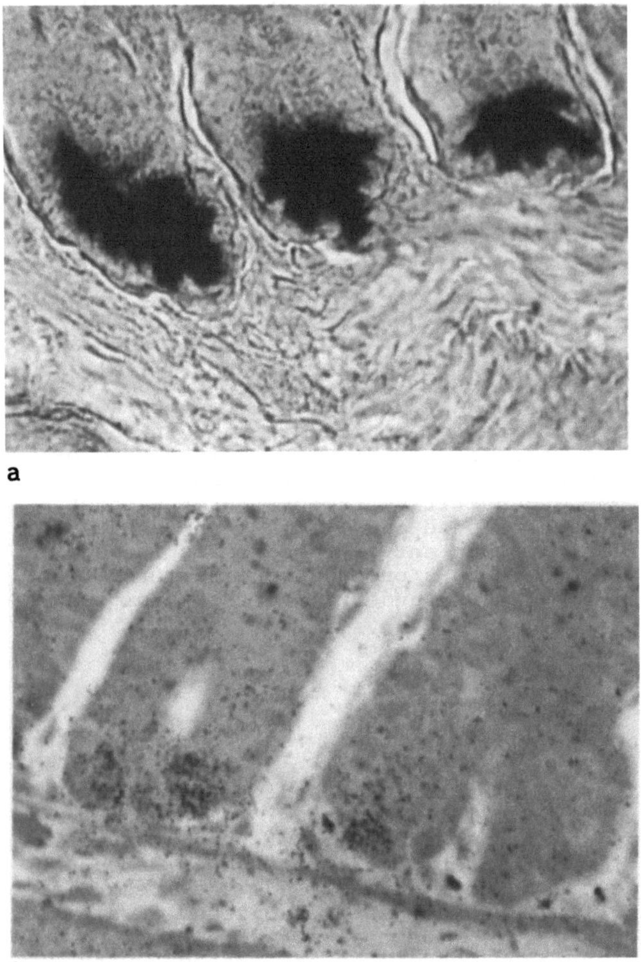

a

b

Fig. 1.(a) Ileal crypts; untreated rat: intense dark staining of the cytoplasm of Paneth cells indicating a high calcium content; nuclei in basal portions of the cells are spared. GBHA reaction. x 250.
(b) Ileal crypts; rat, 2 hours after intravenous application of 0,3 mC ^{45}Ca: Accumulation of activity in the cytoplasm of Paneth cells. Unstained sections; Kodak AR 10 stripping film. x 250

2 hours before being killed. 5 rats served as untreated controls. Fresh ileum was stained in GBHA (KASHIWA, 1966). Paraffin sections of formalin-fixed material were stained by hematoxylin-eosin and Masson-Goldner. Serum calcium was deterimed by flame photometer.

III. Results and Discussion

The GBHA staining was found to be a specific and sensitive method for demonstration of mobile calcium in hard and soft tissues like the ileum of the rat, where it revealed a strongly positive reaction in goblet cells and predominantly in Paneth cells (KASHIWA, 1966). The intense red dye was situated in the cytoplasm of the Paneth cells; the nuclei were spared (Fig. 1a). This finding demonstrates that the Paneth cells of the rat contain high amounts of calcium.

Comparison of the intensity of the staining reaction in differing species revealed that only the guinea-pig, which is closely related to the rat, contains comparable amounts of calcium within the Paneth cells. In contrast, those of the mouse, which are very abundant and well-developed, revealed only a very weak tingation, while those of the Syrian hamster showed a medium intensity of tingation. This experiment demonstrates that the calcium content and the possible role of Paneth cells in intestinal calcium excretion is species-dependent showing the rat in the first position and mouse and man at the end of the list. This result parallels the autoradiographic data for some metals, reported by HALBHUBER et al. (1970). According to this, further experiments for more detailed description of calcium in Paneth cells were undertaken in rats.

At first we stated that prefixation of the tissue in aqueous media before staining completely removed the stainable calcium. Addition of oxalate to the fixation fluid and subsequent GBHA staining retained the calcium in coarse granular form. These findings demonstrate that the calcium in Paneth cells is only loosely bound and not tightly incorporated into the matrix of the secretory granule, which is - in contrast to the stainable calcium - preserved during formalin fixation.

A high calcium content of an excretory cell does not automatically mean a high excretion rate. To investigate the dynamic of calcium incorporation we made an autoradiographic study. After intravenous application of ^{45}Ca we found an accumulation of activity in goblet cells and Paneth cells (Fig. 1b). In Paneth cells the activity was spread over the whole cytoplasm up to the luminal parts of the cells. This pattern could be observed already 10 minutes after injection of ^{45}Ca and was still present after two hours. Obviously accumulation and intracellular spreading of calcium occurs quite rapidly. The correlation of this fact with other autoradiographical data yields interesting aspects. HALBHUBER et al. (1972) observed in Paneth cells of the mouse that 5 hours after application of a triated precursor amino acid nearly all secretory granules are still free of activity. If these data can be transferred to the rat, it seems to be improbable that calcium accumulation is combined with the synthesis of the protein component of the secretory granule, for the protein synthesis is a very slow process, while calcium accumulation occurs rapidly. Two alternatives are possible: 1) Calcium accumulation occurs in structures other than the secretory granule. 2) Accumulation of calcium occurs by rapid addition of calcium to preformed secretory granules.

Ultrastructural studies, employing the pyroantimonate technique for electron dense calcium precipitation, seem to agree with the assumption that both possibilities are existent. Though this technique precipitates some other cations such as sodium as well (KOMNICK and KOMNICK, 1963), the calcium, however, is, according to the intensity of the calcium-specific GBHA reaction, the first candidate for intracellular pyroantimonate precipitation within the Paneth cells. Indeed, especially in these cells we could observe particularly high amounts of precipitates, more than in the surrounding cells. The precipitates were situated at the endoplasmic reticulum, the cell membrane and within the mitochondria (Fig. 2). Important is the different behavior of the prosecretory granules. Some were completely free of precipitates, others were heavily impregnated and others again showed precipitates predominantly in the peripherical space of the granule and at the surrounding membrane systems. There seemed to occur a proceeding ion impregnation of the prosecretory granules before expulsion into the intestinal lumen.

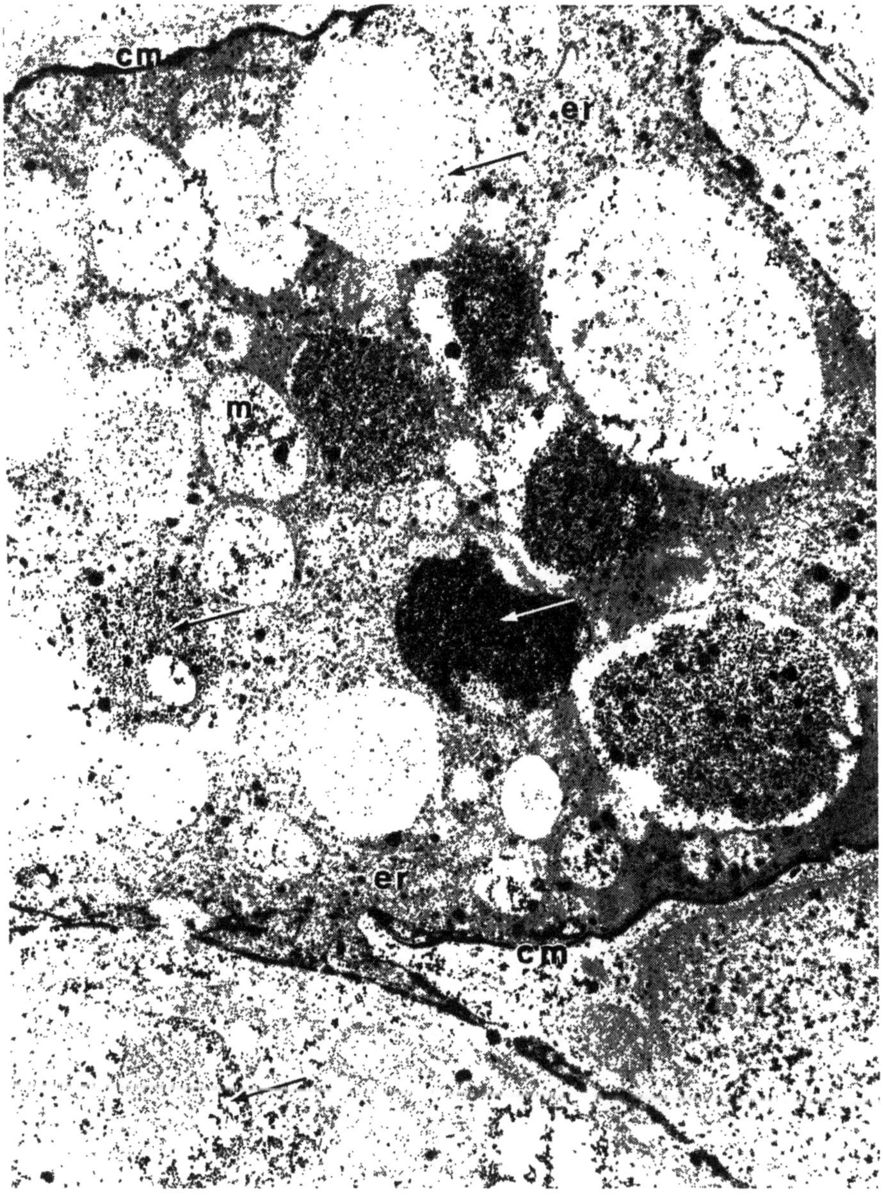

Fig. 2. Paneth cells; untreated rat: precipitates at the cell membrane (cm), the endoplasmic reticulum (er) and within mitochondria (m). Note different stages of ion impregnation in and around prosecretory granules (←). K-pyroantimonate reaction, unstained. x 20 000

These findings together with the presented autoradiographic data and the fact of loose binding and high solubility of the calcium in Paneth cells led us to the following model:

Calcium incorporation into the secretory granule and protein synthesis of the granule matrix are independent processes; after passing the basal cell membrane the calcium spreads within the cell, accumulates in cytoplasmic membrane systems and mitochondria, which are known to be storage organelles for cellular calcium (BORLE, 1971; MATTHEWS et al., 1971), and can at least in part be added in great amounts and with loose linkage to preformed prosecretory granules with subsequent excretion into the intestinal lumen. However, though combining our results, this model is still hypothetical and needs further evaluation.

A further experiment gave evidence that the Paneth cells of the rat may be influenced by high doses of calcitonin or parathyroid hormone. After application of porcine calcitonin the rats developed hypocalcemia with a mean serum calcium of 7,9 (± 0,4) mg%. The Paneth cells showed well developed, dense secretory granules and a particularly intense GBHA reaction. The lumina of the crypts were free of calcium-rich material. After application of high doses of parathyroid hormone the rats developed a marked hypercalcemia (mean serum calcium 18,5 ± 1,0 mg%). The ileum showed a completely different picture. Calcium-containing Paneth cells were decreased in number, magnitude and granulation. Instead of that we found abundant calcium-rich material within the lumina of the crypts, which were often dilated. Within the lumina the Masson-Goldner stain revealed besides amorphous material plugs of desquamated cells, which sometimes contained rests of secretory granules within their cytoplasm and could in some cases be identified as Paneth cells. These findings suggest that after high doses of parathyroid hormone contrary to calcitonin, both, an increased secretion and a desquamation of total, calcium-rich cells, may contribute to an increased loss of calcium into the intestinal lumen.

There is after all strong evidence that in some species the Paneth cells besides lysozyme production contribute significantly to intestinal calcium excretion. Whether they take part in an endocrinologically regulated excretory system cannot be settled at present, because the high doses of parathyroid hormone and calcitonin in our experiment, which were necessary to produce the described light microscopical alterations, do not exclude toxical effects. Probably ultrastructural investigations of Paneth cells by employing lower hormone doses, which are in progress, will give further information about this subject.

IV. Summary

1) Paneth cells of some species contain high amounts of calcium.
2) Calcium contents of Paneth cells are species-dependent. High amounts were found in rats; low amounts were found in mice and human biopsy material.
3) Calcium in Paneth cells is loosely bound. Autoradiographic and electron microscopic findings support the concept that calcium is at least partly added to prosecretory granules by the surrounding membrane systems in later stages of granule development.
4) Light microscopical behavior of Paneth cells of the rat can be influenced by high dosages of parathyroid hormone or calcitonin respectively.

References

BORLE, A.B.: Calcium transport in kidney cells and its regulation. In: Cellular mechanisms for calcium transfer and homeostasis (eds. G. Nichols, and R.H. Wasserman), p. 151-174. New York, London: Academic Press 1971.

HALBHUBER, K.J., STIBENZ, H.-J., HALBHUBER, U., GEYER, G.: Autoradiographische Untersuchungen über die Verteilung einiger Metallisotope im Darm von Laboratoriumstieren. Ein Beitrag zur Ausscheidungsfunktion der Panethschen Körnerzellen. Acta histochem. 35, 307-319 (1970).

HALBHUBER, K.-J., STIBENZ, H.J., GEYER, G.: Autoradiographische Untersuchungen über die Synthese des Prosektrets in den Panethschen Körnerzellen des Dünndarms der Albinomaus. Acta histochem. 42, 143-156 (1972).

KASHIWA, H.K.: Calcium in cells of fresh bone stained with glyoxal bis (2-hydroxyanil). Stain Technol. 41, 49-55 (1966).

KOMNICK, H., KOMNICK, U.: Elektronenmikroskopische Untersuchungen zur funktionellen Morphologie des Ionentransportes in der Salzdrüse von Larus argentatus. Z. Zellforsch. 60, 163-203 (1963).

MATTHEWS, J.L., MARTIN, J.H., ARSENIS, C., EISENSTEIN, R., KUETTNER, K.: The role of mitochondria in intracellular calcium regulation. In: Cellular mechanisms for calcium transfer and homeostasis (eds. G. Nichols and R.H. Wasserman), p. 239-255. New York, London: Academic Press 1971.

OTTO, H.F.: Zur funktionellen Bedeutung der intestinalen Paneth-Zellen. Dtsch. med. Wschr. 98, 220-226 (1973).

SCHÄFER, H.-J., KLÖPPEL, G.: Demonstration of calcium in pancreatic islets. Light-microscopic observations in activated and inactivated B cells of mice. Virchows Arch. Abt. A Path. Anat. 362, 1-12 (1974a).

SCHÄFER, H.-J., KLÖPPEL, G.: The significance of calcium in insulin secretion. Ultrastructural studies on identification and localization of calcium in activated and inactivated B cells of mice. Virchows Arch. Abt. A Path. Anat. 362, 231-245 (1974b).

Chairmen: G.C.H. BAUER and W. REMAGEN

Frictional Ablation – a Neglected Factor in the Mechanisms of Hard Tissue Destruction?

R. F. SOGNNAES)

Ten years ago, in my preface to "Mechanisms of Hard Tissue Destruction" (SOGNNAES, 1963), I discussed the degree to which mechanical friction may be involved in destructive biological mechanisms affecting such solid structures as rocks, corals, shells, antlers, bone, ivory, cementum, dentin, and enamel. Whereas the living culprits of destruction may be identified at the "scene of the crime" - be they rock and shell boring sponges, mollusks, snails, octopods and worms, or osteoclastic giant cells, fungi, algae and bacteria - the precise manner in which these biological systems operate in immediate juxtaposition to the disintegrating surfaces, presents great difficulties in observations, often subject to conflicting interpretations.

I. Oral Hard Tissue Destruction

Among the four principal dental hard tissue destructions that are primarily attributed to oral environmental influences - attrition, abrasion, erosion and caries - no condition has been the subject of more theoretical confusion and practical frustration than the so-called idiopathic dental erosion.

Recent research has shown that not only natural teeth but also various dental restorations, including plastic and metallic fillings, can be prone to erosion-like changes, often superimposed upon other types of pathology, including caries and calculus, and hence not adequately explained by chemical dissolution (SOGNNAES et al., 1972). Erosion-like denture markings, described by BRODIE and SOGNNAES (1973, 1974), have been found both on vulcanite and acrylic materials, affecting labial, buccal, lingual, and palatal surfaces. These show minute substriations when examined under the scanning electron microscope by means of a micro-replication method described elsewhere (SOGNNAES, 1973, 1974). Two special case histories will serve to illustrate unusual wearing patterns, in a form suggestive of idiopathic erosion changes.

CASE no. 1: A very extreme case of intra-oral denture destruction has recently been observed in a 38-year-old mother referred to me by a private practitioner in Watsonville, California, Dr. James A. DUCASSE, D.D.S. This case concerned a maxillary denture with porcelain incisor teeth, but with methylmethacrylate posterior teeth and base (Fig. 1). The incisor teeth were intact whereas the buccal surfaces of the acrylic base and posterior teeth were "eroded" in considerable depth. The patient was found also to have leukoplakia-like changes along the adjacent mucous membrane of her cheeks, now being explored further by the above mentioned indirect method for in vivo microreplication and scanning electron microscopy. This patient, who has worn the denture for four years, has been known to her dentist as a tense, nervous person, employed under considerable pressure and tension as an office manager, "living on sedatives and tranquilizers".

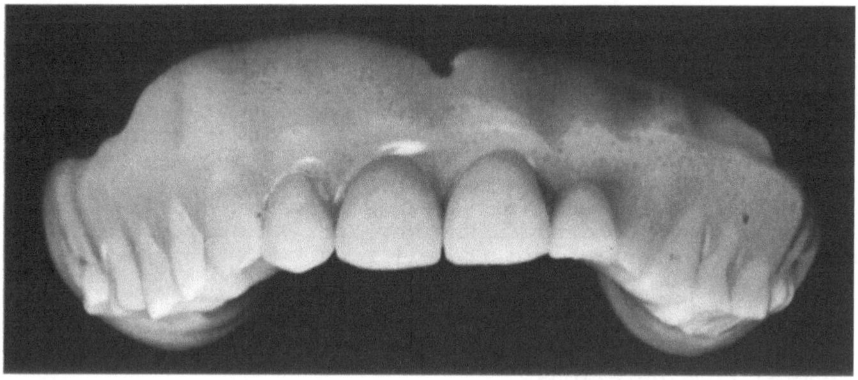

Fig. 1. Maxillary denture worn for four years by 38-year-old woman. The wearing away of the buccal surfaces of the methylmethacrylate posterior teeth and adjacent denture base is thought to be due to "frictional ablation" caused by hyperactive oral soft tissues

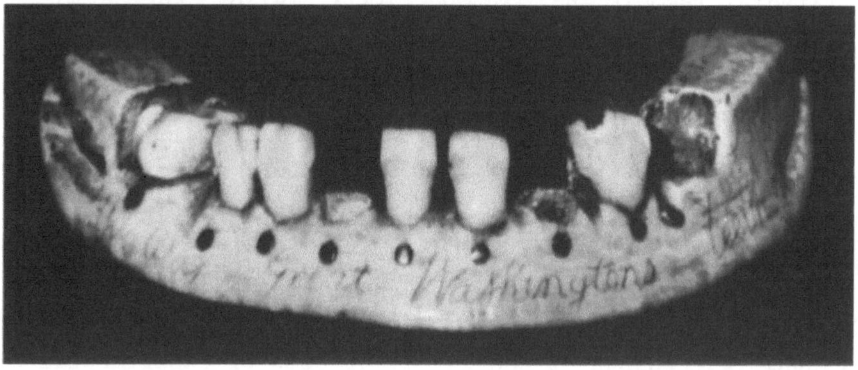

Fig. 2. One of President George Washington's ivory base dentures made **by the** New York dentist John Greenwood in 1789. Note the extensive "wear" of the labial and buccal tooth surface especially on the side where Washingtion's pouting lips pressed the denture against his last remaining tooth, a second premolar accomodated by the hole in the base

One may well ask whether or not this kind of condition is of relatively recent origin, perhaps related to tensions of modern life. However, somewhat similar changes have been observed in a denture worn nearly two centuries ago, belonging to none other than the first President of the United States of America.

CASE No. 2: The oldest known denture attributed to George Washington was made by John Greenwood of New York in 1789 (see Fig. 2). The denture was retained by intermaxillary spiral springs and by Washington's last remaining natural tooth, a lower left premolar. If efforts were made to reduce some of the anterior dental bulk in order to minimize Washington's pouting mouth, the lips would probably continue to press and rub against the anterior teeth with some resistance from the retaining function of the remaining premolar accomodated by a hole in the ivory denture base. Consequently one would expect that the frictional forces would be greatest between the left anterior quadrant and the adjacent lip and cheek surfaces. This is precisely what appears to have been the case, because nearly all of the labial tooth enamel as well as some of the underlying dentin has been "worn" away. In further support of this interpretation, i.e., frictional pressure, wear and tear, is the observation that the remaining lower natural premolar tooth became extensively worn around the neck and ultimately became loose and lost (but not lost to posterity, for it was inherited by Washington's dentist, who had it encased for an adornment of his watch chain, later to be donated to the New York Academy of Medicine, in whose Rare Book Room I found it well preserved). For related details see elsewhere (SOGNNAES, 1972c, 1973).

II. General Considerations

Functional relationships of the oral epithelial lining to the underlying musculature suggest that a possible cause of the erosion-like dental wearing patterns is hyperactivity of the oral soft tissue environment in juxtaposition to the hard structures involved, whether natural or artificial (BRODIE and SOGNNAES, 1973, 1973b). Furthermore, in view of the forceful frictional action which appears to operate in the mouth of some patients, a detrimental thinning effect might even extend to the underlying bone with recession of the gingival supporting tissues and extension of the process to the exposed root surfaces of the teeth, as is being reported elsewhere (SOGNNAES, 1974).

In some types of destructive mechanisms nature has clearly developed biological systems capable of mechanical friction. Thus, readily understood physical forces are at work in the case of destruction by certain large multicellular organisms, e.g., the twisting motion of rock-boring bivalves and the denticle drilling action of the gastropod radula, rasping holes in an oyster shell region, after partial decalcification mediated through chemical action of the snail's accessory boring organ (ABO gland). Indeed, anyone who has observed the cinematographic recordings of the lively process of experimental bone resorption in tissue culture will have a vivid impression that even the osteoclasts - aside from their complex biochemical apparatus - do in fact move around in a slow-motion "twist", rubbing its "pseudopodia" along the presumably pre-softened walls of an eroding Howship's lacuna.

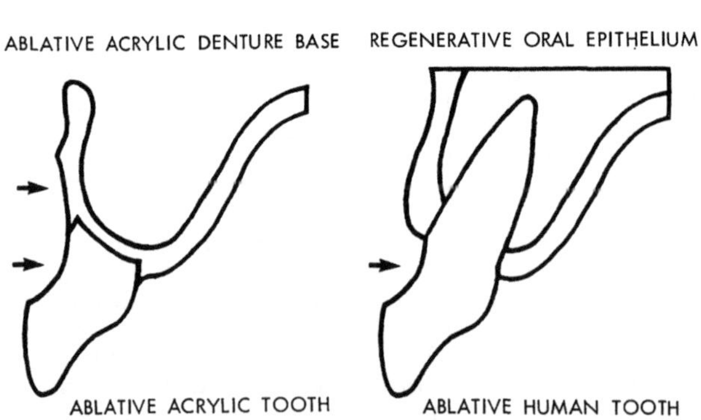

Fig. 3. Diagramatic profile projection of frictional ablation process affecting artificial (left) and natural (right) dentition

III. Suggested Terminology, Summary and Conclusion

A connotation of chemical dissolution rather than mechanical wear is generally attributed to the term erosion in dental and certain other biological contexts. The author wishes therefore to suggest that certain conditions which involve bona fide surface friction, including some of the obscure types of what hitherto has been referred to as idiopathic erosion, may perhaps be more appropriately identified and classified as frictional ablation - a term chosen from space age nomenclature, i.e., the physical surface changes occurring when the nosecone of returning spaceships re-enters the earth's normal environment and undergoes frictional atmospheric dissipation.

In summary, using a diagramatic representation, Fig. 3 illustrates the destructive consequences of frictional ablation, exemplified by the effect on an artificial acrylic denture and on an ablative natural dentition, respectively. Thus, in the context of oral hard tissue destruction, it is hoped that such a designation will help to improve the understanding of and distinction between conditions attributable to distinct chemical erosion mechanisms, vis à vis readily recognized mechanical and masticatory forms of tooth wear, abrasion and attrition (e.g. teeth against teeth), on the one hand, and physical ablation caused by environmental soft tissue friction (e.g. cheeks against teeth), on the other.

In conclusion, returning to the broader introductory aspects of hard tissue destruction, it remains to be determined if further observations on such frictional ablation processes may evolve into a more generalized concept, pertinent to a greater variety of the cells, tissues, organs and organisms which face ablative surfaces involved in mechanisms of hard tissue destruction.

IV. Addendum

After the above was presented, a potentially related physical cell activity has been suggested by scanning electron microscopy of fibroblastic cell types caught (in tissue cultures) at dramatic perambulations of the cell surfaces (J.P. REVEL, Engineer & Sci., Cal. Inst. Tech., Nov.-Dec. 1973, page 5; also TIME, Dec. 31, 1973, p. 41). The physical property referred to as "ruffling" may not merely be reflecting an idle motion but may perhaps in the case of some cell types also contribute to the above mentioned frictional ablation mechanism and facilitate dislodgement of adjacent hard tissue ingredients (Fig. 4).

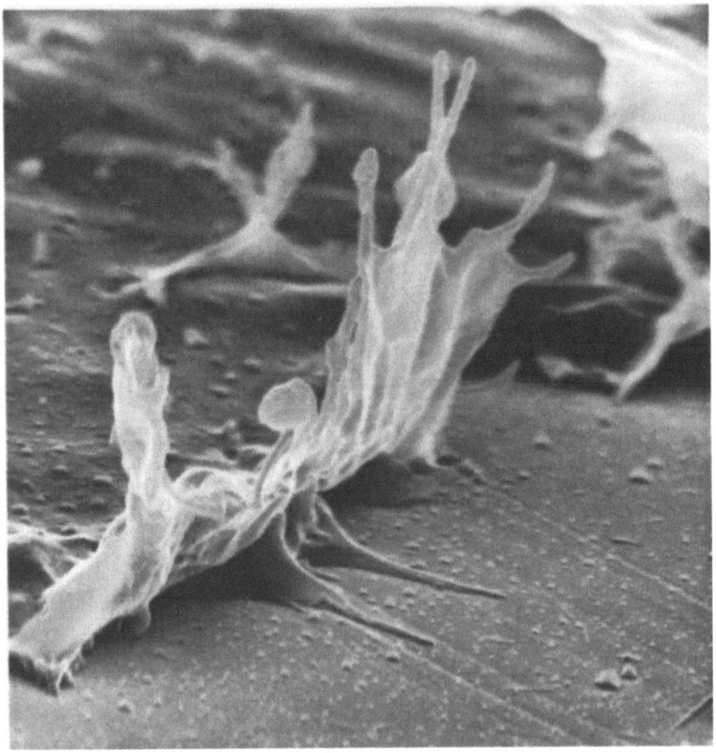

Fig. 4. Fibroblastic cell perambulation by a process of "Ruffling" observed in the scanning electronmicroscope by Professor Jean-Paul Revel at California Institute of Technology. Magnification: 15,000 times (From Engineering & Science, California Institute of Technology, Nov.-Dec. Issue, p. 5, 1973)

References

BRODIE, A.G., SOGNNAES, R.F.: Erosion-like patterns on acrylic dentures attributed to mucosal friction. J. Dent. Res. 52 (Special Issue - Abstract 416), 162 (1973).

BRODIE, A.G., SOGNNAES, R.F.: Erosion-like denture markings possibly related to hyperactivity of oral soft tissues. J. Am. Dent. Assoc. 88, 1012-1017 (1974).

SOGNNAES, R.F. (Editor): Mechanisms of hard tissue destruction. American Association for the advancement of Sience, Publ. No. 75, Washington, D.C. 1963.

SOGNNAES, R.F.: Micro-replication for scanning electron microscopy applied to problems of art, technology and science. Proc. Int. Assoc. Denta. Res. Abstract 128, p. 80 (1972a).

SOGNNAES, R.F.: The use of micro-replication for scanning electron microscopy of dental lesions. Proc. Int. ORCA Conf., Leipzig University; Caries Res. 6, 87-88 (1972b).

SOGNNAES, R.F.: Evolution of dental art and science illustrated by the tale of three cities: President Washington's Washington, Dr. Parkman's Boston, and Hitler's Berlin. Harvard Dental Alumni Bulletin, 32, (No. 2) 52-59 (1972c).

SOGNNAES, R.F.: America's most famous teeth. Smithsonian 3 (No. 11), 47-51 (1973a).
SOGNNAES, R.F.: Some structural observations suggestive of a dento-alveolar ablation mechanism. Presented Sept. 22, 1973, during the 10th Anniversary of the Continental European Division of the International Association for Dental Research, Strasbourg, France. J. Dent Res. 53 (No. 3), 744 (1974).
SOGNNAES, R.F., WOLCOTT, R.B., XHONGA, F.: Dental erosion: I. Erosion-like patterns occuring in association with other dental conditions. J. Amer. Dent. Assoc. 84, 571-576 (1972).

Tumoral Calcinosis with Hyperphosphatemia*

H. G. MORGAN, F. J. DRYBURGH AND E. A. MILLS

Tumoral calcinosis is a rare condition characterized by localized calcific swellings in relation to joints. First named by INCLAN in 1943, the condition appears to be different from metastatic or the better known dystrophic calcifications. Renal function is normal and there is no skeletal demineralization. Most patients are young, of both sexes, and usually of Negro race. Over the past decade or so, some 12 patients with both tumoral calcinosis and a raised plasma phosphorus have been described, often in siblings, but never reported in more than one generation of a familiy (Table 1). We wish to report possibly the first such combination from the United Kingdom or even in Europe.

A girl from East Pakistan, then aged 9, was seen in 1965 with suspected injury to the left foot, which was red and swollen laterally. Radiology showed loculations of opaque material there, and also in the right foot and right hand. She was intelligent, healthy and normal in development. An excision biopsy showed discrete semi-liquid loculi of chalky material in a thin fibrous capsule, not invasive, although passing between muscles. No parasites were seen. The chalky material was composed of tiny spheres of crystals shown to be apatite.

In 1966, the deposit in the right hand was larger, and the left foot was superficially ulcerated. Full radiology showed calcification over each olecranon but no intraabdominal calcification. There was no corneal calcification. The serum phosphorus was 8.0 mg/100ml but all other routine biochemistry was normal, especially calcium 9.8 mg/100ml, alkaline phosphatase 13 KA Units/100ml, and urea 23 mg/100ml. There was no history of drug or excess vitamin D ingestion and no dietary peculiarities. There was no evidence of calcinosis or raised phosphorus in the parents, who are first cousins, or in the only brother. There was no family history in the many relatives.

I. Surgical Progress

Fig. 1 shows the lesion in the right hand in 1966. This lesion had been increasing in size over the past year and continued to do so until excised in 1967. There has been no clinical or radiological recurrence to date. Fig. 2 shows the right elbow before and after surgery in 1971. The lesion seen on the right of Fig. 2 was in fact a recurrence. Excision had been undertaken on two previous occasions, 1967 and 1969. Again there is no evidence, clinically or radiologically, of a further recurrence. The composite diagram (Fig. 3) shows the dates of the many operations. Note that no new sites have developed, although until 1971 fresh masses recurred at operation sites. The lesions in both feet have been most resistant. Despite numerous

+ We express our thanks for help and advice to Professor C.E. DENT, of University College, London, and our most co-operative surgeon, Mr. John WHITE.

‚ical episodes, there is still calcific material in both feet. This :, aged 17, she has no real disability, has developed well, is .er than her mother, and hopes to enter University. Also of inter- are the peculiar localized dental pulp calcifications, with mas- ; swellings of the coronal third of the roots of many teeth. This been reported in detail elsewhere (HUNTER et al.).

.e 1. Patients with Tumoral Calcinosis and Hyperphosphataemia

;enting	Sex	Race	Phosphorus mg/100ml	References
9	M	White	6.5	GHORMLEY, R.K., MACCRARY, W. (1942)
57	F	Negro	5.3	RIEMENSCHNEIDER, P.A., ECCKER, A., (1952)
15	M	Negro	7.8 - 8.1	ANNAMUNTHODO, H. (1960)
13	M	Negro	6.4	BARTON, D.L., REEVES, R.J. (1961)
12	M	Negro	5.5	BARTON, D.L., REEVES, R.J. (1961)
38	M	White	4.5 - 5.8	LAFFERTY, F.W. et al. (1965)
10	F	Negro	6.9 - 7.1	HARKNESS, J.W., PETERS, H.J. (1967)
5	M	Negro	7.3 - 8.7	WILBER, J.F., SLATOPOLSKY, E. (1968)
4	M	Negro	7.1 - 9.8	BALDURSSON, H. et al. (1969)
4	M	Negro	7.0 - 11.5	BALDURSSON, H. et al. (1969)
6	F	Negro	6.1 - 10.0	BALDURSSON, H. et al. (1969)
4	M	Negro	7.7 - 7.8	BALDURSSON, H. et al. (1969)
9	F	Asian	6.5 -10.0	MORGAN, H.G. et al. (1973)

Biochemical Investigations and Treatment

.empts at balance studies were prevented by a mischievous approach our patient. We were helped greatly by Professor C.E. DENT of Lon- .. Good renal function was shown, with no tubular defects other than .al phosphate retention. There was a poor renal response to exogenous

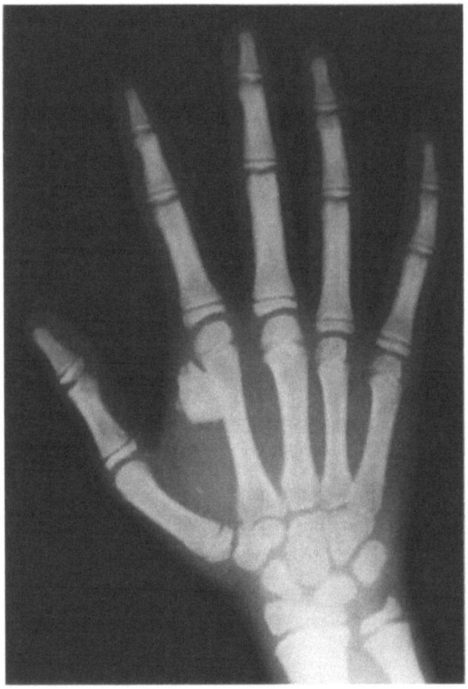

Fig. 1. Right hand, 1966

parathormone (PTH). Blood PTH was not increased in 1973. Minor increases in serum phosphorus levels were caused by dietary phosphate loading, and minor decreases by Aludrox.

The therapeutic principles were to try to lower phosphorus intake by restricting dairy products, to reduce phosphorus absorption by aluminum hydroxide, and to attempt to alter renal handling with Probenecid (BALDURSSON et al.), acetazolamide, bicarbonate, alanine and glycine (MICHAEL et al.). Only the first two measures appeared effective, in that there was a distinct reduction in urinary phosphorus, at times almost to zero, although there was little change in serum phosphorus.

We have persisted with this regime to date, and propose to continue for at least a further year.

Our last hope was that the phosphorus level would fall, although relatively, at puberty, and perhaps modify the rate of recurrences. Her periods began four years ago, close to her 13th birthday, but have been slight and irregular; breast development and hirsutism appear normal. Fig. 4 shows the phosphorus levels over 7 years - the latest value being 6.2 mg/100ml. No persisting change had occured up to 1973, and certainly not at the time of onset of the menarche.

What is the nature of this condition? There are two aspects. First, the occurence of localized nodules in relation to joints. These are rather similar to masses that can occur in renal failure, vitamin D overdosage, or the milk-alkali syndrome, but the racial and familial features are specific, as is the tendency for nodules to occur at pressure sites. Secondly, there is the fixed high serum phosphorus

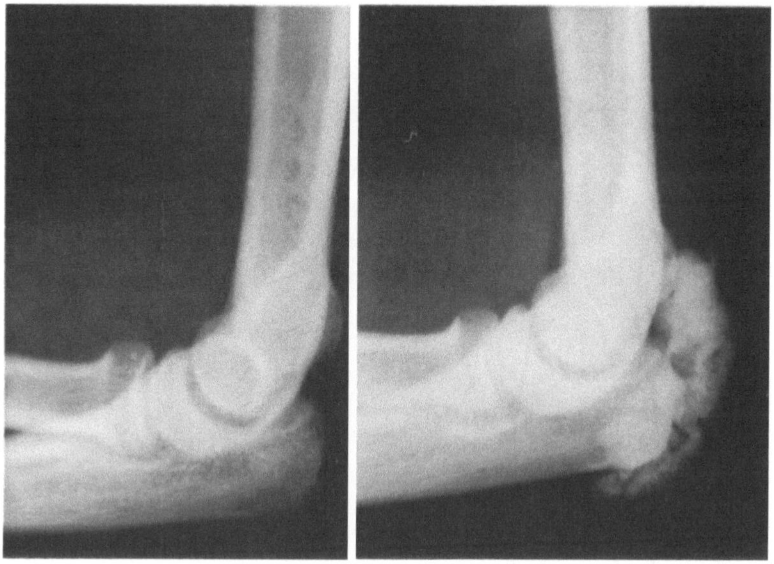

Fig. 2. Right elbow, 1971, before and after operation

level. Other authors have shown a renal phosphate response to calcium and EDTA infusions (presumably an endogenous PTH effect) and to PTH administration (WILBER et al.). A normal blood human growth hormone level has been reported (WILBER et al.).

We are left with a change in phosphate metabolism, often familial, with the renal tubule apparently responding to physiological stimuli, but at a higher level than normal. However, some cases of tumoral calcinosis have a normal serum phosphorus, and there is one instance of a sibling with high phosphorus and no lesions.

Lastly, treatment and prognosis: the consensus of surgical opinion appears to be that early and complete excision of lesions be undertaken where feasible; our patient's experience suggests that this should be combined with a lowered phosphorus intake.

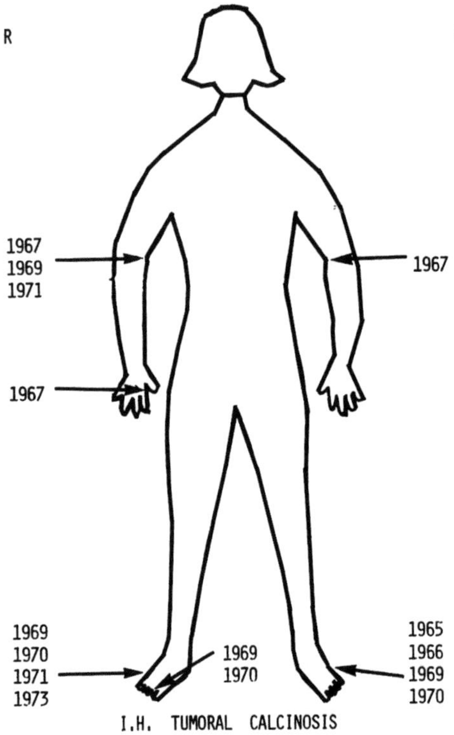

Fig. 3. A composite slide showing sites of calcification and operation dates

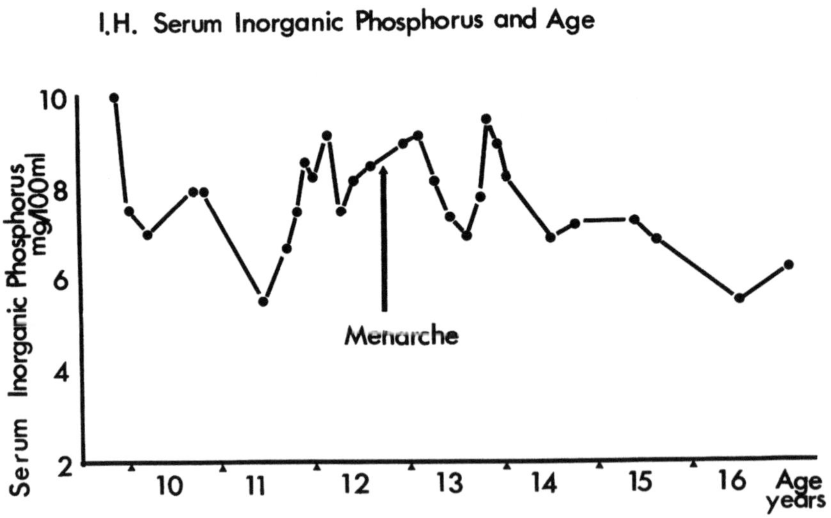

Fig. 4. Plasma phosphate values over 7 years, with probable fall some years after the menarche

References

ANNAMUNTHODO, H.: Calcinosis, Am. J. Surg. 99, 951 (1960).
BALDURSSON, H., EVANS, E.B., DODGE, W.F., JACKSON, W.T.: Tumoral calcinosis with hyperphosphatemia. J.Bone and Jt.Surg. 51A, 913 (1969).
BARTON, D.L., REEVES, R.J.: Tumoral calcinosis. Am. J. Roentgenol. 86, 351 (1961).
GHORMLEY, R.K., MCCRARY, W.: Multiple calcified bursae and calcified cysts in soft tissues. Tr. West. S.A. 51, 292 (1942).
HARKNESS, J.W., PETERS, H.J.: Tumoral calcinosis. J. Bone and Jt. Surg. 49A, 721 (1967).
HUNTER, I.P., MACDONALD, D.G., FERGUSON, M.M.: Developmental abnormalities of the dentine and pulp associated with tumoral calcinosis. Brit. Dent. J. 135, 446 (1973).
INCLAN, A.: Tumoral calcinosis. J.A.M.A. 121, 490 (1943).
LAFFERTY, F.W., REYNOLDS, E.S., PEARSON, O.H.: Tumoral calcinosis: A metabolic diseases of obscure etiology. Am. J. Med. 38, 105 (1965).
MICHAEL, A.F., DRAMMOND, K.N.: Inhibitory effect of certain amino acids on renal tubular absorption of phosphate. Can. J. Physiol. Pharmacol. 45, 103 (1967).
RIEMENSCHNEIDER, P.A., ECCKER, A.: Sciatica caused by tumoral calcinosis. J. Neurosurg. 9, 304 (1952).
WILBER, J.F., SLATOPOLSKY, E.: Hyperphosphatemia and tumoral calcinosis. Ann. Int. Med. 68, 1044 (1968).

Chronic Effects of Parathyroid Hormone Infused to Dogs at Near-Physiological Rates

B. Reit, B. Rafferty and J.A. Parsons

It is well known that hormones have characteristic rates of metabolism and excretion and that their distribution in the body is affected by binding to plasma proteins and membranes. However, one of the consequences of these facts is widely ignored, both in therapy and in the design of experiments. This is that the pattern of concentrations in the blood and at receptors may depend greatly on the rate at which a hormone enters the circulation. Dependence of the pattern of response on entry-rate is likely to be particularly striking when a hormone has a short half-life and affects multiple receptors differing in sensitivity. This is the case with parathyroid hormone (PTH), which probably disappears from the blood with a half-time of 5 or 10 minutes.

As discussed in a recent review (PARSONS and POTTS, 1972), parathyroid hormone appears to act on at least 5 different sets of receptors. The action on osteoclasts is well established and the evidence that it also acts on osteoblasts is impressive. The phosphaturic action has been shown to be exerted on the proximal renal tubule while the calcium-retaining action on the kidney involves the distal tubule (AGUS, et al., 1973). Finally, there is convincing evidence that parathyroid hormone increases intestinal calcium absorption, though the mechanism of this action, which may involve the metabolism of vitamin D, is still not clear.

This paper presents evidence that the pattern of response to parathyroid hormone given in a physiological manner is very different from the effect of giving it by single or repeated injections. We chose to work with large dogs because their calcium metabolism is similar to that of man and their size permits the testing of methods and equipment which we intend to use in a clinical trial. Two indwelling venous cannulae were implanted in an external jugular bein, one for the withdrawal of blood samples, the other to permit uninterrupted infusion, which continued for weeks at a time. Dogs were trained to wear light leather harnesses with miniature motor-driven syringes (developed for us by the Engineering Division of the National Institute for Medical Research), which are the subject of a patent application.

These syringe infusion pumps, powered by small mercury batteries, use disposable plastic syringes to overcome the problem of sterility and will be described in detail elsewhere (SHARPE, LEWIN and PARSONS, 1974). They were set to deliver the contents of one syringe (2 ml) over a 24 hour period.

The following experiments were carried out to determine the minimum rate of PTH infusion that would just cause hypercalcemia and to investigate the mechanisms by which the hypercalcemia was produced. Fig. 1 shows the basis on which a hypercalcemic dose was chosen. Initially, the hormone was infused at rates just below and just above 0.1 U/kg/hr. COPP et al. (1961), using crude Lilly extract, estimated that this was the dose required to maintain plasma calcium in parathyroidectomized dogs kept under anaesthesia.

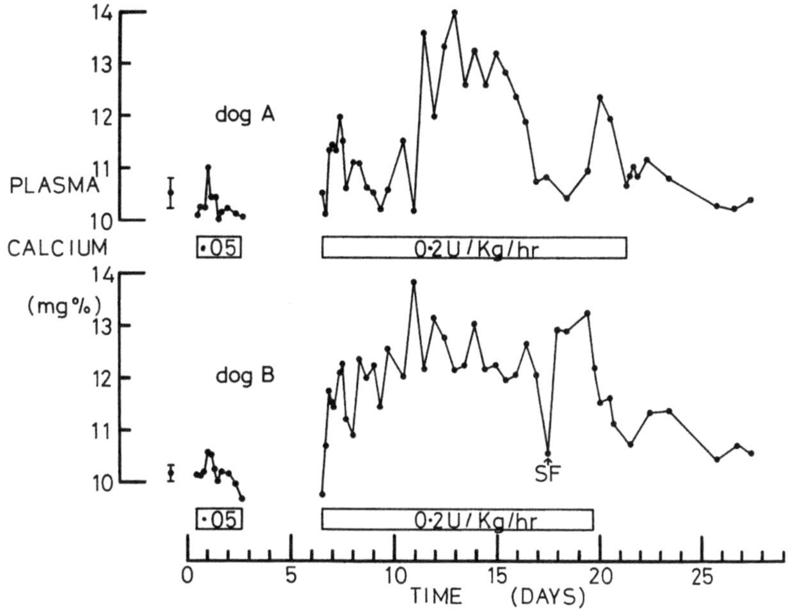

Fig. 1. Plasma calcium levels of dogs A and B during and after infusions of highly purified bovine parathyroid hormone at 0.05 and 0.2 MRC units/kg/hr (25 and 100 ng/kg/hr). Horizontal bars indicate the duration of the infusions, which were separated by an interval of 3 days. The mean and standard deviation of the plasma calcium level during 5 days preceding the first infusion is shown for each dog, at the left of the diagram. The letters SF denote a temporary syringe failure

Results of the present experiments, using Sephadex-purified bovine PTH (2500 M.R.C. units/mg), suggest a figure for the normal endogenous secretion rate which closely agrees with that proposed by COPP et al., in spite of the many differences in method. Though no detectable change in plasma calcium level was caused by the infusion of 0.05 U/mg/hr, the higher rate of infusion (0.2 U/mg/hr) caused gross hypercalcemia in spite of the fact that the animals' own parathyroid secretion was presumably reduced virtually to zero. Although the present results set only an upper limit, infusions in parathyroidectomized dogs should make it possible to estimate the normal endogenous secretion rate more closely and may show that it is even lower than 0.05 U/kg/hr.

The infusion of 0.2 U/kg/hr was repeated in two other dogs for three weeks. Fig. 2 summarizes the plasma calcium changes seen in all four dogs, which were almost identical. The instability of the calcium level seen during these infusions is not methodological and seems to be a real biological effect. Each point is based on three separate measurements by atomic absorption spectrophotometry and the analyses have a coefficient of variation of less than 1%.

Several things can be deduced from these results. Published radioimmunoassay estimates of the half-time of disappearance of intact parathyroid hormone are in the region of 5 or 10 minutes (e.g. SILVERMAN and YALOW, 1973). From this it is possible to calculate the approximate equilibrium concentration of biologically active hormone

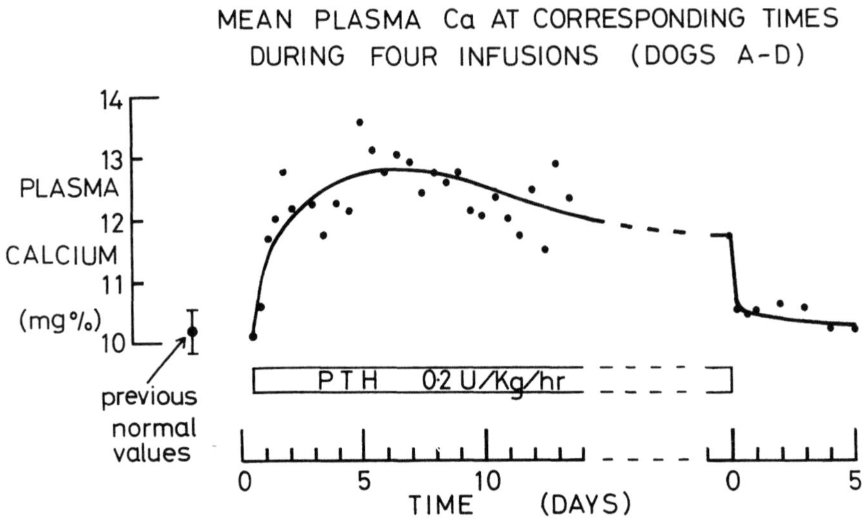

Fig. 2. Mean values of the plasma calcium concentration, at corresponding times during four infusions of parathyroid hormone (0.2 U/kg/hr). Each point is calculated from experiments (dogs A-D) of the type illustrated in Fig. 1, omitting artefacts due to syringe failures. Since two infusions lasted for 2 weeks and two for 3 weeks, the time scale is interrupted and results pooled again at corresponding intervals from the end of the infusions. The solid line is drawn by hand to give an overall impression of the average effect, and to emphasize the rapid fall in calcium level which followed termination of the infusion in all dogs

which an infusion of 0.1 U/kg/hr could support (PARSONS et al., 1974). This calculated concentration (0.01 ng/ml) is lower than the lowest present indications from immunoassay - a discrepancy which is not surprising, because immunoassays can only estimate biological activity very indirectly, for example by comparison of measurements obtained with N- and C-terminal specific systems (PARSONS and POTTS, 1972).

The hypercalcemic infusion of 0.2 U/kg/hr represented a total dose of 4 ug/hr for the largest dog used in this study, weighing 80 lb. This figure and the estimate that the normal circulating concentration is as low as 0.01 ng/ml (about 10^{-10} molar) establish PTH among the hormones of highest biological activity, such as oxytocin, vasopressin and calcitonin, all of which circulate at about this level. Parathyroid hormone used to appear remarkably inactive by comparison with the other hormones when the only figures available for calculation were the doses used in biological assays (corresponding to blood levels of 10^{-7} or 10^{-8} molar). The fact that this paradox disappears when calculating from what we believe is the normal effect of the hormone emphasizes that the acute osteoclastic response used in bioassay simply is not a normal effect.

What can be said about the mechanism of the hypercalcemia caused by these chronic infusions? There was no evidence that the infusions led to an increase in bone breakdown in any of the animals tested. Histological studies carried out by Dr. A. DARBY on iliac crest bone biopsies before and after each infusion showed no increase in osteoclast numbers or in the areas involved in resorption. On the contrary, in one dog the area occupied by osteoid was tripled, suggesting a marked increase in bone formation. As we have said, this dose may be 100 times too low to cause bone breakdown.

The mechanism therefore presumably involves more sensitive responses to parathyroid hormone. Phosphaturia, which is known to be evoked by lower doses than bone breakdown, might have been expected to occur, but was not observed. Because of the difficulty of collecting urine from these large unrestrained dogs, a small phosphaturic effect might have been missed, but plasma phosphate analysis carried out with the same precision as those for calcium showed no significant hypophosphatemia during the infusion periods.

The probable explanation of the hypercalcemia lies in a combination of increased intestinal calcium absorption and decreased renal excretion. The calcium absorption coefficient, measured with a double-isotope method using oral calcium 47 and intravenous calcium 45 (DE GRAZIA et al., 1965) doubled or tripled during each of the hypercalcemic infusions. This was accompanied by a fall in urinary calcium excretion, in spite of the increase in filtered load.

It is clear from Fig. 2 that at the start of the infusions the calcium levels took about 48 hr to rise to a plateau, consistent with what is known about the long latency of the effect on intestinal absorption. When the infusions were stopped, the calcium level fell almost within 4 hours and an equally rapid fall was seen during temporary syringe failures. It therefore appears that neither the intestinal nor the renal effect alone accounts for the hypercalcemia, but the two together can readily do so.

In summary, then, we have found that chronic infusion of parathyroid hormone slightly exceeding the normal secretion rate result in a marked increase in intestinal calcium absorption and concomitant increase in renal calcium retention, with no indication of an increase in bone breakdown.

It seems highly relevant that chronic administration of low doses of PTH has been shown to lead to an increase in bone formation. This was first demonstrated in the rat by PUGSLEY and SELYE (1933) and has been confirmed by the more recent studies of KALU et al. (1970), and WALKER (1971). Particularly interesting is the evidence for increased bone formation in mild clinical states of hyperparathyroidism (DOYLE, 1972; CONNOR et al., 1973).

Considering the clinical implications of these findings, we suggest that the demineralizing condition currently described as hyperparathyroidism may result only when there is gross oversecretion by the parathyroid gland. The dog studies were planned as a prelude to clinical investigation of the possibility, originally suggested on theoretical grounds, that chronic administration of small doses of PTH, alone or in combination with calcitonin, may lead to a positive calcium balance and stimulation of bone formation in osteoporosis (ROBINSON et al., 1972). Their outcome has reinforced our belief in the safety and importance of a limited clinical trial of such therapy.

References

AGUS, Z.S., GARDNER, L.B., BECK, L.H., GOLDBERG, M.: Effects of parathyroid hormone on renal tubular reabsorption of calcium, sodium and phosphate, Am. J. Physiol. 224, 1143 (1973).

CONNOR, T.B., STONER, R.E., FREIJANES, J., MARTIN, L.G., JOWSEY, J.: Osteoclerosis in primary hyperparathyroidism, Transactions of the American Clinical and Climatological Association Vol. 85 (in press, 1973).

COPP, D.H., MOGHADAM, H., MENSEN, E.D., MACPHERSON, G.D.: The parathyroids and calcium homeostasis. In: The Parathyroids (Greep, R.O. and Talmage, R.V., eds.), p. 203. Springfield, Illinois: Thomas 1961.

DE GRAZIA, J.A., IVANOVITCH, P., FELLOWS, H., RICH, C.: A double isotope method for measurement of intestinal absorption of calcium in man. J. Lab. Clin. Med. 66, 822 (1965).

DOYLE, F.H.: Radiological patterns of bone disease associated with renal glomerular failure in adults. Brit. Med. Bull. 28, 220 (1972).

KALU, D.N., PENNOCK, J., DOYLE, F.H., FOSTER, G.B.: Parathyroid hormone and experimental osteoclerosis. Lancet I, 1363 (1970).

PARSONS, J.A., POTTS, J.T. Jr.: The physiology and chemistry of parathyroid hormone. Clin. Endocr. Metab. 1, no. 1, 33 (1972).

PARSONS, J.A., DARBY, A.J., REIT, B.: Anabolic effect of parathyroid hormone studied by chronic infusion to dogs. In: Endocrinology 1973 (Taylor, S., ed.). London: Heinemann Medical Books (in press, 1974).

PUGSLEY, L.I., SELYE, H.: The histological changes in the bone responsible for the action of parathyroid hormone on the calcium metabolism of the rat. J. Physiol. 79, 113 (1933).

ROBINSON, C.J., RAFFERTY, B., PARSONS, J.A.: Calcium shift into bone: A calcitonin-resistant primary action of parathyroid hormone, studied in rats. Clinical Science 42, 235-241 (1972).

SHARPE, J.E., LEWIN, J., PARSONS, J.A.: A miniature syringe pump for continuous administration of drugs and hormones. Lancet (in press, 1974).

SILVERMAN, R., YALOW, R.S.: Heterogeneity of parathyroid hormone: clinical and physiologic implications. J. clin. Invest. 52, 1958 (1973).

WALKER, D.G.: The induction of osteopetrotic changes in hypophysectomised, thyroparathyroidectomised, and intact rats of various ages. Endocrinology 89, 1389 (1971).

The Calcium Orthophosphate Solubilities as Represented by a Model in Three Dimensions

A. R. HAGEN

I. Introduction

The solubility behavior of hard tissues has almost exclusively been assessed by empirical means, e.g. by determining the amount of calcium or phosphate dissolved or by estimating the weight loss or gain. These methods circumvent the very complex solubility problems, and they may be grossly misleading.

In some few studies the solubility product principle has been used. The solubility product is suited for evaluating the behavior of one and the same phosphate under different conditions, but not for comparing different phosphates.

The relative solubilities of any phosphate may be calculated from the n-concept. These relative solubilities may be compared.

The absolute solubilities may also be calculated provided the solubilities are strictly stoichiometric and congruent.

The solubilities of any number of phosphates may be represented on a single diagram as functions of the chemical potentials of $Ca(OH)_2$ and H_3PO_4. This method is convenient, but the experimentally determined variables are not readily visualized on the diagram.

The model to be presented summarizes the concepts developed over the years, and gives a correct and complete representation of the solubility interrelations in calcium orthophosphate systems.

II. Theoretical

The phosphates of biological interest are:

- hydroxyapatite (Hap) or $Ca_5(PO_4)OH$,
- dicalcium phosphate dihydrate (DCPD) or $CaHPO_4 \cdot 2H_2O$,
- octacalcium phosphate (OCP) or $Ca_4H(PO_4)_3 \cdot 2\ 1/2\ H_2O$,
- β-tricalcium phosphate (TCP) or β-$Ca_3(PO_4)_2$.

When solubility equilibrium has been established, the free energy of the solid phase plus the free energy of dissolution must equal the free energy of the ions in solution:

$$\Delta G_f^o(\text{solid}) + \Delta G_d^o = \Sigma v_i \Delta G_f^o(\text{ion})$$

Expressed in terms of the chemical potentials, this equation changes into

$$\Delta G_d^o = \Sigma v_i\ RT\ \ln[\text{activity}]_{ion} = -RT\ \ln K.$$

The constant K is the solubility product.

The solubility product is functionally related to three variables, two of which are independent. These three variables are: 1) the calcium ion activity, 2) the hydronium ion activity, and 3) the activity of the pertinent phosphoric ion. This latter ion is uniquely determined by means of the dissociation constants of phosphoric acid and the pH. In terms of negative logarithms the solubility product is then functionally given by

$$pK = \oint p\left[Ca^{2+}\right] + \oint p[P] + \oint pH \qquad 1)$$

In the ternary system $Ca(OH)_2$-H_3PO_4-H_2O balanced equations may therefore be written for the formation of any phosphate. These equations are given in a coordinate system with $p\left[Ca^{2+}\right]$, $p[P]$, and pH as the three axes in accordance with Equation 1). Choosing the $p\left[Ca^{2+}\right]$ as the dependent variable, the equation for Hap will be:

$$p\left[Ca^{2+}\right] = 0.2\ pK_{Hap} - pKw - 0.6\ pQ_3 - 0.6\ p[P] + 0.2\ pH,$$

where K_w is the dissociation constant of water, and Q_3 is the fraction of total phosphorus existing as PO_4^{3-} ion. Similar equations are written for DCPD, OCP, and TCP.

The limitless number of points calculated from these equations are - for each equation - located on a surface in space. Each of these <u>saturation surfaces</u> has a curvature and a direction different from the other ones.

The electro-neutrality requirements put an additional restriction on the equilibrium conditions. In a pure ternary system the ionic species are connected by the relation

$$2(Ca^{2+}) + (H_3O^+) = (H_2PO_4^-) + 2\ (HPO_4^{2-}) + 3\ (PO_4^{3-}) + (OH^-) \qquad 2)$$

Equation 2) may be transformed and plotted as a <u>neutrality surface</u> in the coordinate system:

$$p(Ca^{2+}) = 0.301 + p\left\{10^{-pK_w+pH} - 10^{-pH} + (Q_1+2Q_2+3Q_3)(P)\right\} \qquad 3)$$

where Q_1 and Q_2 are the fractions of total phosphorus existing as $H_2PO_4^-$ and HPO_4^{2-} ions, respectively. This electro-neutrality surface will intersect all the saturation surfaces and form a line with each of them, a <u>solubility isotherm.</u> Any phosphate can be in solubility equilibrium only at these isotherms, which means that the whole system is uni-variant.

When extraneous salts are added to the ternary system, an additional amount of equivalents will be added to both sides of Equation 2). In the special case, however, where the added salt contains calcium (e.g. $CaCl_2$) or phosphate (e.g. Na_2HPO_4), Equation 3) will not apply without modifications. The calcium concentration calculated from Equation 3) must be increased by the amount of $0.5 \cdot 10^{-3}$ for each milliequivalent of extra calcium added, and decreased by the same amount for each milliequivalent of extra phosphate added. These corrections are termed the <u>electro-neutrality</u> unbalance and is formally defined by

$$U = \pm \sum v_i c_i$$

where the plus sign refers to a surplus of added calcium and the minus sign to a surplus of added phosphate.

III. Construction of the Model

The model is constructed to an ionic strength of 0.05 and a temperature of 37-38°C. To that purpose a general activity coefficient was calculated from the formula of DEBYE and HÜCKEL (1923):

$$\log f_i = - 0.0958 \cdot z_i^2,$$

where z_i is the valency of the ith ion.

The dissociation constants for phosphoric acid were taken from BJERRUM and UNMACK (1929). The activity solubility products used were: 57.43 for pK_{Hap} (MORENO et al., 1968), 46.90 for pK_{OCP} (MORENO et al., 1960), 6.63 for pK_{DCPD} (GREGORY et al., 1970), and 29.66 for pK_{TCP} (MORENO et al., 1970). By these conventions the calcium and phosphate axes in the model are given in concentration units, and the pH axis in activity units.

The region incorporated in the model has pH values from 4.0 to 9.0, and $p(Ca^{2+})$ and p(P) values from 1.0 to 5.0, i.e. they cover a concentration range from 0.01 to 100 mM. All biologic and most inorganic systems will be found within these limits.

1. The Saturation Surfaces

are calculated to:

Hap:
$$p(Ca^{2+}) = -4.6528 - 0.6p(P) + 0.2pH + 0.6\ pY$$
OCP:
$$p(Ca^{2+}) = -5.1024 - 0.75p(P) - 0.25pH + 0.75\ pY$$
DCPD:
$$p(Ca^{2+}) = -3.1056 - p(P) - pH + pY$$
TCP:
$$p(Ca^{2+}) = -4.8716 - 0.67p(P) + 0.67\ pY$$

where

$$Y = 10^{-3pH} + 10^{-2pH-2.0896} + 10^{-pH-8.9645} + 10^{-20.6992} \qquad 4)$$

A schematic representation has been shown in Fig. 1. When the $p(Ca^{2+})$ and p(P) values are plotted in decreasing order, and the pH's in increasing order away from the origo of the coordinate system, any point located above any surface represents a state of supersaturation with respect to that surface. Similarly, any point below a surface represents a state of undersaturation.

2. The Electro-Neutrality Surface

By introducing the appropriate constants valid for an ionic strength of 0.05, Equation 3) becomes

$$p(Ca^{2+}) = 0.301 + p\left\{10^{-14+pH} - 10^{-pH} - \frac{X}{Y}\ (P)\right\}$$

where Y is given by Equation 4) and

$$X = 10^{-2pH-2.896} + 10^{-pH-8.6635} + 10^{-20.2221}$$

When the electro-neutrality unbalance enters, the $p(Ca^{2+})$ must be changed to

$$p\left\{(Ca^{2+}) \pm 1/2\ U.\right\}$$

The electro-neutrality surface will move towards the $p(Ca^{2+})$ axis when calcium is added ("positive unbalance") and towards the p(P) axis when

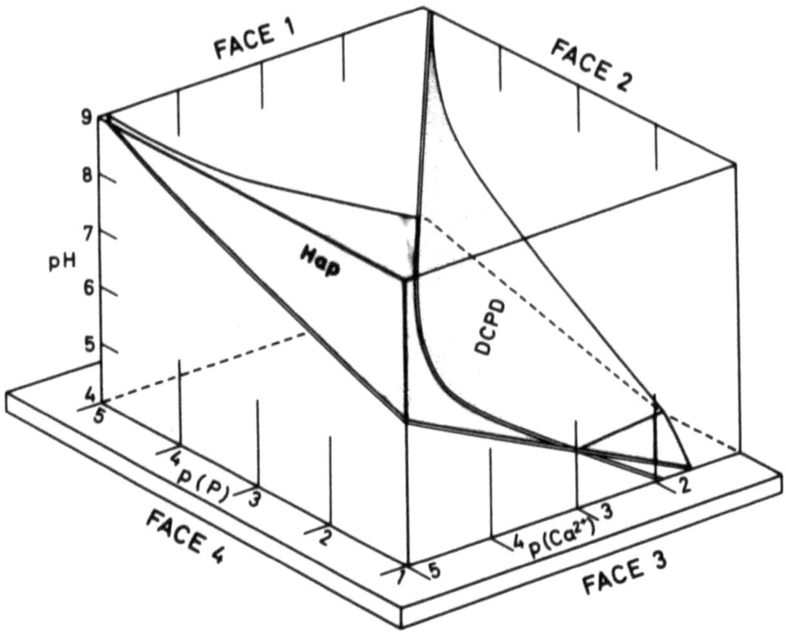

Fig. 1. A schematic drawing of the model

extra phosphate is added ("negative unbalance"). Fig. 2 gives a projection on the basal plane of the model, illustrating the relationship mentioned.

3. The Solubility Isotherms

are formed by the intersections of the saturation surfaces and the electro-neutrality surface. The pertinent calculations give the following equations for these isotherms:

$$\text{Hap: } 0.4702\ p(Ca^{2+}) - 0.1240\ p(P) - 0.2235\ pH + 0.3409 = 0$$

$$\text{OCP: } 0.5774\ p(Ca^{2+}) - 0.1319\ p(P) - 0.2698\ pH + 0.6083 = 0$$

$$\text{DCPD: } 3.1298\ p(Ca^{2+}) - 0.2845\ p(P) - 0.6616\ pH + 2.0756 = 0$$

$$\text{TCP: } 0.8187\ p(Ca^{2+}) - 0.1489\ p(P) - 0.3540\ pH + 0.5446 = 0$$

4. The Singular Lines and the Singular Points

When any pair of the four phosphates is saturated for just the same values of the coordinates, the corresponding saturation state must be described equally well by the equations for the two phosphates in question. This leads to the presence of singular lines in the model (Fig. 2). The equations for the three singular lines are:

$$\text{Hap - DCPD: } 0.5\ p(Ca^{2+}) + 3p(P) + 2pH - 14.3680 = 0$$

$$\text{OCP - DCPD: } 0.5\ p(Ca^{2+}) + 2.3666\ p(P) + 2.4635\ pH - 24.5785 = 0$$

$$\text{TCP - DCPD: } 0.5\ p(Ca^{2+}) - 2.9508\ p(P) + 2.0252\ pH - 17.9352 = 0$$

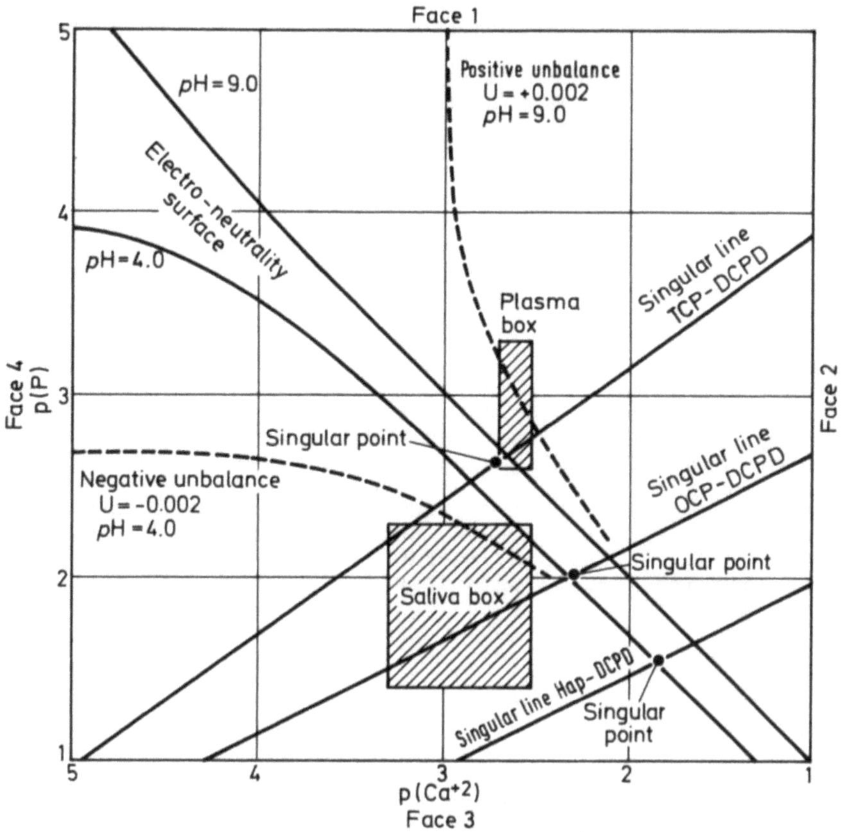

Fig. 2. Projection of the model on the basal plane

When the solubility isotherm intersects a singular line, a singular point will appear. Only at this point can any two phosphates be in equilibrium simultaneously. The coordinates of the singular points in the model are:

Pair of phosphates	pH coordinate	$p(Ca^{2+})$ coordinate	p(P) coordinate
Hap - DCPD	4.3979	1.8222	1.5537
TCP - DCPD	5.3168	2.2308	2.0512
OCP - DCPD	6.8862	2.6797	2.6512

IV. Discussion

The model gives a generally valid representation of the solubilities in calcium orthophosphate systems. The relation of the stability fields to one another is immediately seen. Controversial problems such as the biological metastability and the stoichiometric solubility are solved in rational, physico-chemical terms. Although the model only describes states existing under thermodynamic equilibrium conditions, it also throws light on kinetic questions. In practical respects, the model is a precise instrument for assessing the effect of biologic parameters on bone and dental mineral dynamics.

References

BJERRUM, N., UNMACK, A.: Elektrometrische Messungen mit Wasserstoffelektroden in Mischungen von Säuren und Basen mit Salzen. K. danske Vidensk. Math. fys. Meddel. 9/I, 208 (1929).

DEBYE, P., HÜCKEL, E.: Zur Theorie der Elektrolyte. Phys. Z. 24, 185-206 (1923).

GREGORY, T.M., MORENO, E.C., BROWN, W.E.: Solubility of $CaHPO_4 \cdot 2H_2O$ in the system $Ca(OH)_2$-H_3PO_4-H_2O at 5, 15, 25, and 37.5°C. J. Res. Ntl. Bur. Stand. 74A, 461-475 (1970).

MORENO, E.C., BROWN, W.E., OSBORNE, G.: Stability of dicalcium phosphate dihydrate in aqueous solutions and the solubility of octacalcium phosphate. Soil Sc. Soc. Amer. Proc. 24, 99-102 (1960).

MORENO, E.C., GREGORY, T.M., BROWN, W.E.: Preparation and solubility of hydroxyapatite. J. Res. Ntl. Bur. Stand. 72A, 773-782 (1968).

MORENO, E.C., PATEL, J.M., GREGORY, T.M., BROWN, W.E.: Solubility of whitlockite, β-$Ca_3(PO_4)_2$ IADR, Abstract No. 183 (1970).

Index of Authors

Subject Index